# THE IMMUNE RESPONSE TO VIRAL INFECTIONS

# ADVANCES IN EXPERIMENTAL MEDICINE AND BIOLOGY

## Recent Volumes in this Series

# THE IMMUNE RESPONSE TO VIRAL INFECTIONS

Edited by

## B. A. Askonas
National Institute for Medical Research
London, United Kingdom

## B. Moss
National Institutes of Health
Bethesda, Maryland

## G. Torrigiani
World Health Organization
Geneva, Switzerland

and

## S. Gorini
Fondazione Internazionale Menarini
Florence, Italy

PLENUM PRESS • NEW YORK AND LONDON

Library of Congress Cataloging in Publication Data

International Symposium of the Immune Response to Viral Infections (1988: Florence, Italy)
    The immune response to viral infections / edited by B. A. Askonas . . . [et al.].
       p.    cm. — (Advances in experimental medicine and biology; v. 257)
    "Proceedings of an International Symposium on the Immune Response to Viral In-. fections, held April 28–29, 1988, in Florence, Italy" — T.p. verso.
    Includes bibliographical references.

    1. Virus diseases — Immunological aspects — Congresses. I. Askonas, B. A. II. Title. III. Series.
    [DNLM: 1. Viral Diseases — immunology — congresses. W1 AD559 v. 257 / WC 500 I665i 1988]
RC114.5.I575   1988
616.9′25079 — dc20
DNLM/DLC                                   89-22773
for Library of Congress                              CIP

ISBN-13:978-1-4684-5714-8                e-ISBN-13:978-1-4684-5712-4
DOI: 10.1007/978-1-4684-5712-4

Softcover reprint of the hardcover 1st edition 1989

Proceedings of an International Symposium on
The Immune Response to Viral Infections,
held April 28–29, 1988, in Florence, Italy

© 1989 Plenum Press, New York
A Division of Plenum Publishing Corporation
233 Spring Street, New York, N.Y. 10013

PREFACE

Virus diseases continue to represent serious health problems in most parts of the world. In spite of the fact that diseases such as poliomyelitis and measles have been controlled in the industrialized countries by vaccination, vaccines now in use in tropical countries have proved not to be optimal. Further research is needed to develop new vaccines that will be effective in all countries. To do so we need to understand better the immune response to different viruses so that we may be able to maximize the protective response of new vaccines and minimize their potential immunopathologic effect.

An exciting new discovery which is now being further developed is the possibility of being able to use some viruses (e.g. vaccinia, adenoviruses, etc.), as carriers for other antigens. This may open up the way for the production of vaccines that will be inexpensive and that will confer long-lasting immunity after only one injection.

This meeting has also served to review our present knowledge of virus diseases which are still of great importance such as hepatitis, dengue and influenza.

G. Torrigiani

# CONTENTS

# THE CONTRIBUTIONS OF VIRUS INFECTIONS

# TO OUR UNDERSTANDING OF THE IMMUNE SYSTEM

G.L. Ada

World Health Organization
Geneva, Switzerland

Viruses, on the one hand, symbolise mankind's greatest achievements in controlling infectious diseases yet on the other hand, they still constitute a very great public health risk and provide a scientific challenge to make effective, safe and inexpensive vaccines. Vaccines against smallpox, yellow fever, poliomyelitis, measles and rubella have been or are very successful. Smallpox has been eradicated and transmission of measles reduced to extremely low levels in the USA following the need for parents to provide evidence of vaccination prior to admittance of their children to school. Immunisation with current vaccines to control measles and poliomyelitis is sufficiently effective to suggest that in principle measles could be eradicated and there is a plan to eradicate poliomyelitis in the Americas. In contrast, there are other viral infections of global importance for which either current vaccines are inadequate or too costly or vaccines are unavailable. These include the following viruses - herpes, cytomegalo, dengue, hepatitis A, B and non-A non-B, rota, influenza, parainfluenza, respiratory syncytial and retro viruses, particularly HTLV lll. Viral infections of the G.I. and respiratory tracts are major killers of young children, particularly in third world countries. There is still concern that an influenza re-assortant virus might arise which could cause a pandemic Dengue and the hepatitis viruses are major pathogens in tropical countries, particularly in South-East Asia. Herpes and cytomegaloviruses are continuing problems in all populations especially in immunocompromised people. The HTL viruses now are a major challenge as they may attack cells of the immune system and show great antigenic variation.

There is also increasing evidence that viruses may be the causative agents in a number of common diseases such as cancers eg. Epstein-Barr virus in Burkitt's lymphoma, Hepatitis B virus in primary hepatoma, papilloma virus in cervical cancer and HTLV l in leukaemia. Other viruses have been implicated in diabetes mellitus (Coxsackie virus), multiple sclerosis (HTLV l) and rheumatoid arthritis (E.B. virus).

Viruses may be grouped according to their disease pattern. Thus, some viruses cause acute infections i.e. there is no evidence of virus persistence or post-recovery trauma in non-compromised hosts; examples are influenza and pox viruses. Some viruses may cause chronic and/or persisting infections in apparently normal individuals; examples

1

are measles, cytomegalo and adeno viruses. Other viruses may cause latent infections which may, as seen above, result in neoplasms; examples are herpes, hepatitis B, retro and lenti viruses. The pathology observed during and after primary infection may therefore differ widely.

On the other side of the coin, some viruses have been remarkably effective as tools for unravelling the pathways of cellular metabolism and replication e.g. adenovirus. Others, because of the potential hazard of causing pandemics have been very thoroughly studied by a variety of techniques e.g. the haemagglutinin of the influenza virus is one of the most studied of all proteins. Viruses have also been very important tools for understanding how the immune system works - in fact, as the latter most likely evolved to combat infectious agents of which viruses are an important group, it is not surprising that an important arm of the immune response - effector T cells with cytotoxic activity - evolved mainly if not completely to control many viral infections by aiding in the recovery process. The first demonstration of this phenomenon soon led to a fundamental discovery in Immunology - how the T lymphocyte receptor recognises foreing antigens.

Recently, the "practical" and "academic" aspects of the study of viruses their pattern of infection and the subsequent immune response have come together in a remarkable fashion. The ability to manipulate DNA to form recombinant molecules is having and will continue to have a profound effect on the future directions of scientific research. It is now possible to make recombinant viruses which upon infection of a cell, result in the expression of the inserted "foreign" DNA. Viruses so far used for this purpose include vaccinia, herpes and adeno viruses. As an example, DNA coding for specific antigens of about 10 different infectious, including viruses and parasites, have been incorporated into vaccinia virus. When used to immunise hosts in model systems, some recombinant preparations have protected the host animal from death when challenged with many lethal doses of the agent from which the "inserted" DNA was derived. There is thus a bright prospect for the use of such recombinant viruses as the basis of future vaccines against a variety of infectious agents.

Such recombinant viruses are also proving to be a powerful tool for seeking answers to other questions eg. what antigens of a virus are preferentially recognized by T lymphocytes? They may be vectors of DNA coding for a variety of other substances, such as cellular antigens, growth factors and hormones and so on. Transfection of cells with DNA may achieve the same purpose but recombinant viruses may prove to be a more effective means of achieving desired effects in vivo.

The study of viruses is demonstrably important from several aspects. This Symposium will feature presentations which illustrate many of the more important aspects mentioned in this statment.

# CD4 AS THE RECEPTOR FOR RETROVIRUSES OF THE HTLV FAMILY:

# IMMUNOPATHOGENETIC IMPLICATIONS

M. Carbonari, M. Fiorilli, I. Mezzaroma,
M. Cherchi, and F. Aiuti

Department of Allergy and Clinical Immunology
University of Rome "La Sapienza"
Rome, Italy

## INTRODUCTION

CD4 is the HIV (Human immunodeficiency virus type I) receptor, and the virus seems to bind to it through the Leu3a/OKT4a epitope (Klatzmann et al., 1985). HIV infection may lead to the destruction of CD4+ cells both by direct and indirect mechanisms. Direct mechanisms include the formation of syncitia' among infected cells (Lifson et al., 1986). Indirect mechanisms may involve autoimmune reactions, of both cell mediated and humoral types (Klatzmann and Montagnier, 1986).

A number of autoantibodies are produced by HIV-infected individuals; they include anti-phospholipid, anti-platelet and antilymphocyte antibodies. Some of such autoantibodies seem to induce significant pathological abnormalities, such as thrombocytopenia (Stricker et al., 1985) and, possibly, lymphopenia (Ho et al., 1987). However, the pathogenetic role of antilymphocyte antibodies has been only poorly defined.

Another intriguing point concerning the role of CD4 in the pathogenesis of HIV-related dysorders, is that cells other than T lymphocytes (e.g. macrophages, cells of the central nervous system and EBV-infected B-cells) appear to express this molecule. These non-T CD4+ cells might be both target for HIV-induced destruction and reservoir for the virus.

To further investigate the problems addressed in the above paragraphs, we performed the studies that we are reporting here.

## RESULTS AND DISCUSSION

### Antilymphocyte antibodies in HIV infection

Little is known about the mechanisms which lead to the production of autoantibodies in HIV infection. Polyclonal B-cell activation by HIV antigens is likely to be involved (Pahwa et al., 1985), but other mechanisms might also be of importance. Autoantibodies to CD4+ cells, described by some authors (Dorsett et al., 1985), might be evoked as a part of an antidiotypic circuitry to the CD4-binding region of the viral envelope glycoprotein (Del Guercio and Zanetti, 1987). However, this type of autoantibodies (e.g. to CD4) have been, so far, only very poorly characterized. Therefore, we examined HIV+ sera to evaluate the presence ad clinical significance of antilymphocyte antibodies.

Contrary to previous reports (Dorsett et al., 1985; Tomar et al., 1985), we found no evidence for antibodies reactive with large populations of normal T-cells in any of the sera examined. However, we found that most of the sera had antibodies staining a small (3-5%) subset of peripheral blood lymphocytes from normal donors. Technical details and the representative results from one experiment are described in Fig. 1. A preliminary characterization of the cells stained by HIV+ sera suggested them as being CD3, CD4, CD8-negative, and HLA-DR positive. However, technical considerations (the need for strong compensation of red fluorescence, to avoid leaking of green fluorescence) precluded to us to firmly rule out a (weak) CD3 positivity. Thus, it is likely that we were detecting the autoantibody to the 18 Kd T-cell activation antigen described by Stricker et al., 1987. A major discrepancy between our data and previous reports (Dorsett et al., 1985; Tomar et al., 1985), is that we were unable to identify autoantibodies reacting with large populations of T-cells. This probably depends on technical differences: in fact, we put particular care in reducing, by electronic compensation, the leaking of the strong green fluorescence of polyclonal anti-human Ig antibodies. In fact, a critical re-evaluation of a published report (Tomar et al. 1985) of antilymphocyte antibodies, detected by FACS analysis, suggests that the data may have been biased by insufficient compensation of fluorescence analysis.

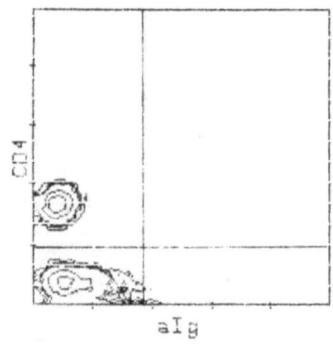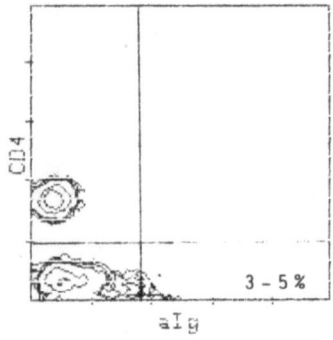

+ normal serum          + HIV-pos. serum

Fig. 1.   HIV-seropositive (right panel) or seronegative (left panel) sera were incubated with normal human peripheral blood lymphocytes; cells were counterstained with Leu3a (red fluorescence) or anti-human IgG antiserum (green fluorescence). Anti-IgG alone was used to set the vertical gating (not shown). In this representative experiment, 3.4% of lymphocytes were stained by HIV+ serum but not by the normal serum. Consistent results were obtained with eight HIV+ (from seropositive asymptomatic, AIDS or AIDS-related complex patients) and five HIV- sera.

Our data, while questioning the existence of antibodies to CD4 or to other major T-cell populations, are compatible with the presence of autoantibodies to a discrete subpopulation of activated T-cells in HIV-infected individuals (Stricker et al., 1987). How these autoantibodies are induced by HIV infection is a matter for speculation. Besides polyclonal B-cell activation (Pahwa et al., 1985), a model for this autoimmune response might be the following. HIV-infected individuals are known as having large numbers of activated T-cells in vivo (i.e., T-cells expressing class II MHC products) (Fig. 2). We examined the distribution of HLA-DR positive cells among CD4+ and CD8+ T-cell subpopulations, and found grossly the same distribution among these two subsets (Fig. 2).

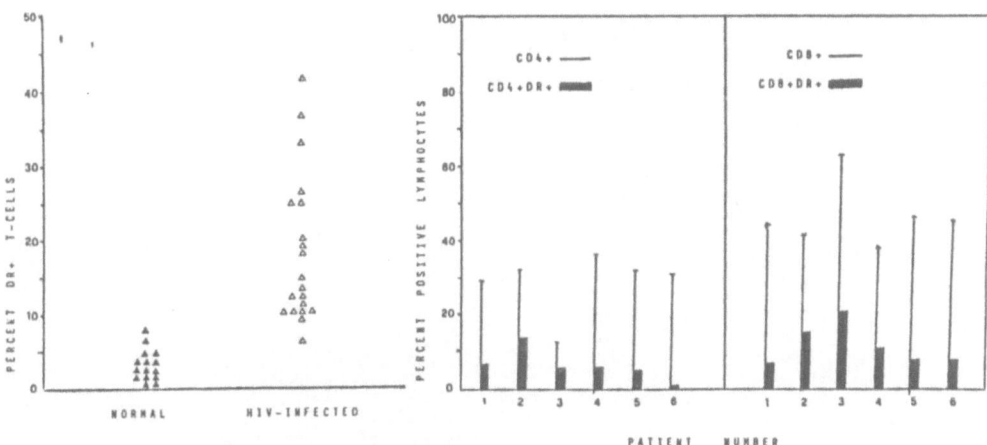

Fig. 2. Percentage of HLA-DR positive T-cells (CD3+)
in normal and HIV-infected subjects (left panel),
and their distribution between CD4+ and CD8+ subsets
(left panel).

Thus activated, MHC class II-positive, CD4-positive cells are abundant in HIV+ individuals. These cells might be also coated with HIV antigens (either produced endogenously by an infected cell, or absorbed as soluble antigen from body fluids by uninfected cells). Viral antigen(s) present on the cell surface might, in turn, bind some membrane structure(s) (e.g. the 18 Kd molecule expressed by T cells during activation) and render it (auto)immunogenic by steric modification. This modified autoantigen would then be efficiently self-presented, owing to te presence of class II MHC products on the cell surface.

Expression of CD4 outside T lymphocytes

A point of major significance concerning the role of CD4 in the expression of HIV infection, is its presence on cells other than T lymphocytes. So far, monocyte/macrophages, some cells of the central nervous system, and some EBV-transformed B-cells have been found to express CD4. While macrophages appear to bear the same CD4 molecule of T-lymphocytes (Stewart et al., 1986), cells of the nervous system (glial cells and some neurones) express a truncated form of CD4 mRNA, which is perhaps not even translated into protein (Maddon, 1986; Gorman et al., 1987). The CD4 expressed by some EBV-transformed B-cell lines (Salahuddin, 1987) has not been, so far, characterized. As the role of CD4 in terms of infectability of EBV-B cell lines by HIV is unclear (Salahuddin, 1987), we undertook a molecular characterization of the CD4 molecule expressed by an EBV-transformed lymphoblastoid cell line obtained in our laboratory from the B cells of a normal donor. This cell line (named blym-2) has 50-80% CD4-positive cells (Fig. 3). Immunoprecipitation studies and FACS analysis showed that it bears a typical 55 Kd rMW CD4 species, carrying all of the following epitopes: Leu3a, OKT4, OKT4a, OKT4b, OKT4d, OKT4e. Thus, some EBV-infected B cells bear the same CD4 as T cells, including the putative attachment site for

HIV (i.e. the Leu3a epitope). These cells may therefore represent, in vivo, a potentially important reservoir for HIV, expecially considering their proliferative capacity. This might account for the recently recognized role for EBV coinfection on the expression of HIV-related pathologic manifestations (Ragona et al., 1986).

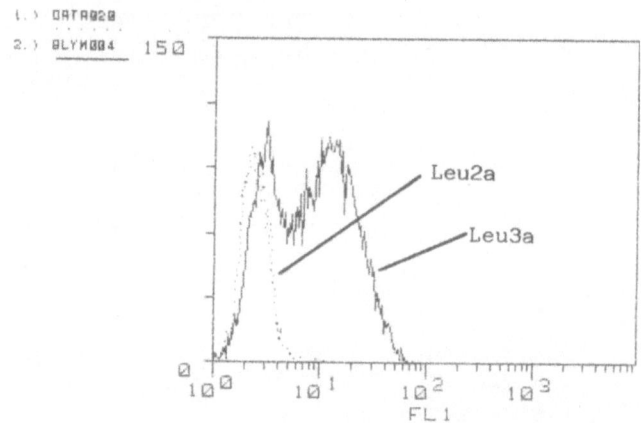

Fig. 3.  Leu3a positivity of an EBV-transformed
B cell line (blym-2).

It is unclear why only some EBV-transformed B-cell lines (about 15%) express CD4. On one hand, normal circulating B cells do not appear to express detectable CD4 (our unpublished data). On the other hand, EBV superinfection does not seem to induce CD4 on CD4-negative B-cell lines (Salahuddin, 1987). Understanding of the relationship between EBV infection, induction of CD4 on transformed B-cells, and expression of HIV infection, might highlight a significant co-factor for the development of AIDS.

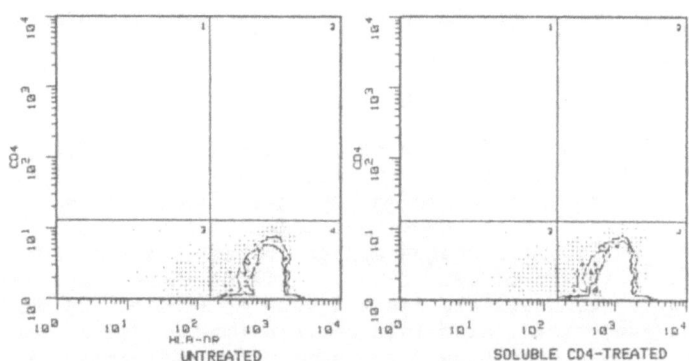

Fig. 4.  Soluble CD4 does not bind surface class II
MHC products. Cells from an EBV-transformed B
cell line (HLA-DR positive, CD4 negative) were
incubated at one million/ml with soluble CD4
(kindly provided by E. Reinherz) at 2 ug/ml final
concentration (or with saline as control), for
one hour in the cold. Then, cells were washed
and stained with FITC-labelled anti-CD4
(Leu3a+Leu3b, Multiclone, Becton-Dickinson) and
with anti-HLA-DR monoclonal antibodies.

In addition, since CD4 in soluble form has recently been demonstrated capable of inhibiting viral binding to CD4+ lymphocytes and, thus, of neutralizing HIV-1 infectivity, its use in vivo as a therapeutical agent has been hypothesized (Weiss, 1988). However, a potential side effect might be its interference with the functioning of the immune system via its interaction with MHC class-II molecules. We, therefore, performed experiments to assess whether CD4 might bind to MHC class-II products expressed by a (CD4-negative) EBV-transformed B cell line.

Preliminary results (Fig. 4) suggest that soluble CD4 does not bind HLA-DR; thus, it probably would not interfere with normal antigen presentation. Whether this might occur in vitro using other test systems, or upon in vivo administration of soluble CD4, has still to be defined.

REFERENCES

Del Guercio P, Zanetti M. Immunol. Today 8, 204, 1987.

Dorsett B, Cronin W, Churma V, Jochain H. Autoantibodies to helper T lymphocytes in AIDS patients. Am. J. Med. 78, 621, 1985.

Gorman SD, Tourvieille B, Parnes JR. Structure of the mouse gene encoding CD4 and unusual transcript in brain. Proc. Natl. Acad. Sci. USA 84, 7644, 1987.

Ho DD, Pomeranz JR, Kaplan JC. Pathogenesis of infection with human immunodeficiency virus. N.Eng. J. Med. 317, 278, 1987.

Klatzmann DE, Champagne E, Chouret S, Grunest J, Guetart D, Hercend T, Gluckman JC, Montagnier L. T lymphocyte T4 molecule behaves as the receptor for human retrovirus LAV. Nature 312, 767, 1985.

Klatzmann D, Montagnier L. Approaches to AIDS therapy. Nature 319, 10, 1986.

Lifson JD, Feinberg MD, Rabin SR, et al. Induction of CD4 dependent cell fusion by HTLVIII/LAV envelope glycoprotein. Nature 323, 725, 1986.

Maddon PJ, Dagleish AG, McDouglas JS, Clapham DR, Weiss RA, Axel R. The T4 gene encodes the AIDS virus receptor and is expressed in the immune system and the brain. Cell 47, 333, 1986.

Pahwa S, Pahwa R, Saxinger C, Gallo RC, Good RA. Influence of HTLVIII/LAV on functions of human lymphocytes: evidence for immunosuppressive effects and polyclonal B cell acrivation by banded viral preparations. Proc. Natl. Acad. Sci. USA 82, 8198, 1985.

Ragona G, Sirianni MC, Soddu S, Vercelli B, Sebastiani G, Picchi M, Aiuti F. Clin. exp. Immunol. 66, 17, 1986.

Salahuddin SZ, Ablashi DV, Hunter EA, Gonda MA, Sturzenegger S, Markham PD, Gallo RC. HTLV-III infection of EBV genome positive B lymphoid cells with or without detectable T4 antigens. Int. J. Cancer 39, 1987.

Shaw GM, Harper ME, Hahn BH. HTLV-III infection in brains of children and adults with AIDS-encephalopathy. Science 227, 177, 1985.

Stewart SJ, Fujimoto J, Levy R. Human T lymphocytes and monocytes bear the same Leu3(T4) antigen. J. Immunol. 136, 3773, 1986.

Stricker RB, Abrams DI, Corash L, Shuman MA. Target platelet antigen in homosexual men with immune thrombocytopenia. N. Engl. J. Med. 313, 1375, 1985.

Stricker RB, McHugh TM, Moody DJ, Morrow WJW, Strikes DP, Shuman MAS, Levy JA. An AIDS-related cytotoxic autoantibody reacts with a specific antigen on stimulated CD4+ T cells. Nature 327, 710, 1987.

Tomar RH, John PA, Hennig AK, Kloster B. Cellular targets of antilymphocyte antibodies in AIDS and LAS. Clin. Immunol. Immunopathol. 37, 37, 1985.

Weiss RA. Receptor molecule blocks HIV. Nature 331, 15, 1988.

# T-CELLS IN RESPIRATORY SYNCYTIAL VIRUS (RSV) INFECTION

B.A. Askonas, P. Openshaw and M. Cannon

National Institute for Medical Research
The Ridgeway, Mill Hill
London NW7 1AA

## INTRODUCTION

T-cells are an important arm of our anti-viral defense, and generally are required to clear virus once a virulent infection has occurred. In certain virus infections T-cells can cause immunopathology. For development of more rational vaccines it becomes important to acquire knowledge about the role of T-cell subpopulations in infection, their viral recognition as well as the induction requirements of protective T-cell responses.

RSV is a paramyxovirus belonging to the genus Pneumovirus. RSV infection occurs regularly in winter and causes serious disease in infants and also in old people. It is the most common single cause of admission to hospital of otherwise normal infants in their first year of life (for example review by Stott & Taylor[1]). Most RSV deaths are in children with cardiac abnormalities and these children are more susceptible to the most severe bronchiolitis. In view of its seriousness for infants, RSV has been designated for vaccine development by WHO and there is a special WHO program for respiratory paramyxovirus infections of childhood. Maternal antibodies appear to protect against RSV only for the first 6-8 weeks and this contrasts far longer term protection against many other viral infections; although high levels of certain types of neutralising antibodies can be protective[2], overall there is no correlation between antibody titres and protective immunity. Moreover, early trials with formaldehyde inactivated vaccines (which induced neutralising serum antibodies) failed to protect and resulted in exacerbated disease and high morbidity when vaccinated children were exposed to natural infection[3].

Adults get reinfected with RSV every few years although neutralising serum antibodies are present.

There are indications that T-cells are important in recovery from RSV infection in as far as T-cell deficient individuals suffer from severe RSV infection, and the virus infection persists in athymic nude mice[4]. Isaacs et al.[5] find that infants with severe bronchiolitis show very low anti RSV CTL activity in culture from PBL. On the other hand the strong lymphoid cell infiltration observed in the lung in bronchiolitis may well contribute to the immuno-pathology.

These problems prompted us to study T-cell mediated immune responses, their role in RSV infection as well as the viral recognition patterns by cytotoxic T-cells (CTL) and T-helper cells ($T_H$). Although it is possible to study recognition of viral proteins by T-cells in vitro in both mouse and man, we still require animal models to examine the function of T-cell subsets in vivo in infection. Knowledge in this respect is essential to permit development of safe and protective vaccines against RSV. Since the age group of children most susceptible to bronchiolitis is 8-24 weeks, it is unlikely that vaccination using any attenuated live virus preparation will fill the bill. We therefore need an animal model in order to define protective immune responses, to test the immunogenic potential of vaccines, and the immune recognition of the virus and its components. We have used a mouse model for our studies; mice support RSV replication in the lungs after intranasal infection with some human isolates of RSV (e.g. the A2 strain), with histological changes not dissimilar to human pulmonary infection[6]. The infected mice show little or no overt illness, and lethal RSV infection in normal mice has not been demonstrated even with high doses of infectious virus. After RSV infection of BALB/c mice, virus titres in the lung peak days 4-5; imunocompetent mice clear the virus by day 10[6], while in athymic or irradiated mice a persistent infection is established.

RSV specific CTL are detected in lung by day 5 of infection[7], at an earlier time than virus specific antibodies which appear from day 9 onwards. By 2 weeks of infection memory $T_H$ and CTL cells can be recovered from the spleen. Memory CTL in spleen can be stimulated in vitro with RSV infected syngeneic spleen cells as antigen presenting cells and after 5 days  culture T-cell mediated cytotoxicity is generated and the ability to lyse MHC compatible RSV infected target cells (tumour cells and persistently infected fibroblasts) can be assayed[8]. T-cells from normal mice do not mount a primary RSV-specific CTL response _in vitro_, hence the CTL response generated following infection of hosts reflects the level of memory CTL. $T_H$ memory cells can also be primed by i.n. infection of mice, and antigen stimulation _in vitro_ induces the release of lymphokines such as IL-2 or IL-3. The amount of lymphokine secreted can be assayed on IL dependent cell lines and reflects the level of $T_H$ cell memory[9]. We have used these methods to examine T-cell memory and the RSV proteins that can be recognised by primed T-cells.

VIRAL RECOGNITION PATTERNS OF CTL and $T_H$ CELLS IN MOUSE AND MAN

We collaborated with Professor G. Wertz, Dr. A. Ball and their associates who had constructed recombinant vaccinia viruses (VV) with inserted genes coding for individual RSV proteins[10,11]. It is very difficult to purify RSV components in sufficient quantity for the required assays. Moreover purified proteins are poor in forming target cells for CTL, and with some exceptions antigen presentation and induction of CTL require intracellular formation of the viral proteins after virus infection or infection with recombinant VVs resulting in expression of the RSV protein they code for. Fig. 1 shows the cytotoxicity of class I MHC restricted CTL from BALB/c mice vaccinated with the different recombinant VV containing A2 RSV genes[12]; the spleen cells from these mice were stimulated _in vitro_ with RSV infected spleen cells and cytotoxicity assayed using BCH4 target cells (BALB/c fibroblasts) persistently infected with Long strain RSV (kindly donated by Dr. B. Fernie). In the mice, FVV results in a far stronger CTL responses than NVV. In restimulated human CTL lines, the reverse is true. Our overall results for RSV can be summarised as follows: the highly glycosylated attachment glyco-protein (G) and 1B proteins are not significantly recognised by CTL in mouse[12] or man[13,14] and only marginally by $T_H$ cells. The RSV nucleoprotein (N) and fusion (F) protein are target antigens for CTL and are also recognised by $T_H$ cells[9]. Of the other RSV VV tested IA and IC give a marginal T-cell response. We realise that CTL can recognise additional RSV proteins, since we have CTL lines and one CTL clone specific for RSV, and yet these CTL show no recognition of VV`s tested (negative for F, N, G, IA, IB, & IC and partial matrix proteins). Thus further target antigens need to be defined. A major proportion of memory CTL primed by infection do not discriminate between the subgroups of RSV distinguishable by serology[15]. It should be pointed out that this paramyxovirus, with its strong fusion protein, unlike influenza virus, can induce CTL memory _in vivo_ even following u.v. irradiation of the virus. Similarly u.v. irradiated RSV is able to form target cells for CTL[8] as had been shown previously for Sendai, another paramyxovirus[16].

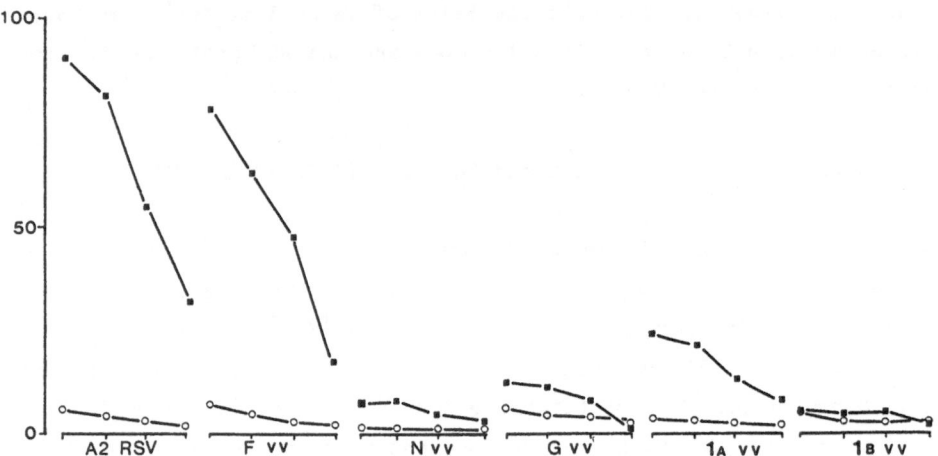

Fig. 1. CTL recognition of RSV-infected targets from mice immunised with
recombinant vaccinia viruses.
Mice were primed by i.n. infection with A2 RSV or one of the
recombinant vaccinia viruses by scarification at the base of the
tail. One month later, spleen cells were stimulated in vitro for 5
days with A2 RSV infected spleen cells. The generation of CTL was
tested using RSV persistently infected BCH4 cells[24] as targets (■)
or uninfected BALB/c fibroblasts (O) as control target cells.
Horizontal axis, from left to right: K:T ratios for each culture
80:1, 40:1, 20:1, 10:1.
Vertical axis: % specific lysis of $^{51}$Cr labelled targets in a 4
hour assay.

CYTOTOXIC T-CELLS CAUSE PATHOLOGY IN RSV INFECTED MICE

We have shown[4] that intranasal (i.n.) RSV infection of athymic nu/nu
mice or mice immunosuppressed by irradiation results in persistence of RSV
infection. Transfer of T-cells (primed T-memory cells) from immunocompetent mice that had recovered from RSV infection resulted in the
clearance of RSV from the lungs of persistently infected hosts[4]. We
therefore wished to know which subpopulation of T-cells was responsible

for this viral clearance. We selected CTL lines and cloned RSV specific
CTL to examine their effect on lung virus replication by transferring the
CTL i.v. into irradiated mice at the time of RSV infection . We were
surprised to find that RSV-infected mice given $3 \times 10^{6}$ or more CTL became

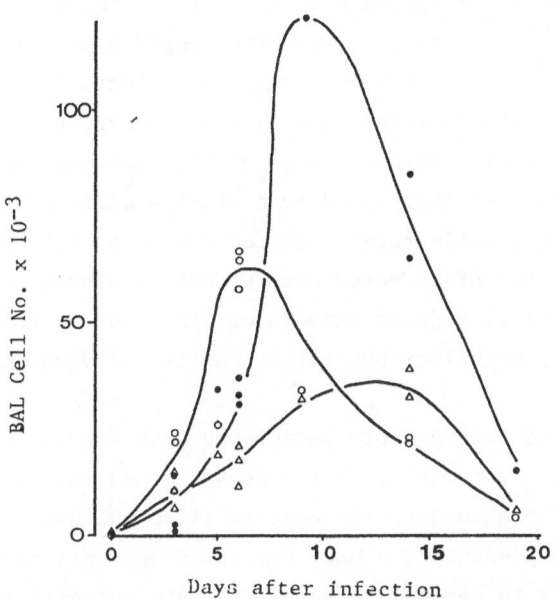

Fig. 2. Phenotype and number of lymphocytes ($X10^{-3}$) recovered
per mouse at various times after infection i.n. with
A2 strain RSV. Broncho-alveolar lavage was performed
on groups of 4 mice days 3,5,6,9,14 & 19 and cells
pooled. The recovered cells were stained for CD4
(anti-L3T4, mAb GK1.5 directly coupled with
phycoerythrin) and CD8 (anti-Lyt2, mAb 3.168 directly
coupled to FITC) and analysed by 2 colour flow
cytometry. Cells which are CD4⁻8⁻ peak at day 5 (O);
these are surface Ig⁻ and may possibly have NK like
activity. Thereafter, CD8⁺4⁻T cells (●) outnumber
CD4⁺8⁻T cells (△).

very ill with respiratory distress and many died[17]. No virus was detected
in the lungs of surviving mice tested on day 5. Similar clinical effects
but lower mortality were noted in RSV-infected immunocompetent recipients
of CTL.

Bronchial lavage (BAL) fluid samples showed the presence of RBC, indicating the development of haemorrhagic pneumonitis after transfer of CTL. In some of the T-cell recipient mice, polymorphonuclear cells could be recovered in large numbers in BAL fluid. CTL transfer into normal mice has no effect on lung pathology and the PMN and RBC appear in BAL only after the transfer of RSV specific CTL into RSV infected mice. Intranasal RSV infection alone results in increases of lymphoid cells in BAL (Fig.2), with a particularly striking efflux of CD8[+] T-cells from the lung[18] but does not cause lung haemorrhage. The T-cell efflux presumably reflects earlier inflammatory infiltration into the infected lung tissue. Transfer of fewer CTL ($\leqslant 10^6$) also cleared the virus, although more slowly, and the animals did not show clinical symptoms. We thus noted quantitative differences in the induction of immunopathology, while relatively few CTL resulted in the clearance of virus from the lungs of infected mice without enhancing pathology or respiratory distress, greater numbers of CTL exacerbated the clinical symptoms and lung pathology but still resulted in viral clearance[17].

These results contrast our experience with a mouse model of influenza type A infection. Our earlier experiments showed that i.v. transfer of influenza specific cloned CTL is beneficial to the host. Although infection is not prevented, cloned CTL speed up viral clearance from the lungs and trachea in immunocompetent influenza infected host mice[19]; the virus induced lung pathology recovers far more rapidly (almost normal by day 4) than in the absence of the CTL transfer (Taylor, McKenzie and Askonas, manuscript in preparation). This requires $6 \times 10^6$ cloned CTL, or more.

What is the cause of this interesting difference in the role of CTL in two respiratory virus infections? The most likely explanation relates to the difference in the spread of the two infectious viruses in the hosts. While influenza infects the larger air passages, RSV spreads to the lining of the alveoli and bronchioli - where any cell damage would be expected to have serious consequences because of their small dimensions. Bronchio-litis occurs frequently in RSV infection but is not a feature of influenza. Overall, mice look ill after influenza and human influenza virus replicates efficiently in mice. RSV shows a lower level of infectious virus in lung, the mice show no symptoms of sickness unless CTL are transferred. Thus although RSV is a low grade infection in mice, our experiments show that it provides a model for the study of RSV and T-cell induced pathology. RSV specific CD4[+]T-helper cells have not been cloned as yet and their function in RSV infection awaits study.

It has become clear that each virus infection is a law unto itself and protective immune responses need to be examined in each case. Strong immune responses also can result in immunopathology mediated by CTL and/or $T_H$ cells or antibodies, depending on the site of infection, cell damage and perhaps other factors. CTL specific for LCMV cause brain pathology and death following intracerebral infection, but are effective in clearing virus peripherally[20, 21, 22]. Local HSVI infection of skin, requires herpes specific $T_H$-cells for clearance, while Lyt$2^+$ CTL are important in preventing herpes infection of the nervous system[23]. It is not possible to generalise from one virus infection to another as to which types of immune responses are inducible or desirable following vaccination.

CONCLUSIONS

We have been able to examine the recognition by T-cell subpopulations of the components of RSV. By infecting target cells or APC with recombinant vaccinia virus, into which RSV genes coding for individual viral proteins are inserted, or priming mice with these recombinant vaccinia viruses, cytotoxic T-cells (CTL) in mouse and man were shown to recognise the RSV nucleoprotein and fusion protein with a very minor part of the repertoire seeing other recombinants presently available. Other virus target antigens still await definition. The highly glycosylated attachment G protein is not recognised. A similar pattern of recognition was found for murine T-helper cells, while $T_H$ cells in humans have not yet been analysed. Since fusion protein can be recognised by CTL, $T_H$ and B-cells, it may become a vaccine candidate if it can be prepared in undenatured form.

To examine whether CTL are responsible for immunopathology or protective effects, we have selected long term mouse CTL lines (about 80% Lyt$2^+$ and 20% CD$4^+$) and 2 CTL clones specific for RSV. These Lyt$2^+$ cells were transferred into RSV infected, immunodepressed or immunocompetent host BALB/c mice. CTL in small numbers ($\sim 3 \times 10^5$) clear the virus gradually and completely (by d 10) and do not cause respiratory distriss. However, larger numbers of CTL ($10^6$ or more) lead to severe respiratory symptoms in infected mice and some mortality. Bronchial lavage shows the presence of RBC and (in some of the T-cell recipients) PMN infiltration. In survivors, RSV is cleared from the lung within 5 days despite induction of immunopathology.

These results contrast our previous studies with influenza, where large numbers of CTL limit virus replication _in vivo_ and reduce virus pathology. Thus CTL behave very differently in two respiratory infections and this is likely to be attributable to the spread of infection by RSV to the narrow air passages of the lungs. The quantitative differences in the effects of CTL on RSV infection also show the importance of achieving a balanced immune response in vaccination. We do not know yet the role of T-helper cells _in vivo_ in RSV infection and this awaits analyses to see whether $T_H$ cells are able to clear virus or not and whether these also induce immunopathology.

## ACKNOWLEDGEMENTS

We are most grateful to Professor G. Wertz and Dr. A. Ball and their associates for their collaboration and the RSV-recombinant vaccinia constructs.

## REFERENCES

1. E.J. Stott, and G. Taylor, Respiratory syncytial virus, _Arch. Virol._, 84:1 (1985).
2. G.A. Prince,V.G. Hemming, R.L. Horswood, P.A Baron, and R.M. Chanock, Effectiveness of topically administered neutralizing antibodies in experimental immunotherapy of respiratory syncytial virus infection in cotton rat, _J. Virol._, 61:1851 (1987).
3. P.F. Wright, R.B. Belshe, L.P. Kim, L.P. Van Voris and R.M. Chanock, Administration of a highly attenuated live respiratory syncytial virus vaccine to adults and children, _Infectn. Immun._, 37: 397 (1982).
4. M.J. Cannon, E.J. Stott, G. Taylor and B.A. Askonas, Clearance of persistent respiratory syncytial virus infections in immunodeficient mice following transfer of primed T cells, _Immunol._, 62:133 (1987).
5. D. Isaacs, C.R.M. Bangham and A.J. McMichael, Cell-mediated cytotoxic response to respiratory syncytial virus in infants with bronchiolitis, _Lancet_, 2(8562):769 (1987).
6. G. Taylor, E.J. Stott, M. Hughes and A.P. Collins, Respiratory syncytial virus infection in mice, _Infectn. Immun._, 43:649 (1984).
7. G. Taylor, E.J. Stott and A.J. Hayle, Cytotoxic lymphocytes in the lungs of mice infected with respiratory syncytial virus, _J. Gen. Virol._, 66:2533 (1985).
8. C.R.M. Bangham, M.J. Cannon, D.T. Karzon and B.A. Askonas, Cytotoxic T-cell response to respiratory syncytial virus in mice, _J.Virol._, 56:55 (1985).
9. P.J.M. Openshaw, R.M. Pemberton, L.A. Ball, G.W. Wertz and B.A. Askonas, Helper T-cell recognition of respiratory syncytial virus in mice, _J. Gen. Virol._, 69:305 (1988).
10. L.A. Ball, K.K.Y. Young, K. Anderson, P.L. Collins and G.W. Wertz, Expression of the major glycoprotein G of human respiratory syncytial virus from recombinant vaccinia virus vectors, _Proc. Nat. Acad. Sci. USA._, 83:246 (1986).
11. G.W, Wertz, E.J.Stott, K.K.Y. Young, K. Anderson, and L.A. Ball, Expression of fusion protein of human respiratory syncytial virus from recombinant vaccinia virus vectors and protection of vaccinated mice, _J. Virol_. 61:293 (1987).

12. R.M. Pemberton, M.J. Cannon, P.J.M. Openshaw, L.A. Ball, G.W, Wertz and B.A. Askonas. Cytotoxic T cell specificity for respiratory syncytial virus proteins: fusion protein is an important target antigen, J.Gen. Virol., 68:2177 (1987).

13. C.R.M. Bangham and A.J. McMichael, Specific human cytotoxic T cells recognise B cell lines persistently infected with respiratory syncytial virus, Proc. Nat. Acad. Sci. USA., 83:9183 (1986).

14. C.R.M. Bangham, P.J.M. Openshaw, L.A. Ball, A.M.Q. King, G.W. Wertz and B.A. Askonas, Human and murine cytotoxic T cells specific to respiratory syncytial virus recognise the viral nucleoprotein (N), but not the major glycoprotein (G) expressed by vaccinia virus recombinants, J. Immunol., 137: 3973 (1986).

15. C.R.M. Bangham and B.A. Askonas, Murine cytotoxic T cells specific to respiratory syncytial virus recognize different antigenic subtypes of the virus, J. Gen. Virol., 67:623 (1986).

16. U. Koszinowski, M.J. Gething and M. Waterfield, T cell cytotoxicity in the absence of viral protein synthesis in target cells, Nature (Lond.)., 267:160 (1977).

17. M.J. Cannon, P.J.M. Openshaw and B.A. Askonas, Cytotoxic T-cells clear virus but augment lung pathology in mice infected with respiratory syncytial virus, J. Exp. Med., 168: (1988) in press.

18. P.J.M. Openshaw, Flow cytometric analysis of pulmonary lymphocytes from mice infected with respiratory syncytial virus, Clin. Exp. Immunol., (1988) in press.

19. P.M. Taylor and B.A. Askonas, Influenza nucleoprotein specific cytotoxic T-cell clones are protective in vivo, Immunol., 58:417 (1986).

20. J.A. Byrne and M.B.A. Oldstone, Biology of cloned cytotoxic T lymphocytes specific for lymphocytic choriomeningitis virus, J. Immunol., 136:698 (1986).

21. J. Baenziger, H. Hengartner, R.M. Zinkernagel and G.A. Cole, Induction or prevention of immunopathological disease by cloned cytotoxic T cell lines specific for lymphocytic choriomeningitis virus, Eur. J. Immunol., 16:387 (1986).

22. T.P. Leist, S.P. Cobbold, H. Waldmann, M. Aguet and R.M. Zinkernagel, Functional analysis of T lymphocyte subsets in antiviral host defense, J. Immunol., 138:2278 (1987).

23. A.A. Nash, A. Jayasuriya, J. Phelan, S.P. Cobbold, H. Waldmann and T. Prospero, Different roles for L3T4+ and Lyt2+ cell subsets in the control of an acute herpes simplex virus infection of the skin and nervous system, J. Gen. Virol., 68:825 (1987).

# NERVES AND NEUROPEPTIDES IN THE REGULATION OF MUCOSAL IMMUNITY

John Bienenstock, Ken Croitoru, Peter B. Ernst, and
Andrzej M. Stanisz

Department of Pathology and the Intestinal Disease
Research Unit, McMaster University Health Sciences Centre
Hamilton, Ont., Canada L8N 3Z5

The subject of mucosal immunity continues to generate considerable interest. It is clear that the presentation of antigen to the mucosa, especially in the form of replicating virus or live attenuated organisms produces a local mucosal immune response. This response is better than if the antigens are presented in other forms, and certainly better than if they are presented via other routes[1-3]. Thus, the experiments performed some fifty years ago on immunization of volunteers against dysentery by oral immunization[4] with Shigella organisms, which produced incomplete but definite protection amongst those vaccinate, clearly hold up.

With the understanding that IgA was the predominant immunoglobulin of secretions have come attempts to explain this phenomenon. Not surprisingly, since 90% of the B cells synthesizing immunoglobulin, e.g. in the intestinal tract, appear to be making IgA, emphasis has been placed on this immunoglobulin molecule. Most studies involved in characterizing immune responses at local mucosal surfaces have focused on IgA, and it has become dogma that this is the most important mechanism of protection at mucosal surfaces. This indeed may be true, but it has not been categorically proven. Indeed, it would be difficult to do so. There are very few clear experiments which support the thesis, by passive transfer of immunoglobulin or other means, that IgA is the predominant form of protection at mucosal surfaces. It is obvious that IgA must play an important role, but whether it is the most important role in terms of potentially invasive organisms, still remains to be established. For example, the cellular component of mucosal immunity has received considerably less attention, and it is less well known and understood how cells may involve themselves in provision of protection, at mucosal surfaces. Furthermore, the interaction between immune cells and the epithelium, and the regulation of so-called non-specific protective mechanisms is both poorly studies and understood.

In view of the fact that mucosal immunity and its principles have been well reviewed and discussed elsewhere, and such reviews are readily available to the interested reader, we will review some potentially interesting more recent developments in this field.[1-3,5,6]

It is well established that lymphoid tissues are innervated, and that large amounts of neuropeptides are found in mucosal tissues.[7] Interest has been elicited by observations of increased levels of neuropeptides such as vasointestinal polypeptide (VIP), in the involved tissue, in inflammatory bowel disease.[8] We ourselves have been interested in a relationship between neurotransmitters, nerves and immunity in mucosal tissues.[9,10]

In this short review, we will attempt to cover this aspect of the potential regulation of mucosal immune events, since it is an area which has received relatively little attention over the years, and is one which may play an important part in protection. Furthermore, with the recent observations by Ader and Cohen, and subsequently by others, it has become[11] clearer that the central nervous system (CNS) might influence immunity. Other experiments have suggested that the immune system itself might influence the CNS, but we will not discuss this important area herein.

In order to put our studies into a better perspective, while the reader should go elsewhere for details, we will very briefly summarize some principles of mucosal immunity before discussing the possible role of nerves and neuropeptides in this system.

PRINCIPLES OF MUCOSAL IMMUNITY

## IgA

As stated before, IgA is the predominant immunoglobulin found in mucosal secretions. It is found in the so-called secretory forms, usually covalently coupled to secretory component, which acts as the transport protein for the dimeric IgA secreted by plasma cells in the tissues, across the epithelial cells to their apical surfaces. At the cell surface, the molecule is cleaved leaving a portion of the secretory component behind in the membrane, and releasing the secretory IgA molecule. The molecule in this form is highly resistant to proteolysis and therefore particularly suited to the potentially degradative surroundings in which it finds itself. The function of the IgA molecule is primarily to interfere with binding to surface sites, and thus prevent bacterial colonization and viral attachment. Once bound, IgA also seems able to prevent complement lysis and furthermore may cooperate with cells (particularly activated macrophages and/or cells of the suppressor/cytotoxic T-cell lineage) to[12] produce antibacterial antibody-dependent cell-mediated cytotoxicity.

## Lymphoid Aggregates

The initiation of the IgA response is either in the aggregates of lymphoid tissue in the gut (GALT) or the lung (BALT) or in the draining lymph nodes. Evidence has been adduced to suggest that the epithelium itself, especially if it expresses Ia molecules, as it does when involved in inflammatory events, may process antigen so as to present it in a classical form, to T cells and B cells.[13] The extent of processing by the epithelium as compared to the lymphoid aggregates is undetermined at the present time. There are two not antagonistic views which state that initiation of the immune response therefore can occur in either the lymph node or the lymphoid aggregate, but it is in the lymphoid aggregates particularly that amplification of the system occurs (secondary immune response).

The M cells overlying the follicular aggregates have the selective capability of phagocytosing antigen and passing this into the follicles. Lymphocytes migrate from the lymphoid tissue to the draining lymph nodes and through the lymphatics to the blood. There seems to be a selective localization of lymphocytes, derived from mucosal tissues, within mucosal tissues, especially those from which they were originally derived.[14] This has produced the theory of 'a common mucosal immune system' in which cells, for example, of the IgA lineage from the mesenteric lymph node, have a tendency to selectively populate mucosal tissues such as respiratory and GI tracts distal to their point of origin.[15] More recently, it seems that this selective mucosal localization is on the basis of molecules expressed on lymphocyte cell surfaces, with complementary receptors expressed on the[16] endothelium of post-capillary vessels in mucosal tissues. What regulates the expression of these molecules is currently under intensive examination.

## T cell Regulation

The lymphoid system of mucosal tissues is under intense immune regulation. On the positive side, this is through the generation of $T_7$ helper cells which promote isotype-specific immunoglobulin synthesis,[17] or other T cells known as switch cells,[18] which promote the switch at the gene level from expression of IgM to IgA. A number of cytokines are now known to be involved in this process of IgA synthesis involving signalling between T cells and B cells, and IL-4 and IL-5 are particularly involved in this promotion of IgA synthesis.

A third set of T cells appear to be involved in promotion of IgA immune responses. These are a part of a cellular network known as the contra-suppressor pathway.[19] These T cells induce suppressor cells which in turn suppress suppression of IgA, thus rendering help.

Obviously, this positive regulation of the mucosal immune system must be carefully balanced by a negative effect, and down-regulated. This local suppressor concept extends to the whole concept of oral tolerance in which antigen presented orally promotes depression or suppression of the immune response in peripheral systemic tissues, while preserving mucosal responses.[20] While it is generally agreed that this is a very complex event, T suppressor cells generated in the mucosal lymphoid tissue are clearly involved.[21]

## Epithelial Leukocytes

Within the epithelium lie a population of leukocytes most of them granulated.[22] These cells are above the basement membrane and have been termed intraepithelial leukocytes or lymphocytes (IEL). They represent an enormous compartment of lymphoid tissue which, in the intestine alone, occupy a volume the size of an average spleen. Many of the cells bear the CD8 phenotype most commonly associated with suppressor/cytotoxic T cells and lack pan-T markers in mice, rats and humans. It is possible that many of these leukocytes are indeed not T cells and some evidence has recently come forward suggesting that they may be more related to the myeloid lineage than to T cells (Croitoru and Ernst, unpublished data). Nevertheless, within this population are lymphocytes which have been shown to possess natural killer activity against tumors and viruses, contain precursor cells for cytotoxic T lymphocytes as well as the mature cells themselves, and mitogen activated cytotoxic cells.[23-25] They also contain a population of progenitor cells which under the influence of IL-3 differentiate into mast cells and have the highest mast cell precursor frequency in the body, higher even than that found in bone marrow.

## Mucosal Nerves

The nervous network within mucosal tissues is very extensive.[26] It has been calculated that the number of nerve cell bodies present in the gastrointestinal tract is at least equivalent to that found in the spinal cord. Neurotransmitters are found in very large amounts in these tissues, particularly substance P (SP), VIP and somatostatin (SOM). In addition, nerves predominate in T cell zones of lymphoid aggregates, where they contain both neuropeptides, and especially sympathetic neurotransmitters, such as noradrenaline. The blood vessels are surrounded by nerve plexuses, particularly at the level of post-capillary venules which are sites of lymphocyte traffic out of the circulation. Recently our own group has demonstrated that intestinal mucosal mast cells appear to be associated with enteric nerves lying in very close apposition to them.[27] In view of the network of enteric nerves and the fact that twigs from these nerves pass into the epithelium in respiratory and GI tracts, it is obvious that neuropeptides released at these nerve endings could greatly influence the function and activity of lymphocytes.

## Neuropeptide Effects on Lymphocytes

Many different neuropeptides appear _in vitro_ to have functional effect on immune cells such as B lymphocytes, T lymphocytes, natural killer cells macrophages and mast cells.[28-32] The neuropeptides which have received the greatest attention are SP, SOM, and VIP.

_In vitro_ these appear to have so-called bidirectional effects, having different effects at very low concentration to those found at higher concentrations, but still within physiological limits. Furthermore, the culture conditions, dose, time of exposure, organ derivation, etc., all appear to exert effect on the responses of such cells to neuropeptides. What is more, these effects appear to be cell-cycle dependent, and this may account for the developing confusion which is currently found when delving into this literature. It should also be noted that if any effects of neuropeptides are found _in vivo_, which reflect those found _in vitro_, this does not necessarily mean that the same events are occurring as _in vitro_, since _in vivo_ a multiplicity of signalling events obviously occurs between many organs and cells, which is only reflected, as the balance of these events, in the read-out system.

## Neuropeptide Receptors

Specific receptors for SP are seen on human lymphocytes, particularly T helper cells. In the mouse however, high affinity receptors for SP are found both on B and T cells although lymphocytes from Peyer's patches have from 5 to 7 times as many receptors as those from spleen.[31] Receptors for SOM are seen on both murine T and B cells but again a greater percentage of cells from Peyer's patches bind SOM, and here the phenotype is predominantly Lyt2[+], indicating the predominance of the suppressor cytotoxic phenotype of the SOM-bearing cell.

## Neuropeptide Effects in vitro

Somatostatin inhibits the thymidine uptake of both B and T cells when stimulated by mitogen (Con A, PHA or pokeweed mitogen).[30] This is interesting in view of the Lyt2[+] phenotype-bearing cell predominance described above and may explain the mode of action of SOM. In terms of immunoglobulin synthesis, SP tends to stimulate antibody synthesis, while SOM inhibits it, as measured either by a radioimmunoassay or reverse plaque forming assay. Most interestingly, SP produces an apparent organ-specific (Peyer's patches) synthesis of IgA in seven day cultures in the presence of Con A.[30]

## Neuropeptide Effects on Cytotoxicity

VIP has been shown _in vitro_ to increase human NK activity.[33] This again was bidirectional since preincubation with VIP caused stimulation, whereas coincubation caused an inhibitory effect, not dependent on adenylate cyclase activation. Our own recent observations have shown that SP stimulated the NK activity of IEL, while producing a minimal effect on NK activity of splenic lymphocytes.

## Neuropeptide Effects in vivo

In order to obtain some information on whether these _in vitro_ phenomena had any _in vivo_ significance, we have performed experiments in which substance P was administered for seven days via subcutaneously implanted mini-osmotic pumps. This allowed serum levels of substance P to be attained, at a level two to three-fold higher than normal.[6] Cells freshly isolated from Peyer's patches and spleen from such animals, as opposed to controls, showed an increased proliferative capacity, and furthermore synthesized more immunoglobulin, appropriate to the organ derivation as we had seen

previously in vitro under the influence of Con A.  On the other hand, SOM,
when infused in a similar fashion had an opposite effect to that previously
seen in vitro, namely it now caused proliferation of cells and increased
antibody production.  This suggests agains that neuropeptides may have dual
or bidirectional effects as have been noted before, and extensively documented in the endocrine literature (for example, stimulation versus suppression of secretion).

Infusion of SP appeared to regularly promote the increased NK activity
of cells isolated from the IEL compartment compatible with the in vitro
findings.

NEUROPEPTIDES, NERVES AND MAST CELLS

Since we and others have shown that neuropeptides can cause degranulation of mast cells,[34] and since mast cells are involved in axon reflexes
in the generation of so-called neurogenic inflammation,[35] we have examined
relationships between mast cells, nerves, and neuropeptides.  Extensive
work has shown that only substance P causes degranulation of isolated mucosal
mast cells, which differ in a number of ways from their conventional connective tissue counterparts.  A very careful morphometric study furthermore
has revealed that mucosal mast cells lying immediately under the epithelium
are apparently selectively associated with enteric nerves and that this is
not a random finding.[9]  Ultrastructural observations have shown very
frequent intimate membrane/membrane contact between nerves and mast cells,
and similar but not so careful studies have demonstrated such contacts before
in a variety of tissues.[37]

We have now performed very extensive observations in vitro on mast
cell/nerve interactions involving Ussing chambers.  Electrophysiological
measurements have shown that mast cells are involved with nerves in the
physiological regulation of epithelial cell function and integrity.  Other
observations have led us to the conclusion that this seems to apply also
in vivo since capsaicin treatment at birth, which depresses or ablates the
sensory afferent system which contains substance P, has profound effects in
these model systems.  We have concluded elsewhere that mast cells and nerves
are involved as regulatory units in a variety of local regulatory events in
the mucosal tissues of the lung and intestine.[9,10]

CONCLUSIONS

There is a large body of literature which suggests that there are
neuroendocrine influences on the immune system and this idea is by no means
new.[10,38]  Considerable evidence exists for hormonal influence on many
different aspects of the immune response.  Corticosteroids have been well
characterized in this respect, and it is known that sex hormones, independent
of the gender of the animal, appear to have a significant influence on the
synthesis and secretion of IgA in the eye.[39]  Secretion of intestinal IgA
has been shown in other experiments to be modulated by cholecystokinin (CCK)
infusions which were blocked by atropine or CCK antagonists.[40]  The well
known effects of some of these hormones and neuropeptides on post-capillary
venules,[14] on lymphoblast migration[41] and expression of secretory component
in sex hormone-dependent situations should also be emphasized.[42]

It should hardly surprise us therefore that neuropeptides have profound
effects on lymphocytes, as well as their circulation and migration.  It now
becomes even more clear that an examination of how neuropeptides regulate
these events, and how these can be harnessed to promote mucosal immunity,
must become the target for further investigation.  It is reasonable to hope
at the present time that these types of studies will allow the development
of some new approaches to the regulation of mucosal immunity.

REFERENCES

1. J. Bienenstock and A.D. Befus, Mucosal immunology, _Immunology_ 41: 249-270 (1980).
2. J. Bienenstock and A.D. Befus, The gastrointestinal tract as an immune organ, _in_: "Gastrointestinal Immunity for the Clinician", R.G. Shorter and J.B. Kirsner, eds., Grune and Stratton, New York, pp 1-22 (1985).
3. _In_: "Local Immune Responses of the Gut", T.J. Newby and C.R. Stokes, eds, CRC Press, Florida, pp 1-248 (1984).
4. A. Besredka, Local Immunization, Williams and Wilkins, Baltimore, MD, p 1 (1927).
5. C.O. Elson, M.F. Kagnoff, C. Fiocchi, A.D. Befus and S. Targan, Intestinal immunity and inflammation: recent progress, _Gastroenterology_, 91: 746-768 (1986).
6. J. Mestecky and J.R. McGhee, Immunoglobulin A (IgA): Molecular and cellular interactions involved in IgA biosynthesis and immune response, _in_: "Advances in Immunology" Vol. 40, F.J. Dixon, ed., Academic Press, pp 153-245 (1987).
7. D.L. Felten, S.Y. Felten, S.L. Carlson, J.A. Olschowka and S. Livnat, Noradrenergic and peptidergic innervation of lymphoid tissue, _J. Immunol._, 135: 755-765s (1985).
8. A.E. Bishop, J.M. Polak, M.G. Bryant, S.R. Bloom and S. Hamilton, Abnormalities of vasoactive intestinal polypeptide-containing nerves in Crohn's disease, _Gastroenterology_ 79: 853-860 (1980).
9. R. Stead, J. Bienenstock and A.M. Stanisz, Neuropeptide regulation of mucosal immunity, _Immunol. Rev._ 100: 333-359 (1987).
10. J. Bienenstock, M. Perdue, A. Stanisz and R. Stead, Neurohormonal regulation of gastrointestinal immunity, _Gastroenterology_ 93: 1431-1434 (1987).
11. R. Ader and N. Cohen, CNS-immune system interactions: conditioning phenomena, _Behavioural and Brain Sci._ 8: 379-395 (1985).
12. A. Tagliabue, L. Nencioni, L. Villa, D.F. Keren, G.H. Lowell and D. Boraschi. Antibody-dependent cell-mediated antibacterial activity of intestinal lymphocytes with secretory IgA, _Nature_ 306: 184-186 (1983).
13. L. Mayer and R. Shlien, Evidence for function of Ia molecules on gut epithelial cells in man, _J. Exp. Med._ 166: 1471-1483 (1987).
14. J. Bienenstock, A.D. Befus, M. McDermott, S. Mirski and K. Rosenthal, Regulation of lymphoblast traffic and localization in mucosal tissues with emphasis on IgA, _Fed. Proc._ 42: 3213-3217 (1983).
15. M.R. McDermott and J. Bienenstock, Evidence for a common mucosal immunologic system. I. Migration of B-immunoblasts into intestinal respiratory and genital tissues, J. Immunol. 122: 1892-1898 (1979).
16. E.C. Butcher, The regulation of lymphocyte traffic, _in_:"Current Top. Microbiol. Immunol." Springer-Verlag Berlin-Heidelberg, vol. 128, pp 85-122 (1986).
17. C.O. Elson, J.A. Heck and W. Strober, T-cell regulation of murine IgA synthesis, _J. Exp. Med._ 149: 632-643 (1979).
18. H. Kawanishi, L.E. Saltzman and W. Strober, Mechanisms regulating IgA class-specific immunoglobulin production in murine gut-associated lymphoid tissues. I. T cells derived from Peyer's patches that switch sIgM B cells to sIgA B cells in vitro, J. Exp. Med. 157: 433-450 (1983).
19. I. Suzuki, K. Kitamura, T. Kurita, D.R. Green and J.R. McGhee, Isotype specific immunoregulation: Evidence for a distinct subset of T contrasuppressor cells for IgA responses in murine Peyer's patch, _J. Exp. Med._ 164: 501 (1986).
20. P.B. Ernst, R. Scicchitano, B.J. Underdown and J. Bienenstock, Oral immunization and tolerance, _in_: "Immunology of the Gastrointestinal Tract and Liver", A.L. Jones et al., eds., In press, Raven Press, N.Y. (1988).

24

21.  G.L. Asherson, M. Zembala, M.A.C.C. Perera, B. Mayhew and W.R. Homas, Production of immunity and unresponsiveness in the mouse by feeding contact sensitizing agents and the role of suppressor cells in the Peyer's patches, mesenteric lymph nodes and other lymphoid tissues, Cell. Immunol. 33: 145–155 (1977).

22.  P.B. Ernst, A.D. Befus and J. Bienenstock, Leukocytes in the intestinal epithelium: an unusual immunological compartment, Immunol. Today 6: 50–55 (1985).

23.  S. Targan, L. Britvan, R. Kendal, S. Vimadalal and A. Soll, Isolation of spontaneous and interferon inducible natural killer like cells from human colonic mucosa lysis of lymphoid and autologous epithelial target cells, Clin. Exp. Immunol. 54: 14–22 (1983).

24.  P.S. Carman, P.B. Ernst, K.L. Rosenthal, D.A. Clark, A.D. Befus and J. Bienenstock, Intraepithelial leukocytes contain an unique subpopulation of NK-like cytotoxic cells active in the defense of gut epithelium to enteric murine coronavirus, J. Immunol. 136: 1548–1553, 1986.

25.  J.R. Klein and M.F. Kagnoff, Non-specific recruitment of cytotoxic effector cells in the intestinal mucosa of antigen-primed mice, J. Exp. Med. 160: 1931–1936 (1984).

26.  H.J. Cooke, Neurobiology of the intestinal mucosa, Gastroenterology 90: 1057–1081 (1986).

27.  R.H. Stead, M. Tomioka, G. Quinonez, G.T. Simon, S.Y. Felten and J. Bienenstock, Intestinal mucosal mast cells in normal and nematode infected rat intestine are in intimate contact with peptidergic nerves, Proc. Natl. Acad. Sci. USA 84: 2975–2979 (1987).

28.  D.G. Payan, J.P. McGillis and E.J. Goetzl, Neuroimmunology, Adv. Immunol. 39: 299–323 (1986).

29.  M.S. O'Dorisio, Neuropeptides and gastrointestinal immunity, Am. J. Med. 81: 74–82 (1986).

30.  A.M. Stanisz, D. Befus and J. Bienenstock, Differential effects of vasoactive intestinal peptide, substance P and somatostatin on immunoglobulin synthesis and proliferation by lymphocytes from Peyer's patch, mesenteric lymph node and spleen, J. Immunol. 136: 152–156, (1986).

31.  A.M. Stanisz, R. Scicchitano, P. Dazin, J. Bienenstock and D.G. Payan, Distribution of substance P receptors on murine spleen and Peyer's patch T and B cells, J. Immunol. 139: 749–754 (1987).

32.  C.A. Ottaway and G.R. Greenberg, Interaction of vasoactive intestinal peptide with mouse lymphocytes specific binding and the modulation of mitogen responses, J. Immunol. 132: 417–423 (1984).

33.  M. Rola-Pleszczynski, D. Bolduc and S. St. Pierre, The effects of vasoactive intestinal peptide on human natural killer cell function, J. Immunol. 135: 2569–2573 (1985).

34.  F. Shanahan, J.A. Denburg, J. Fox, J. Bienenstock and D. Befus, Mast cell heterogeneity: effects of neuroenteric peptides on histamine release, J. Immunol. 135: 1331–1337 (1985).

35.  J. Foreman, Peptides and neurogenic inflammation, Brit. Med. Bull., In press (1988).

36.  J. Bienenstock, M. Tomioka, H. Matsuda, R.H. Stead, G. Quinonez, G.T. Simon, M.D. Coughlin and J.A. Denburg, The role of mast cells in inflammatory processes: Evidence for nerve/mast cell interactions, Int. Archs. Allergy appl. Immunol. 82: 238–243 (1987).

37.  G. Skofitsch, J.M. Savitt and D.M. Jacovowitz, Suggestive evidence for functional unit between mast cells and substance P fibres in the rat diaphragm and mesentery, Histochemistry 82: 5–8 (1985).

38.  D.J.J. Carr and E. Blalock, A molecular basis for bi-directional communication between the immune and neuroendocrine system, in: Progress in Immunology VI, B. Cinader and R.G. Miller, eds., Academic Press, Florida, pp 619–28 (1986).

39. D.A. Sullivan, Endocrine control of the ocular secretory immune system, in: "Hormone and Immunity", I. Berczi and K. Kovacs, eds., MTP Press, pp 54 (1987).

40. S. Freier, M. Eran and J. Faber, Effect of cholecystokinin and of its antagonist, of atropine and of food on the release of immunoglobulin A and immunoglobulin G specific antibodies in the rat intestine, Gastroenterology 93: 1242-1246 (1987).

41. P. Weisz-Carrington, M.E. Roux, M. McWilliams, J.M. Phillips-Quagliata, and M.E. Lamm, Hormonal induction of the secretory immune system in the mammary gland, Proc. Natl. Acad. Sci, USA 75: 2928-2932 (1978).

42. D.A. Sullivan and C.R. Wira, Hormonal regulation of immunoglobulins in the rat uterus: uterine response to multiple estradiol treatments, Endocrinology 114: 650-658 (1984).

# POSSIBLE FACTORS CONTRIBUTING TO PERSISTENCE OF CYTOMEGALOVIRUS INFECTION

R.V. Blanden

Division of Cell Biology
John Curtin School of Medical Research
Australian National University
PO Box 334
Canberra City, ACT 2601
Australia

## INTRODUCTION

The purpose of this article is to discuss factors which may contribute to the persistance of cytomegalovirus (CMV) infection in the face of host responses. In general, available evidence suggests that cell-mediated rather than humoral immunity is important for the control of CMV infection. For example, immunosuppressed humans die of progressive HCMV infection in the presence of normal or high antibody levels (Rasmussen et al., 1982). Furthermore T cell deficient mice are more susceptible to MCMV infection than normal controls (Grundy and Melief, 1982) and transferred T cells confer protection (Grundy and Melief, 1982; Starr and Allison, 1977; Ho, 1980; Reddehase et al., 1985). Therefore I intend to focus on factors which may influence the expression and effectiveness of cell-mediated mechanisms against CMV infection.

## THE CHRONOLOGY OF CELL-MEDIATED IMMUNE RESPONSES IN MURINE CMV (MCMV) INFECTION

Within 2 days after infection with MCMV, there is an augmentation of natural killer (NK) cell activity in infected mice (Quinnan and Manischewitz, 1979). Presumably this occurs as a consequence of the production of class I interferon by infected cells; interferon then activates NK

cells (Quinnan and Manischewitz, 1979). There is strong
evidence that genetically resistant mice show more
augmentation of NK cell activity following MCMV infection
than do susceptible strains (Bancroft et al., 1981). This
evidence suggests a cause and effect relationship, i.e. that
NK cells contribute in some way to resistance to MCMV
infection. The mechanisms which would influence these events
are discussed below. Following the early NK cell response
there is a response of cytotoxic T lymphocytes (Tc cells)
(Sethi and Brandis, 1979; Quinnan et al., 1978, 1980; Ho and
Ashman, 1979; Ho, 1980, Reddehase and Koszinowski, 1984;
Sinickas et al., 1985). It is likely that this response also
contributes to the resolution of MCMV infection. First, as
stated above, T cell deficient animals do not recover from
infection. Second, cell transfer experiments have
established that T cells transferred into infected recipients
can reduce titres of MCMV in the tissues of the recipients
(HO, 1980), and that the phenotype of the cells responsible
is Lyt2$^+$, L3T4$^-$ (Reddehase et al., 1985) and class I MHC-
restricted (Ho, 1980; Sissons et al., 1986).

The evaluation of the efficiency of these cell-mediated
immune mechanisms requires some discussion of the nature of
the ligands recognized by the two classes of killer cells.

NATURE OF THE LIGAND RECOGNIZED BY NK CELLS

NK cells are known to have two intriguing inverse
relationships with Tc cells. First, Tc cells are present in
normal animals and require a thymus gland for their
development. In contrast, in athymic, nude mice Tc cells are
not present but NK cells are present in increased numbers.
This suggests a possible origin of both cell lineages from a
common precursor but further speculation is irrelevant in the
present context.

The second reciprocal relationship bears on the nature
of the ligand required on the target cell for recognition and
tiggering of cytotoxic function by either Tc cells or NK
cells. In the case of Tc cells, increased concentration of
class I MHC antigen on target cells leads to increased
tiggering of cytotoxic function (O'Neill and Blanden, 1979;
Shimonkevitz et al., 1985; Blanden et al., 1987). In

contrast, if MHC class I antigen concentration is increased
by, for example gamma interferon, there is decreased
susceptibility to lysis by NK cells (Pointek et al., 1985;
Storkus et al., 1987; Müllbacher and King, 1989).

There are two potential classes of explanation for this
phenomenon.  One explanation would be that whereas a positive
signal is delivered to Tc cells as a consequence of MHC
antigen binding, a negative signal is delivered to NK cells.
If this is the case, NK cells should bind to MHC and
consequentially target cells with increased MHC expression
should be superior competitors to normal target cells for
binding to NK cells.  There is no evidence for such increased
binding (Müllbacher and King, 1989).

An alternative class of explanation is that the ligands
bound by Tc cells or NK cells are different and are
reciprocally regulated in target cells.  This line of
reasoning led Müllbacher and King (1989) to test the
possibility that the ligand for NK cells is $\beta_2$ microglobulin.
Normally $\beta_2$ microglobulin is expressed on the cell surface
noncovalently attached to class I MHC molecules and in most
cases is required for the surface expression of class I MHC.
Therefore, while MHC and $\beta_2$ microglobulin genes must be
coregulated to some extent it is possible that there could be
variations in the ratios of the two different molecules in
different cell classes and that upregulation by gamma
interferon may not uniformly affect both genes.  This leads
to the possibility that "free" $\beta_2$ microglobulin, unassociated
with class I MHC (Solheim and Thorsby, 1974) could be
available on some cells to act as a ligand for NK cells.
Susceptibility to NK cell-mediated lysis may then depend upon
the amount of "free" $\beta_2$ microglobulin expressed.  The nature
of the linkage of "free" $\beta_2$ microglobulin to the cell surface
is unknown but possibile precedents exist in proteins of the
immunoglobulin super gene family (Stroynowski et al., 1987).

Müllbacher and King (1989) have produced evidence
consistent with the above hypothesis.  First, target cells
which are unable to express class I MHC antigens but which
express $\beta_2$ microglobulin are sensitive to NK cell-mediated
lysis and furthermore, interferon gamma treatment increased
$\beta_2$ microglobulin expression and concurrently increased

29

susceptibility to NK cells.  A cell line which expresses neither $\beta_2$ microglobulin nor class I MHC is not susceptible to NK cells.

Target cells expressing both $\beta_2$ microglobulin and class I MHC are vulnerable to lysis by both NK cells and allo-reactive Tc cells.  If monoclonal antibodies against $\beta_2$ microglobulin or class I MHC are applied to the target cell, antibody against $\beta_2$ microglobulin blocks NK cell-mediated lysis, not Tc cell-mediated lysis, whereas antibody against class I MHC block Tc cell-mediated lysis, not NK cell mediated lysis.  In summary this evidence suggests that the ligand for NK cells is $\beta_2$ microglobulin unassociated with class I MHC and that the concentration of cell surface $\beta_2$ microglobulin is a determinant of target cell susceptibility to NK cell mediated lysis.

THE LIGAND FOR Tc CELLS

It is now clear that the antigen-specific receptors of antiviral Tc cells bind to a ligand formed by two polymorphic helices in the $\alpha_1$ and $\alpha_2$ domains of class I MHC (Bjorkman, et al., 1987a, 1987b) plus a virus peptide which binds to a cleft between the helices of the MHC molecule (Townsend et al., 1986a).  Available evidence suggests that the complex of MHC and viral peptide is produced inside infected cells (Townsend et al., 1986b).  Thus the timing of host and viral protein synthesis and their local concentrations will be among factors that determine the amount of complex formation. This is illustrated by work with recombinant vaccinia viruses encoding other antigens.  The use of different vaccinia promoters has allowed the investigation of the effect of changing the timing of antigen production in relation to viral infection.  Thus when influenza haemagglutinin (HA) was expressed from recombinant vaccinia under the control of the p7.5 promoter which allows the initiation of protein synthesis within 2 hours after infection, class I MHC restricted Tc cells specific for HA could recognize target cells infected by the recombinant vaccinia (Coupar et al., 1986).  However when a late promoter (PL11) was used (with the synthesis beginning only after viral DNA replication), even though as much HA was produced in vaccinia-infected cells as with the p7.5 promoter, HA-specific Tc cells did not lyse the infected targets (Coupar et al., 1986). This could be interpreted to mean that if target cell MHC synthesis has been turned off as a consequence of vaccinia infection before there is substantial viral protein synthesis, then the concurrent, intracellular concentrations of MHC and peptide derived from HA will be unable to allow formation of a sufficient concentration of MHC-HA peptide complexes on the target cell surface for triggering Tc cell lytic function.

Similar reasoning can be applied to data obtained by using recombinant vaccinia viruses encoding murine class I MHC. We have such viruses which encode either H-2K$^d$ alone or double constructs which encode both H-2K$^d$ and influenza nuclear protein (NP) (Andrew et al., 1987). This has enabled us to test target cells infected with these viruses for susceptibility to Tc cells restricted by H-2K$^d$ and specific for NP. When target cells were infected with the double construct encoding both H-2K$^d$ and NP they were susceptible to lysis by H-2K$^d$ restricted, NP specific Tc cells (Andrew, et al., 1987). In contrast, when double infection was employed in which one vaccinia virus encoded H-2K$^d$ and another encoded NP, despite simultaneous expression of both K$^d$ and NP on the doubly-infected target cells such target cells were not susceptible to lysis by H-2K$^d$ restricted NP-specific Tc cells (Andrew et al., 1987). Controls indicated that the doubly-infected target cells expressed similar concentrations of both H-2K$^d$ and NP to the targets infected with the double construct. We interpret these contrasting results as an indication of the effect of local intracellular concentrations of H-2K$^d$ and NP peptide within infected cells. In the case of the double construct, synthesis of H-2K$^d$ and NP might occur in close physical proximity to each other. This should allow ready association between H-2K$^d$ and peptide derived from NP. In the case of double infection, H-2K$^d$ might be synthesised in one location and NP in a different location within the doubly infected cells. This should reduce the probability of intracellular association between H-2K$^d$ and NP peptide and, in turn, result in a lower concentration of H-2K$^d$-NP peptide complexes on the infected cell surface, even though total intracellular concentrations of the individual proteins would be similar to the case of the double construct.

## LIGAND CONCENTRATION ON STIMULATOR-TARGET CELLS DETERMINES THE EXTENT OF KILLER CELL ACTIVATION

Evidence has accumulated over the last decade that triggering of responses or effector function in Tc cells is dependent upon ligand concentration on stimulator-target cells (reviewed by Blanden et al., 1987). The experimental evidence ranges from the effects of genetic regulation of class I MHC expression (O'Neill and Blanden, 1979) to the effects of the concentration of purified MHC protein added to cell sized beads (Goldstein and Mescher, 1987). There is also abundant evidence regarding increased susceptibility to Tc cell mediated lysis of target cells expressing increased amounts of class I MHC as a consequence of treatment with gamma interferon (Shimonkevitz et al., 1985; Blanden et al., 1987). Similar evidence has been obtained with respect to increased $\beta_2$ microglobulin expression after gamma interferon treatment and susceptibility to NK cell lysis (Müllbacher and King, 1989) as alluded to above. In the case of Tc cells and comparable evidence for class II restricted helper T (Th) cells, it seems that a certain signal threshold as a consequence of receptor binding to ligand is required to activate T cell function whether it be proliferation, release of lymphokine, or cytotoxicity (Blanden et al., 1987). We have discussed in detail elsewhere (Blanden et al., 1987) the idea that signal strength received by a T cell will be determined by the concentration of T cell receptors and accessory molecules such as CD4 or CD8, by the concentration of ligand on stimulator-target cells, and by the affinity of the interactions between receptors and ligands. A similar dependence upon ligand concentration is in evidence with NK cells with respect to $\beta_2$ microglobulin (Müllbacher and King, 1989) as discussed above.

## VIRAL STRATEGY IN RELATION TO HOST CELL-MEDIATED MECHANISMS OF VIRAL CLEARANCE

The mechanisms by which cell-mediated immunity may clear viral infection are potentially numerous and are generally ill-defined *in vivo*. However evidence exists which supports the following possibilities. The first requisite is that the effector cells must be present in or near the focus of infection if they are to operate efficiently against the virus (Blanden, 1971). Therefore in non-lymphoid target organs the effector cells must be available in the circulation for recruitment into foci of infection. This recruitment may be mediated by factors emanating from the infectious site or may be immunologically specific, as in the case of ligand binding by T cells (Blanden, 1974). Once the effector cells are assembled in the focus of infection, a number of possible mechanisms could operate. First, the lysis of virus-infected cells before the viral replication cycle is completed could make a significant contribution to the control of infection. Second, the secretion of lymphokines such as gamma interferon and interleukin-2 could also contribute both directly and indirectly. Apart from the direct antiviral effects of gamma interferon, its capacity to up-regulate MHC expression could be an important factor in ensuring that ligand concentrations on the surfaces of infected cells will be sufficient to trigger T cell or NK cell function (Blanden et al., 1987). IL-2 could boost T cell numbers at the site of infection and could contribute to NK cell activation in the focus. Lymphokines could also activate macrophages to augment their phagocyctic and intra-cellular destructive capabilities. The roles of various lymphokines and interplay between them may well be defined by future research.

It is axiomatic that if viruses are capable of causing chronic, persisting infection in an immuno-competent host then they must have evolved strategies to evade cell-mediated immune mechanisms. One factor concerns the intrinisic chemotatic properties of infected foci. For example, in mice infected with ectromelia virus, foci of infection in the liver do not attract inflammation; only when virus-specific T cells enter the lesion for immunologically specific reasons (i.e. specific ligand binding), does a major inflammatory influx occur (Blanden, 1971). In contrast, in the case of vaccinia virus, a pox virus very closely related to ectromelia virus, liver lesions are invaded by inflammatory cells immediately after infection (Karupiah and Blanden, unpublished observations). This may well be a factor in the different virulence of the two viruses for mice, ectromelia being capable of high virulence, whereas vaccinia virulence is very low. Furthermore, there is evidence that genetically determined resistance to MCMV is reflected in the extent of inflammation in lesions in target organs in the early days after infection (Grundy et al., 1981); resistant strains of mice exhibit significantly more inflammation then susceptible strains. Thus, the assembly of certain cell-mediated mechanisms in foci of infection may be more rapidly established in resistant strains.

Another mechanism by which viruses may reduce the efficiency of cell-mediated immunity is by reduction of ligand concentration on the surfaces of infected cells (Gardner et al., 1975). It was of particular importance in developing assays for MCMV-specific Tc cells. When mouse embryo fibroblasts were used as MCMV-infected target cells their susceptibility to Tc cells could be markedly improved by treatment with gamma interferon to up-regulate MHC class I expression prior to MCMV infection (Sinickas et al., 1987).

The mechanisms by which cytopathic viral infection reduces the expression of MHC on cell surfaces has not been precisely defined. In some cases it may be no more than a reflection of the fact that host protein synthesis is shut down after viral infection and the infected cell membrane becomes increasingly dominated by the expression of virus-specified proteins. However in the case of CMV some very intriguing possibilities have emerged from recent research. First, it has been shown that HCMV binds $\beta_2$ microglobulin (Grundy et al., 1987). Two viral proteins have been identified as being responsible for this property (Grundy et al., 1987). While extracellular virus is known to be coated with $\beta_2$ microglobulin when present in urine (McKeating et al., 1987) it has not been determined whether either or both of the viral proteins could bind to $\beta_2$ microglobulin inside infected cells. If this were the case, it could result in such a reduction of the concentration of intracellular $\beta_2$ microglobulin, which is required for the cell surface expression of most class I MHC molecules, so that simultaneously the concentrations of class I MHC and "free" $\beta_2$ microglobulin would be reduced on the cell surface. In consequence, the ligands required for triggering NK cell and Tc cell function would be reduced, possibly below the thresholds required for effective triggering.

A recent paper by Beck and Barrell (1988) has shown that the DNA of HCMV contains a complete coding sequence for a class I MHC-like molecule. There are no introns, implying that this DNA was derived by reverse transcription of cellular mRNA at some point of the evolution of the virus. While there was sufficient homology with class I MHC sequences to leave no doubt that the gene was obtained from cellular sources, the viral sequence is distinguished by additional potential glycosylation sites which would cover the polymorphic $\alpha$-helices that constitute an essential part of Tc cell ligands, thus preventing T cell recognition of the molecule (Wiley, 1988). This additional glycosylation could result in increasing the apparent molecular weight of the viral protein to 65,000 (Wiley, 1988), which is compatible with one of the viral proteins identified by Grundy et al. (1987) as responsible for binding to $\beta_2$ microglobulin.

There are, of course, unanswered questions. Is this MHC-like protein actually expressed in HCMV-infected cells? Does it bind to $\beta_2$ microglobulin inside these cells, thus reducing the availability of ligands for NK cells and Tc cells? Does this protein bind to CD8? If so, does this binding cause the infection of the CD8[+] Tc cells which could contribute to recovery from HCMV infection? Is there a similar gene in MCMV? Research to answer these and related questions promises an exciting future in the evaluation of CMV strategy.

REFERENCES

Andrew, M.E., Coupar, B.E.H., Boyle, D.B. and Blanden, R.V., 1987, Eur J. Immunol. 17:1515.
Bancroft, G.J., Shellam, G.R. and Chalmer, J.E., 1981, J. Immunol. 126:988.
Beck, S. and Barrell, B.G., 1988, Nature 331:269.
Bjorkman, P.J., Saper, M.A., Samraoui, B., Bennett, W.S., Strominger, J.L. and Wiley, D.C., 1987a, Nature 329:506.
Bjorkman, P.J., Saper, M.A., Samraoui, B., Bennett, W.S., Strominger, J.L. and Wiley, D.C., 1987b, Nature 329:512.
Blanden, R.V., 1971, J. Exp. Med. 133:1090.

Blanden, R.V., 1974, Transplant. Rev. 19:56.

Blanden, R.V., Hodgkin, P.D., Hill, A., Sinickas, V.G. and Müllbacher, A., 1987, Immunol. Rev. 98:75.

Coupar, B.E.H., Andrew, M.E., Both, G.W. and Boyle, D.B., 1986, Eur. J. Immunol. 16:1479.

Gardner, I.D., Bowern, N.A. and Blanden, R.V., 1975, Eur. J. Immunol. 5:122.

Goldstein, S.A.N. and Mescher, M., 1987, J. Immunol. 138:2034.

Grundy (Chalmer), J.E. and Melief, C.J.M., 1982, J. Gen. Virol. 61:133.

Grundy (Chalmer), J.E., Mackenzie, J.S. and Stanley, N.F., 1981, Infect. and Immun. 32:277.

Grundy, J.E., McKeating, J.A. and Griffiths, P.D., 1987, J. Gen. Virol. 68:777.

Ho, M., 1980, Infect. Immun. 27:767.

Ho, M. and Ashman, R.B., 1979, Aust. J. Exp. Biol. and Med. Sci. 57:425.

McKeating, J.A., Griffiths, P.D. and Grundy, J.E., 1987, J. Gen. Virol. 68:785.

Müllbacher, A. and King, N.J.C., 1989, Scand. J. Immunol., in press.

O'Neill, H.C. and Blanden, R.V., 1979, J. Exp. Med. 149:724.

Pointek, G.E., Taniguchi, K., Ljunggren, H., Gronberg, A., Kiessling, R., Klein, G. and Karre, K., 1985, J. Immunol. 135:4281.

Quinnan, G.V. and Manischewitz, J.E., 1979, J. Exp. Med. 150:1549.

Quinnan, G.V., Manischewitz, J.E. and Ennis, F.A., 1978, Nature 273:541.

Quinnan, G.V., Manischewitz, J.E. and Ennis, F.A., 1980, J. Gen. Virol. 47:503.

Rasmussen, L., Kelsall, D., Nelson, R., Carray, W., Hirsch, M., Winston, D., Preikaitis, J. and Merigan, T., 1982, J. Infect. Dis. 145:191.

Reddehase, M.J., Weiland, F., Munch, K., Jonjic, S., Luske, A. and Koszinowski, U.H., 1985, J. Virol. 55:264.

Reddehase, M.J. and Koszinowski, U.H., 1984, Nature 312:369.

Sethi, K.K. and Brandis, H., 1979, Arch. Virol. 60:227.

Shimonkevitz, R., Luescher, B., Cerottini, J.-C. and MacDonald, H.R., 1985, J. Immunol. 135:892.

Sinickas, V.G., Ashman, R.B. and Blanden, R.V., 1985, J. Gen. Virol. 66:747.

Sinickas, V.G., Ashman, R.B. and Blanden, R.V., 1987, Immunol. and Cell Biol. 65:173.

Sissons, J.G.P., Borysiewicz, L.K., Rodgers, B. and Scott, D., 1986, Immunology Today 7:57.

Solheim, B.G. and Thorsby, E., 1974, Tissue Antigens, 4:83.

Starr, S.E. and Allison, A.C., 1977, Infect. Immun. 17:458.

Storkus, W.J., Howell, D.N., Salter, R.D., Dawson, J.R. and Cresswell, P., 1987, J. Immunol. 138:1657.

Stroynowski, I., Soloski, M., Lowe, M.G. and Hood, L., 1987, Cell 50:759.

Townsend, A.R.M., Rothbard, J., Gotch, F.M., Bahadur, G., Wraith, D. and McMichael, A.J., 1986a, Cell 44:959.

Townsend, A.R.M., Bastin, J., Gould, K. and Brownley, G.G., 1986b, Nature 324:575.

Wiley, D., 1988, Nature 331:209.

# THE IMMUNE RESPONSE OF HUMANS TO LIVE AND INACTIVATED

# INFLUENZA VACCINES

Yuri Ghendon

World Health Organization
Geneva, Switzerland

Influenza has long been recognized as a problem in both developed and developing nations. Influenza is not a trivial disease. For example, in the USA in 1957 the Asian strain of influenza virus caused an estimated 70 000 deaths; the Hong Kong strain of influenza virus that appeared in 1968 caused about 30 000 deaths[39]. Even in years not associated with antigenic shift many people die as a result of influenza infection, so 10 000 or more excess deaths have been documented in the USA during each of 18 different epidemics from 1957 to 1985[39]. It was estimated also[35] that in the USA, during an influenza outbreak, about 70 million people catch influenza at a cost of about $4.5 billion.

For influenza prophylaxis two sorts of influenza vaccines are now available: inactivated concentrated and purified for parenteral administration and live attenuated cold-adapted recombinant vaccines destined for instillation or pulverization into the upper respiratory paths.

Most of the information on the mechanism of immunity to influenza was obtained in animal models which suggests that resistance to the disease can be correlated with the presence of antibody to the surface proteins of the virus and that cell-mediated immunity plays an important role in recovery from influenza infection[1,21]. There is much less information available about the mechanism of protection of humans against influenza viruses.

It is known that previous infection of humans with an identical strain of virus confers immunity upon later challenge. It should be noted that immunity to influenza can be both solid and long-lasting as shown by the resistance of the population older than 20-25 years to H1N1 influenza virus infection on the reappearance of virus of this subtype in 1977 after 20 years of the last outbreak induced by H1N1 influenza virus.

It has been generally accepted that high titres of haemagglutinin inhibiting antibody are significantly associated with protection against a virus with a closely related haemagglutinin (HA)[37]. It was also shown that neuraminidase inhibiting antibody in serum contributed to immunity to influenza[16,32].

The antibody response to influenza virus or vaccine depends on the recipient's prior antigenic experience. During primary infection of humans with influenza virus IgM and IgG in serum occur regularly and IgA less frequently; in nasal secretions IgA followed by IgM was the dominant response[34]. Previously seropositive subjects produced antibodies of the IgG and IgA class more frequently than previously seronegative persons; in contrast, IgM antibodies occur more frequently in unprimed subjects than in primed ones.[5]

Resistance to influenza infection in humans has been correlated with HA antibody in nasal washings[10,11,17]. It should be noted that resistance to wild type influenza A infection has been demonstrated in adults with nasal wash neutralizing antibody, but without detectable serum antibody[33]. On the other hand a contribution of serum antibody alone has been inferred from studies of the correlation of resistance to illness caused by influenza A virus with the level of maternally transferred antibody in neonates[40]. Perhaps in humans antibody present in either the local or systemic compartment can contribute to resistance to illness caused by influenza virus, but it is possible that to be protective, HA antibody must be present at the mucosal surface, having been produced either locally or derived from serum.

As to recovery from influenza infection it is possible to assume that antibody is not essential because patients with agammaglobulinaemia recover from influenza infection. Data available at present show that cell-mediated immunity is a host factor responsible for the recovery process from influenza infection[1,21].

Since local and serum antibodies and also cytotoxic T-cells (Tc-cells) appear to be mediators of immunity to influenza infection it is important that influenza vaccines induce all these components of immunity.

Inactivated influenza vaccines in primed individuals induced a protective level of serum HA antibody in over 85% of recipients[38]. The induction of a secretory antibody response to inactivated influenza vaccines is dependent both on the route of administration and on the recipient's prior antigenic experience. In unprimed recipients local antibody responses are of low magnitude and occur infrequently after both parenteral and intranasal administration of vaccine. Parenteral administration of inactivated vaccine produced a local IgG response in 94% of primed recipients, whereas local IgA responses developed in only 38%[12]. In contrast after intranasal administration of inactivated vaccines local IgA response developed in the majority of primed recipients[47].

Murphy et al[34] found in studies with ca live influenza vaccines that young children vaccinated with these vaccines (H3N2 or H1N1) had serum IgG, IgM and IgA antibody response. In nasal washes of most of the vaccinees were found IgA and IgM antibody and in 50% of vaccinees, IgG antibody. Most of the IgA and IgM HA antibody was actively secreted locally, whereas only some of the IgG HA antibody could be shown to be actively secreted into the respiratory tract. These date indicate that intranasal vaccination of susceptible children with ca live influenza A vaccines efficiently stimulates both systemic and local antibody responses.

It should be noted that ca live influenza vaccine (H3N2) can induce local IgA antibody not only in the majority of seronegative but also seropositive vaccinees[12].

A substantial difference between the response to natural infection and inactivated vaccines is seen when the dynamics of the antibody response are compared[38]. Serum and anti-HA titres gradually decrease over the first six months after infection and may then persist for several years, possibly due to subsequent infections by related virus strains. The duration of serum anti-HA after vaccination with inactivated vaccine varies - primed subjects retain protective levels of antibody for at least one year, whereas antibody levels decline rapidly in unprimed subjects.

In comparative studies of Johnson et al[24] of antibody responses of young children vaccinated with intranasal ca live vaccines (H3N2 or H1N1) or intramuscular inactivated influenza vaccine (H3N2), it was found that six weeks after vaccination the titres of HAI antibody were more or less the same. But in several other studies it has been reported that parenteral vaccination with inactivated influenza vaccine stimulated systemic antibodies in humans more efficiently than does intranasal vaccination with live attenuated vaccine[8,9,11,29,51].

Feldman et al[22] also found that after vaccination of seronegative children with ca live H1N1 influenza vaccine, only 57% of vaccinees showed serological responses. But a natural H1N1 challenge which occurred shortly after completion of the vaccination showed that most vaccinees were protected against infection and symptomatic illness despite its failure to stimulate high levels of serum HAI antibody.

It has been suggested that local antibody may play an important role in protection of vaccinees with ca live influenza vaccine. In fact, Clements and Murphy[13] also found that inactivated influenza vaccine induced serum IgA and IgG in most vaccinees in comparison with ca live vaccine and induced higher titres of serum antibodies than did live vaccine. But in contrast only 38% of inactivated virus vaccinees had local IgA responses compared with 83% of vaccinees immunized with ca live vaccine. The same results were obtained by Zahradnic et al[51] who showed that parenteral inactivated vaccine was relatively ineffective in stimulating neutralizing secretory antibodies when compared with intranasal ca live vaccine.

Johnson et al[24] also found in comparative studies of inactivated (intramuscular) and ca live (intranasal) vaccines on young children that nasal secretory IgA developed almost exclusively in vaccinees with live vaccines and persisted for up to one year (vaccine H3N2) and after vaccination with ca live vaccine H1N1, nasal IgA was demonstrable as long as 30 months. Persistent nasal secretory IgG was detected in vaccinees vaccinated with both inactivated and live vaccines but the titre of this antibody was higher in vaccinees with ca live vaccine.

In subsequent studies by Johnson et al[25] 59 young children were divided into four groups based on prior exposure to influenza A(H3N2) virus, natural infection, live ca vaccine given intranasally, inactivated vaccine given intramuscularly, and no previous exposure. Virus challenge with homologous live ca vaccine occurred 12 months after vaccination or natural infection. It was found that prechallenge local IgA detected almost exclusively in subjects naturally infected or vaccinated with ca live vaccine was associated with protection against shedding. Any detectable nasal IgA (⩾ 1log2) suppressed viral shedding. Effect of nasal IgG was not as sharply defined in this study; however, at higher levels (> 4log2) shedding was reduced.

It should be noted that although inactivated vaccine failed to produce significant local IgA during the primary response, it primed infants to a better response to both nasal IgA and IgG after challenge with ca live virus.

Clements et al[15] investigated the role of serum and nasal wash antibodies in resistance of humans vaccinated with ca live or inactivated influenza A vaccines to experimental challenge with influenza A wild-type virus. Protection of vaccinees receiving inactivated vaccine for infection or illness correlated with the level of haemagglutinin - inhibiting (HAI) antibody and neuraminidase-inhibiting antibody in serum and local anti HA IgG (but not IgA) antibody. Protection of vaccinees receiving ca live vaccines against infection correlated with local anti-HA IgA antibody and neuraminidase-inhibiting antibody in serum, but not with HAI antibody in serum. The authors suggest that live vaccine-induced immunity may involve different compartments of the immune system but sufficient antibody in either serum or nasal secretion is capable of conferring resistance.

The origin of nasal-wash HA antibodies induced by influenza vaccines is not completely known. In the case of intranasal vaccination with live influenza vaccine there is evidence that secretory IgA and to some extent IgM and IgG are synthesized by nasal epithelial cells and actively secreted locally[6,7,34]. It should be noted that peripheral blood lymphocytes obtained from children immunized with ca live vaccine produced in vitro IgG but not IgA antibody[18]. The absence of IgA-producing B-cells in the peripheral blood of children vaccinated with live vaccine may reflect that these cells belong to local mucosal sites.

The studies of persistence of the serum and nasal-wash IgA, IgG and IgM antibody have shown that 12-24 months after vaccination with ca live vaccine had significantly less decay of serum HAI and IgG antibody in contrast to the antibody induced by inactivated influenza vaccine[9,24,48] which declines significantly in the first six months after vaccination. Serum IgM antibody persisted for one year after vaccination with inactivated vaccine[48].

As to secretory IgA antibodies it was found that in adults such antibody induced by live influenza vaccines is relatively short-lived[7,13,48]. However, data from studies in seronegative children have indicated that nasal IgA antibody could be detected for one or more years in about 50% of naturally infected or vaccinated with live influenza vaccine children but in only 5% of inactivated vaccine recipients[24].

The results presented above show that inactivated and live influenza vaccines stimulate both systemic and secretory antibodies in children and adults and that inactivated vaccine is more potent in inducing serum antibodies but live vaccines are superior in stimulation of secretory antibodies that may play an important role in host protection.

As noted above the cell-mediated immunity also can be an important component of immunity to influenza especially at the stage of recovery from influenza infection.

In several laboratories in studies on mice it was found that Tc lymphocyte response to influenza infection is required for recovery from influenza pneumonia[1]. In experiments on mice it was found that inactivated influenza vaccines can stimulate Tc cell responses[19,41,45].

Studies on human volunteers support the contention that Tc cells are important in recovery from influenza virus infection. Mitchell et al[30A] have shown that Tc cell memory in peripheral blood lymphocytes correlated with the rapid clearing of administered virus in individuals, some of whom lacked specific antibody to HA or NA.

There are very few studies on Tc response after immunization of humans with influenza vaccines.

McMichael et al[30] have demonstrated that inactivated vaccines in primed humans stimulated a cross-reactive Tc cell response. However this response is not universal and appears to be related to the pre-immunization level of Tc memory.

Ennis et al[20,21] found that in primed volunteers immunized with live or inactivated influenza vaccines (H1N1) both types of vaccines induced HLA-restricted T-lymphocyte responses specific for influenza A virus. But by six months after vaccination, both the memory of Tc cell activity and the directly detected Tc cell activity had returned to pre-immunization levels. It should be noted that in these studies there was no absolute correlation between antibody responses and an increase in  Tc cell activity: there were several volunteers who had antibody responses without increase in specific Tc cell activity and vice versa.

The ability of influenza vaccines to stimulate the Tc cell responses in unprimed humans has not been determined but in mice ca live influenza vaccine can induce a primary Tc cell response and can sensitize the lungs for a secondary Tc cell response[27,28]. It was found that the dose of a live ca vaccine strain required to induce the same level of Tc response was 100 to 1000 times greater than that of the parenteral wild strain[27]. The difference in dosage required for priming could be overcome by giving two small doses of ca virus three weeks apart[43]. Using this approach ca live vaccine may induce in mice cross-protection against different subtype viruses[44]. It was found also that stimulation of Tc cell response in mice by live virus infection was superior to inactivated virus[45].

Several authors found that human and murine Tc cells can recognize not only glycoproteins of outer membranes of influenza virions as HA and neuraminidase but also other viral-specific proteins with common antigenic specificity such as NP, M1, PA, PB1, PB2 and NS1 (see 50). Although Tc cells cannot per se prevent infection, infection in primed persons may be modified early in its course. As Tc cells recognize the common internal viral proteins this protective effect can be broader than that of antibody response. It was found recently that local injection of mice with purified NP of influenza virus primes for influenza A virus cross-reactive Tc memory cells and leads to protection of hosts against intranasal infection with a lethal dose of influenza virus[49]. But at the moment there are no data on the possibility of Tc cells recognizing internal proteins of influenza virus to take part in the immunity of humans to influenza infection.

In spite of convincing data obtained on mice showing that influenza vaccine stimulates the Tc cell response the ability of inactivated and live influenza vaccines to induce cell-mediated immunity in unprimed and primed humans should be determined.

The protection afforded by influenza vaccines was evaluated in volunteers and in field trials.

One of the first comparative studies of protective effects of live and inactivated influenza vaccine (H1N1) was done by Clark et al[8,9]. In a study of short term immunity young adults were vaccinated with ca live vaccine (H1N1) or inactivated vaccine. One month after vaccination, protection against challenge with homologous ca live vaccine was equivalent in the two groups as assessed by rises in titres of HAI antibody. The same data was obtained in a study of long term immunity when groups of young adults were challenged eight months after vaccination. But it should be noted that this study did not include investigation of local antibody and viral shedding data.

Clements et al[11] in comparative studies on seronegative adult volunteers who were vaccinated with ca live vaccine (H3N2) or inactivated vaccine showed that after challenge with the homologous wild-type virus five to eight weeks after vaccination, recipients of live vaccine were completely protected against illness compared with a 72% efficacy in the inactivated vaccine recipients. Wild-type virus was recovered from only 13% of live vaccine vaccinees compared with 63% of inactivated vaccine vaccinees. The few infected vaccinees immunized with ca live vaccine shed 1000 times less wild-type virus than did infected inactivated vaccine vaccinees or unvaccinated control. This striking reduction in virus shedding suggests that influenza transmission may be more efficiently interrupted with live than with inactivated vaccine.

In subsequent studies by Clements et al[14] of resistance of vaccinees to challenge with influenza wild-type virus seven months after vaccination it was found that vaccine efficacy, measured by reduction in febrile or systemic illness in vaccinees, compared with that in controls was 100% for ca live H3N2 vaccine, 87% for inactivated H3N2 vaccine, 79% for ca live H1N1 vaccine and 67% for inactivated H1N1 vaccine. The authors concluded that ca live influenza vaccine induced significantly greater resistance to wild-type influenza virus one to two months after vaccination but at seven months post-vaccination, the resistance induced by live vaccines is only slightly greater than that induced by inactivated vaccine. But it should be noted that in the studies of Johnson et al[25] and Wright et al[48], vaccinees immunized with ca live (H3N2) or inactivated vaccine and challenged with homologous live ca vaccine or natural challenge with wild-type influenza virus 12 months after vaccination, live ca vaccine significantly reduced ca virus shedding after challenge compared with inactivated vaccine.

As to the efficacy of inactivated or live influenza vaccines it should be noted that at the moment there are no publications on comparative studies on the efficacy of the two types of vaccines during the same field trials. But as many studies have shown, adults who have experienced one or more infections by influenza virus and have been immunized with inactivated vaccines, may have protection against antigenically homologous influenza virus for at least one to two years. In unprimed individuals, usually young children, this immunization is less protective, probably because of the poor ability of parenterally administered inactivated virus to prime for a local humoral or Tc cell response[1].

The efficacy of live influenza vaccines in primed adults is more or less similar to the efficacy of inactivated vaccines. Efficacy studies of ca live vaccines on unprimed children has shown[3,4,46] that children receiving a H3N2 ca live vaccine appeared to be protected against subsequent natural infection with related strains of influenza virus. Ca live influenza A vaccine was studied in a controlled field trial among more than 16 000 children between three and 15 years of age vaccinated

with bivalent vaccine (H1N1 + H3N2)[2].   Protective efficacy of both
components of the vaccine was developed in two time decrease in morbidity
of vaccinees during outbreaks of influenza A/H1N1 and A/H3N2.

In comparative studies by Couch et al (unpublished data) on the
efficacy of inactivated and ca live influenza vaccines it was found that
for adults a greater effectiveness for inactivated vaccine was suggested,
but among children a greater effectiveness for ca live vaccine than
inactivated vaccine was suggested, particularly for children less than 10
years of age.

Another important problem related to the efficacy of influenza
vaccines is the protective effect of vaccine against drift variants of
influenza virus.   Hoskins et al[23] showed that the protective effect
of inactivated influenza vaccine was limited to schoolchildren who were
vaccinated for the first time with the inactivated vaccine made from the
prevailing strain of influenza virus.   Revaccination with inactivated
vaccine produced from the later prevailing strain did not provide
protection against a new drift variant.   On the other hand natural
infection with live influenza virus afforded almost complete protection
during successive outbreaks involving drift variants for more than four
years.   As the specificity of the antibody response after vaccination
with inactivated vaccine and after infection with live virus is similar,
differences in the extent of cross-reactive Tc responses to vaccination
and infection may account for the lesser degree of heterotypic protection
seen after vaccination with inactivated vaccine.   Taking the above into
account it is possible to suggest that live influenza vaccine may have the
advantage over inactivated vaccines in protection against new drift
variants of influenza viruses.

It was found recently that adaptation of influenza viruses to growth
in embryonated eggs resulted in selection of variants which were
antigenically and biologically distinguishable from viruses isolated from
the same source in mammalian cell line MDCK[36,42].   It was also found
that haemagglutination-inhibiting or virus neutralizing antibodies in
human sera can be detected more frequently, and to a higher titre, in
tests employing virus grown exclusively in MDCK cells than in tests with
virus adapted to growth in embryonated eggs[36].

Taking into account that the substrate for production of influenza
vaccines - inactivated or live - is embryonated chicken eggs, these
findings have raised concern regarding the suitability of eggs for
cultivation of influenza viruses used in vaccine production.

Results obtained recently[26,31] show that antigenic differences
between viruses grown in eggs and MDCK cells did not influence their
ability to protect if live viruses were used for immunization:   ferrets
infected with either live egg-grown or MDCK-grown virus were protected
equally well from challenge with virus grown in either host cell type.
On the other hand, if ferrets were vaccinated with formalin-inactivated
viruses, there were differences in protection related to the host cell
used to prepare the vaccine:   egg-grown vaccine is not fully efficacious
against challenge with MDCK-grown virus.

Of course the data obtained on animal models and results on humans may
be different but the experiments on ferrets has shown a certain advantage
in live influenza vaccine.

In conclusion it should be noted that most of our knowledge on
immunity to influenza in humans is very weak and based on the data
obtained in mice, but the pathogenesis of influenza infection in mice and

humans is not the same.   Much more information is needed on the real mechanisms of protection of humans against influenza, in particular the role of different serum and local antibodies, and participation of Tc cells, especially those which can recognize the internal proteins of influenza virus, in immunity to influenza, and on the mechanisms responsible for long-lasting immunity to influenza in humans.

As to the vaccines against influenza which are available at the moment, it is possible to say that the inference that live intranasal vaccination may provide better and more durable protection than inactivated intramuscular vaccination remains unproven, and the factors responsible for protection should be defined more clearly.   Nevertheless, with our present knowledge it is possible to suggest that local administration of live attenuated influenza vaccines can mimic natural influenza infection and may provide more natural and cross-reactive immunity to this disease than vaccination with inactivated vaccine especially in protection of seronegative children.

REFERENCES

1.   Ada, G.L., Jones, P D:   Curr. Top. Microbiol. Immunol., 1986, 128:1-55.
2.   Alexandrova, G.I. et al.:  Vaccine, 1986,4:114-118.
3.   Belshe, R B, Van Voris, L P:  J. Infect. Dis., 1984,149:735-740.
4.   Belshe, R B, et al:  J. Infect. Dis., 1984,150:834-840.
5.   Beyer, W E et al:  J. Hyg. Camb., 1986,96:513-522.
6.   Bienenstock, J., Befus, A D:Immunol, 1980,41:249-270.
7.   Butler W T et al:  J. Immunol. 1970,105:584-591.
8.   Clark, A et al:  J. Hyg. Camb., 1983,90:351-359.
9.   Clark, A et al:  J. Hyg. Camb., 1983,90:361-370.
10.  Clements M L et al:  Infect. Immun., 1983,40:1044-1051.
11.  Clements M L et al:  Lancet, 1984,i:705-708.
12.  Clements M L et al:  J. Clin. Microbiol., 1985,21:997-999.
13.  Clements, M L, Murphy, B R:  J. Clin. Microbiol., 1986,23:66-72.
14.  Clements, M L et al:  J. Clin. Microbiol, 1986,23:73-76.
15.  Clements, M L et al.:  J. Clin. Microbiol., 1986,24:157-160.
16.  Couch, R B et al.:  J. Infect. Dis., 1974,129:411-419.
17.  Couch, R B et al, in:  Nayak D P, Fox C F (eds) Genetic variation among  influenza viruses.ICN-UCLA Symposium on Molec. and Cell. Biol., 1981,XXI:535-546.
18.  Edwards, K.M. et al.:  Vaccine, 1986,4:50-54.
19.  Ennis, F A et al.:  Nature, 1977,266:418-419.
20.  Ennis, F A et al.:  Lancet, 1981,ii:887-891.
21.  Ennis, F A et al.:  Arch. Virol., 1982,73:207-217.
22.  Feldman, S et al.:  J. Infect. Dis., 1985,152:1212-1218.
23.  Hoskins, T W et al.:  Lancet, 1979,i:33-35.
24.  Johnson, P.R. et al.:  J. Med. Virol., 1985,17:325-335.
25.  Johnson, P R et al.:  J. Infect. Dis., 1986,154:121-127.
26.  Katz, J M et al.:  Virology, 1987,156:386-395.
27.  Mak, N K et al.:  Infect. Immun., 1982,38:218-225.
28.  Mak, N K et al.:  Immunol., 1984,51:407-416.
29.  Mann, J. et al.:  J. Immunol, 1968,100:725-735.
30.  McMichael, A.J. et al.:  Clin. Exp. Immunol., 1981,43:276-284.
30A. Mitchell, D M et al.:  Br. Med. Bull, 1985,41:80-85.
31.  Molecular  epidemiology of influenza viruses:  memorandum from a WHO meeting.   Bull. WHO, 1987,65:161-165.
32.  Murphy, B R et al.:  N. England J. Med., 1972,286:1329-1332.

33. Murphy, B R et al.: J. Infect. Dis., 1973,128:479-487.
34. Murphy, B R et al.: Infect. Immun., 1982,36:1102-1108.
35. New Vaccine Development, Establishing Priorities. Part I - Washington, D.C. Natl. Acad. Press, 1985:55.
36. Oxford, J S et al.: Bull. WHO, 1987,2:181-187.
37. Potter, C W, Oxford, J.S.: Br. Med. Bull., 1979,35:69-75.
38. Potter, C: In: Basic and Applied Influenza Research, ed. Beare A S, CRC Press, Florida 1982:119-156.
39. Prevention and control of influenza. Morbid. Mortal. Weekly Rep., 1986,35:317-325.
40. Puck, J M et al.: J. Infect. Dis., 1980,142:844-849.
41. Reiss, C S, Schulman, J L: J. Immunol., 1980,125:2182-2188.
42. Schild, G C et al.: Nature, 1983,303:706-709.
43. Tannock, G A et al.: Infect. Immun., 1984,43:457-462.
44. Tannock, G A et al.: Arch. Virol., 1987,92:121-133.
45. Webster, RG, Askonas, B A: Eur. J. Immunol., 1980,10:396-401.
46. Wright, P F et al.: J. Infect. Dis., 1982,146:71-79.
47. Wright, P F et al.: Infect. Immun., 1983,40:1092-1095.
48. Wright, P F et al.: In: Options for the control of influenza. Alan R. Liss, 1986:243-253.
49. Wright, P F et al.: J. Gen. Virol., 1987,68:433-440.
50. Wraith, P.: Immunol. Today, 1987,8:239-246.
51. Zahradnic, J.M. et al.: J. Med. Virol., 1983,11:277-285.

PHYSIOLOGICAL MECHANISMS OF PRODUCTION AND ACTION OF INTERFERONS IN
RESPONSE TO VIRAL INFECTIONS

F. Dianzani and G. Antonelli

Institute of Virology, University "La Sapienza"
Rome, Italy

The Interferons (IFNs) are the members of a family of proteins
produced by animal cells in response to a variety of stimuli. Although
originally described as an extraordinarily potent antiviral agents, they
were subsequently found to affect many other cellular functions and now
they are considered pleiotropic hormonal effectors. There are several
experimental and clinical evidences that IFNs represent an active
defense against viruses in vivo: i. in many viral infections a strong
correlation has been established between IFN production and natural
recovery; ii. inhibition of IFN production or action enhances the
severity of infection; iii. treatment with IFN usually protects animals
from several viral infections.

So far, three known types of IFN have been characterized: IFN-α (or
leukocyte), IFN-β (or fibroblastic), IFN-Υ (or immune). They show
different chemical and biologic characteristics and it is tempting to
speculate that they may have different physiological roles.

CLASSES OF IFN

Historically, the first type of IFN (fibroblastic IFN or IFN-β) was
discovered by Isaacs and Lindenmann during virus infection of cells
(1957). Viral and other foreign nucleic acids are the inducers of IFN-
production by most body cells (i.e., fibroblasts, epithelial cells, and,
as far as IFN- 2 is concerned, macrophages) (Marcus, 1987). Up now,
there are two known subtypes of IFN-β: IFN-β1 and IFN-β2 (Seghal, 1980).
While IFN-β1 is still considered only an antiviral protein with modest
immunomodulatory activity, IFN-β2 is emerging as an important mediator
of interaction between immunocompetent cells (Garman et al., 1987).

In 1961 a second type of IFN (IFN-α) was discovered that was
antigenically, genetically, and structurally different from IFN-β . This
IFN is produced by B lymphocytes, NK cells and macrophages induced by
viruses, foreign cells, virus infected cells, tumor cells, and bacterial
cells. Mitogens for B cells may mimic this induction (Dianzani and
Capobianchi, 1987). More recently another type of IFN-α has been found
in the serum of patients with AIDS or with autoimmune diseases (Chadha,
1987). This IFN appears to be antigenically identical to regular
IFN- but it is inactivated to acid pH as opposed to regular IFN-
that is stable. For this reason it is named IFN-α acid labile.

Up now, there are at least 18 subtypes of IFN-α; some of them are
not species-specific.

The third type of IFN (variously named immune IFN, type 2 IFN, or

IFN-γ) differs molecularly and antigenically from IFN-αand -β. IFN-γ is produced (along with other lymphokines) by T-specifically sensitized lymphocytes induced by their antigen. Mitogens for T cells may mimic this induction (Dianzani and Antonelli, 1987).

Several authors, in other systems, reported that also NK cells in appropriate conditions are able to produce IFN-γ . Only one type of IFN-γ has been isolated.

Important progress has been made in the fields of biochemistry and molecular genetics of the IFNs (Petska,1987; Joklik, 1985). The molecular characterists of IFNs and their gene location are shown in table 1.

Table 1. BIOCHEMICAL AND GENETIC CHARACTERISTICS OF IFNs.

| Characteristics | Types of IFN | | |
|---|---|---|---|
| | α | β | γ |
| Molecular Weight | 20K | 20K | 45K (20+25) |
| n. of genes        at least | 14 | 2 | 1 |
| n. of aminoacids | 166 | 166 | 145 |
| Presence of sugars | No | Yes | Yes |
| Exposure to pH2 | stable" | stable | unstable |
| Gene(s) located in Chromosome n. | 9 | 9,2,5 (?) | 12 |
| Presence of introns | No | No | Yes |
| Produced by | PBL | Fibroblast | T cells |

" Except acid labile IFNα .


"PHYSIOLOGICAL" PRODUCTION OF IFN

Generally the first cells to be infected in a viral infection are epithelial or fibroblastic cells which are IFNβ -producing cells. This event depends on penetration and uncoating of viral genome. The actual mechanism whereby viral RNA induces IFN β is still to be defined. This type of IFN appears to remain in the site of infection since blood levels are hardly detected even after exogenous administration. If the viral replication is not blocked by IFN β production, the virus can spread over linfatic or blood vessels thus interacting with mononuclear cells which are IFNα -producing cells. This event does not require virus penetration into the cell. On the contrary, it seems due to incorporation of viral glycoproteins into the plasma membrane of the cells followed by interaction and induction of B cells or macrophages. IFN-α is steadily released into the blood-stream and gives a relatively high plasma level that declines with time. IFN-γ is produced later in the infection, when the specific immune response causes the development of T lymphocytes sensitized against viral antigens. However, since many viruses have mitogenic properties, it is tempting to speculate that they may induce IFN also in unsensitized lymphocytes.

In order to establish the antiviral state, the IFNs have to interact with cells by binding to specific receptors present on the cell surface. IFNs inhibit viral replication by inducing the synthesis of antiviral proteins. The antiviral state, which is reversible, begins to develop within a few minutes and becomes maximum in about 5-8 hours to then declain in about 24 hours. Therefore the outcome of a viral infection depends on a sort of a race between the rate of virus replication and the rate of IFN production and action. Much depends on the effectiveness of the virus to stimulate IFN production and on the susceptibility of the virus to the antiviral action of IFN. Since IFN and the effector (antiviral) proteins must be produced de novo by the infected cells, it is easy to understand that rapidly replicating viruses, able to block cell protein synthesis at a very early stage of the infection, will escape the IFN defense more effectively than slow replicating viruses with little effect on the host macromolecular synthesis.

Along with the general mechanism of IFN induction in vivo there are several observations on the mechanism of IFN production at cellular level. It is worth noting that almost all types of viruses, both RNA- and DNA viruses, are able to induce IFN-β in the appropriate cells (Marcus, 1987). In general RNA viruses are good IFNβ -producers, while DNA viruses are poor IFN-β inducer. The mechanism underlying this phenomenon is still unclear. However it is known that the induction process requires virus internalization and exposure of viral nucleic acid. There is evidence that dsRNA upon infection is the IFN inducer moiety and this event can be triggered by one molecule of ds RNA per cell. At this regard, they way through DNA viruses can induce IFN as well, is still entirely unknown. A clou may be offered by the observation that single stranded "plus" RNA viruses can induce IFN without undergoing RNA replication and that IFN production is activated also by single stranded syntetic polyribonucleotides. Thus, IFN induction by DNA viruses might occur after synthesis of viral mRNA. After the induction process several cellular events leading to production of mRNA for IFN are triggered but the whole sequence of events leading to the derepression of IFN gene is not known.

The production of IFN alfa is activated by the interaction of several types of white blood cells with a number of viral and non viral stimuli. The triggering event seems to be a membrane interaction of the responder cells with the inducer, without the requirement of its engulfment (Dianzani et al., 1980; Capobianchi et al., 1985). The interaction is facilitated by high cell density and, at least in the case of viruses, requires the integrity of sugar network on the cell membrane. Furthermore this interaction is most efficient when the lymphocytes contact virus-infected cells rather than free virions. This can provide an efficient host defense mechanism against spread of viral infections through virus-infected cells present in the bloodstream or in the endothelial layers of the blood vessels; on the other hand increasing viral resistance of these latter cells enhances the effectiveness of natural barriers such as blood-brain, blood lung, etc.

Recently in our laboratory (Capobianchi et al., 1988), we demonstrated that the IFN induced by HIV-infected cells although sharing with IFN-α both antigenic properties and M.W., was strongly inactivated by treatment at pH lower than 4 (IFN- acid labile), while free infectious HIV induces only the classical IFN-α . The ability to induce acid-labile IFN-α was exerted both by the chronically-infected cell line H9/HIV and by PBMC acutely infected in vitro with HIV. The capacity of inducing acid labile IFN-α is shared by cells infected with other human retroviruses, and not by cells infected with other enveloped viruses, such a HSV, VSV, NDV, etc. Although the actual mechanism of this process is still to be defined, these findings suggest a role of this type of IFN in the progression of some disease such as AIDS, since the presence of acid labile IFN is usually reported only in the serum of patients affected with autoimmune disorders. Alternatively they could also suggest the responsibility of retroviruses in such disorders.

As regard to IFN-γ the data reported indicate that this type of IFN is produced by T lymphocytes probably following a calcium influx through the membrane (Dianzani et al., 1980) and that, for mitogenic and antigenic induction, oxidation of galactose residues on macrophage membrane is critically needed (Dianzani et al., 1982; Antonelli et al., 1985). In the presence of T cell mitogens or oxidizing agents macrophages elaborate, along with IL-1, which alone is not capable of inducing IFN-γ, another soluble factor named MBF, that is able of inducing production of IFN-γ and IL-2 by resting T cells. This latter event leads to lymphocyte differentiation and then to the modulation of the immune response. On the other hand the agents which induce an increase in intracellular calcium concentration can bypass the macrophage requirement since they act directly on T lymphocyte (Antonelli et al., 1985 and 1988). It remains to be established whether also specific antigens lead to production of IFN-γ through a similar sequence of events. Recent results from our laboratory indicate that macrophages can specifically participate to the process. In fact, when these cells are derived from immunized animals, they cooperate with T lymphocytes much more effectively than macrophages derived from unimmunized animals (Antonelli et al., 1986). In addition primed macrophages produce a monokine different from IL-1 and similar or equal to MBF, which is capable of inducing IFN-γ production in unprimed T cells (Dianzani et al. in preparation). The whole process of IFN-production is modulated by membrane components, such as β2 microglobulin (Antonelli et al., 1988),and soluble mediators such as prostaglandin E 2 and leukotrienes B 4 (Antonelli et al., in preparation).

## MECHANISM OF ANTIVIRAL ACTION OF IFN

As already stated, the IFNs do not directly inactivate the virus. Indeed, they must interact with specific receptors on the cell surface to induce an antiviral state in the treated cells. The receptors for the various types of IFN are different. In particular, IFN-α and β share the same receptor (coded for by chromosome 21) while IFN-γ binds to a different receptor (coded for by chromosome 6) (Branca and Baglioni, 1981; Zoon, 1987). It has been clearly shown that the induction of the antiviral state does not require internalization of IFN. However after IFN binding, the receptors are subject to diffusion, aggregation on the cell surface, endocytosis, and probably migration to the nucleus (Grossberg, 1986). It has not been established whether other IFN actions, such as anticellular, immunomodulatory, etc. require IFN uptake and internalization. These events lead to the establishment of an antiviral state which begins to develop within a few minutes and becomes maximum in about 5-8 hours (Dianzani and Baron, 1975).

Many IFN-induced proteins have been identified but only for some of them it is clear the association with the antiviral state (Fig.1).

It is worth nothing that treatment with IFN-α and IFN-γ enhanced the synthesis of 12 and 28 proteins, respectively and some of these proteins were induced in common by other cytokines (IL-1 and TNF) (Beresini et al., 1988).

The first mechanism to explain the antiviral activity of IFN involves the induction of the 2'-5' oligo A synthetase (Kerr et al., 1978; Baglioni and Maroney, 1980; Merlin et al.,1983). This enzyme induces the formation of a series of oligoadenylates which in turn activate a ribonuclease, RNAse L, that is normally present in a latent form in the cytoplasm. This latter enzyme requires, to be activated, also dsRNA and, once activated, it cleaves viral mRNA and inhibits viral protein synthesis.

A second step of the antiviral action of IFN is the activation of protein kinases (Lengyel, 1982; Petska et al.,1987). Like in 2'-5'oligo A system, the IFN-induced protein kinase is present in the mammalian cells, but only IFN is able to activate it, thus promoting the phosphorilation of different proteins (67K and 72K proteins) and the small fraction of initiation factor eIF2. Under these conditions formation of the tertiary initiation complex (GTP, eIF2, and aatRNA) does not take place (Revel, 1979; Samuel et al. 1979; Petska et al., 1987).

While the consequences of the phosphorilation of both protein (67K and 72K) are completely unknown, it is known that the phosphorilation of the factor eIF2 causes a dramatic fall of protein synthesis, thus blocking the viral protein synthesis as in the case of 2'-5' oligo A synthetase. It remains to be established the mechanism through which these enzymes can specifically block viral messages and not cellular messages.

Another protein which is present in IFN-treated cells is Mx protein (Horisberger, 1983). This protein has been shown to confer a cellular resistance specific for the influenza virus. Up now it is known that Mx protein inhibits transcription and translation of influenza mRNA but both the exact mechanism of this phenomenon and the reason why it acts only on influenza virus are still to be defined.

IFNs can establish an antiviral state through several other mechanisms.

First IFN can modify the structure and the fluidity of cell membrane (Joklik, 1985). In fact there are several evidences that IFN can, in some case, inhibit the penetration of VSV or the release of retroviruses, through an increase in membrane rigidity. Additionally, it has been demonstrated that in IFN-treated cells, some viruses are produced with some deficiencies in the envelope structure. Secondly, it has been reported that IFN can alter the mechanism of normal methylation of mRNA. For example mRNA of some viruses produced in IFN-treated cells, fail to shown cap methylation and then to associate with ribosomes.

IFNs can also interfere with the transformation of cell by tumor viruses. Although the studies are still in progress, it is noteworthy that IFNs markedly reduced the efficiency of transformation by virus such as SV40, KiSV, and RSV. In some case IFN acts on early viral gene expression (SV40), in other they act on inhibiting the synthesis or the integration of proviral DNA (retroviruses) (Joklik, 1985). Recently, it has been proposed that IFNs can also inhibit the cellular trasformation through an inhibition of several oncogene expression (Friedman, 1987).


IMMUNOMODULATORY ACTIVITY OF IFNs

Along direct antiviral activity, all types of IFNs can exert also immunopotentiating activity towards cellular functions which are normally induced in the response to viral infections. It is worth noting that, at the same antiviral unitage, IFN-γ has been reported to be more active than the other types of IFN in modulating immunity compared to direct inhibition of virus replication.

Studies on the action of IFNs on immunity have revealed a complex behavior of these cytokines. On one hand enhancement of the immune response may occur in some conditions, on the other hand cell-mediated immunity and humoral immunity can be inhibited by high doses of IFNs.

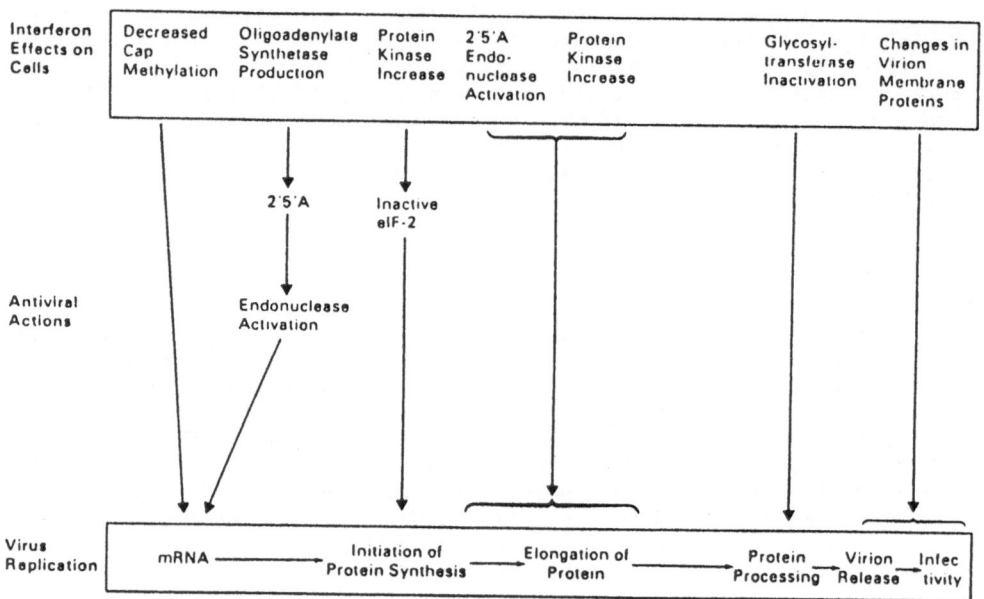

Figure 1

## Effect on macrophage activities

Macrophage activated by T cell-derived lymphokines play an important role in the defense against viral infections, as well as in cancer and bacterial diseases, and IFN- plays a major part in the activation of monocytes and macrophages. In fact macrophages stimulated with IFN- increase the expression of Ia or DR antigens (Steeg et al., 1982), show a higher oxidative metabolism and phagocytosis, inhibit the replication of many intracellular microorganism, such as Mycobacterium tuberculosis, Leishmania donovani, Toxoplasma gondii and show a higher expression of Fc receptor (Imanishi et al., 1975; Degre' et al., 1980; De Maeyer Guignard, 1987).

Moreover it has been reported that gamma IFN has a strong enhancing effects on the interleukin 1 (IL-1) secretory potential of human monocytes and that it is more efficient than IFN- or B at reversing the loss of endotoxin-induced IL-1 secretion observed in aged cultures (Arenzana-Seisdedos, 1985).

Macrophages have been reported also to be able to suppress some lymphocyte reactions, such as mitogen-induced blastogenesis, and this suppressive activity is reduced by IFN treatment. However while IFNα and β cause such impairement by reducing the release of prostaglandins (PGE) and O2, impairment of suppression by IFN-γ appears to be due to a decrease of prostaglandins production and induction of IL-1 by macrophages (Boraschi et al., 1984).

## Action on B lymphocytes

IFNs appear to act directly on B cells and can either inhibit or stimulate mitogen induced immunoglobulin production (Baron et al., 1987; De Mayer, 1987). The effect depends on the relative dosage or timing of exposure of immunocompetent cells to IFNs or to the antigen. Namely, IFN suppresses production of antibodies when added at high doses and at the same time as the antigen; on the contrary, it enhances the number of antibody-producing cells or the amount of immunoglobulin produced when added at low doses or after the antigen. All these conclusions ar derived from studies conducted in vitro. Less clear is the behavior of IFNs in vivo.

## Effect on cytotoxic cells

IFNs have been shown to be able to enhance spontaneous natural killer (NK) cell cytotoxicity directed against tumor or virus-infected cells in vitro and in vivo (Herberman, 1984).

The action of IFN could be explained both in terms of promoting the proliferation of NK progenitors, and in terms of enhancing the activity of already differentiated NK cells.

Earlier studies suggested that partially purified IFN-γ was more efficient in enhancing NK cell cytotoxicity than the other types of IFN. However, recent data have shown that recombinant IFN-γ is much less efficient than IFN-α and β in this respect. Differences in the target cell used, contamination of effector cells with cells of different phenotypes and/or partial inefficacy of recombinant IFN-γ as compared to the natural type, may explain such discrepancies.

As to specific T cell cytotoxicity, some reports indicate that IFNs can also aid in the generation of alloantigen-specific cytotoxic cells (Herberman, 1984). In fact IFN γ produced during T cell responses has been considered by some reports necessary for the differentiation and maturation of antigen specific T cells. However others authors suggest that the production of IFN-γ and generation of cytotoxic T lymphocytes are two indipendent and concomitant expression of the same type of immune response. In addition, the influence of IFN-γ has been found also in the development of the antibody-dependent cell-mediated cytotoxicity (ADCC) (Herberman, 1984).

## Enhancement of lymphocyte surface antigens

Several studies have shown that IFNs can play a role in enhancing the expression of lymphocyte surface antigens. The altered expression of such antigens could be part of the mechanism by which IFN modulate lymphocyte function. In fact it is well known that in the human system IFN-γ induces the appearance of new surface markers or receptors associated with differentiation and enhances expression of both class I- and class II-MHC antigens. This function is not unique for IFN-γ , since also IFN-α and β can have some effect on the expression of surface antigens in lymphocytes (Ameglio et al., 1983). However IFN-γ shows both wider range of effects, as compared to and B types, and stronger activity in terms of antiviral units requested. Recent studies have shown that IFN-γ induces mRNA for class II antigens. In particular such an effect is greater for HLA-DR and SB loci rather than for HLA-DC locus (Trinchieri, Perussia, 1985).

In the mouse system IFNs have been demonstrated to be the inducer of Ia antigen expression on murine macrophages. Increased Ia expression in response to IFN-γ has been proposed as a mechanism by which antigen presentation, mixed leukocytes reactions, and other inflammatory manifestations are effected and amplified.

Recently it has been reported that after exposure to IFN-γ not only lymphocytes but also other types of cells, such as amnion cells, are capable of expressing detectable levels of class I antigens and β2-microglobulin (Hunt, 1986).

## Effect on hematopoietic cell differentiation

The abnormal production of IFNs in patients with aplastic anemia, and the blood count depression associated with IFN therapy suggest that these molecules may have a role in regulating normal hematopoiesis (for a review see Klimpel, 1987). In fact several reports have showed that: IFN-α/β can inhibit in some circumstances erythroid differentiation (Rossi et al. 1977); IFN-γ and α show cooperative effects in the suppression of murine hematopoietic colony formation (Klimpel et al., 1982); IFNγ competes with colony-stimulating factor (CSF) in murine and human systems; moreover, IFN-γ is able to induce myeloid cells to express a series of surface markers, enzymes and functional activities characteristic of cells differentiating along the monocytic pathway (Perussia et al., 1983).

## Hormonal effect

There is some evidence suggesting that IFN may act on the cells with an hormonal-like activity (Inglot, 1983; Smith and, Blalock, 1987; In fact IFNs: i) act at distal as well as proximal sites; ii) has high specific activity; iii) does not need to be internalized but acts through cell surface receptors. However, differently from classical hormones, which are essentially tissue-specific but not species-specific, IFNs behave as species-specific hormones.

When applied to cultures of mouse myocardial cells, mouse IFN-α and causes a noradrenaline-like increase in their beat frequency. These observations are consistent with the finding that noradrenalin induces an IFN like antiviral state in the mouse myocardial cells. This suggests on one hand that the effects of both substances on the beat frequency and antiviral activity occur through a common mechanism, and on the other hand that, in addition to their known functions, classical hormones can also protect specific tissues against viruses. Furthermore there is evidence that IFN-α and β show common receptors with the two glycoprotein hormones thyrotropin (TSH) and human chorionic gonadotropin (HCG), and that biogen amines, such as histamine, serotonin and mexamine, can induce, in some conditions, circulating IFN in mice.

GENERAL CONSIDERATION ON IFN SYSTEM AND APPLICATION IN VIRAL THERAPY

The importance of the IFN defense against viral infections in vivo has been established by many studies in which theoretical or experimental considerations have been proposed to demonstrate the absolute defensive role of IFN. However new and more detailed work is necessary for definitive elucidation of mechanisms underlying production and action of IFN in vivo. In fact, along with many reports which demostrate the critical defensive role of the IFN system, there are some studies in which it is demostrated that IFNs play a role in the pathogenesis of viral infection. For example, Wabuke-Bunoti et al. reported that treatment with either anti-IFN-α or IFN-β ameloriated the disease and reduced mortality in mice lethally infected with influenza virus (1986). In addition, despite the poor ability to induce IFN in vitro, picornavirus infections in vivo cause a quick IFN response in the host (Dianzani et al., 1988). IFN response can be modulated by several host-determined factors and the immune mechanisms are certainly involved in the amplification of IFN effects either by ameliorating or by worsening the course of the disease according to what arm of the immune system is actually interacting with IFN. There are other systems in which pathogenesis of viral infections is correlated with the production of IFN (Gresser et al, 1976).

The whole mechanism is still under study but it is tempting to speculate that IFN produced in vivo may contribute to the cell damage, through activation of cytotoxicity (NK, T cells, LAK, etc.) towards infected cells.

Clinical application of IFN to viral diseases, already started in the sixties, has still to be considered at a preliminary stage. Several considerations support this conclusion. First, pharmakokinetics and pharmakodinamics of the IFNs is not yet fully defined; secondly, the trials conducted over the years are mostly unhomogenous as far as type or purity of the IFNs used, dosage, skedule of administration, selection of patients, evaluation criteria are concerned; thirdly, it is becoming increasingly clear that only long term infections are suitable for treatment, since during an acute viral infection the amount of endogenous IFN produced, especially during the early critical phases after onset, largely exceed the amount which can be given exogenously; finally most of the subtypes which compose IFN-α and one of the subtypes of IFN-β are not yet available for clinical trials.

So far best results in clinical trials have been obtained in infections by papovavirus and hepadnavirus and some efficacy has been reported in infections by herpesviruses and, profilactically, by rhinoviruses (table 2) (Merigan et al., 1973; Quesada et al., 1987; Rizzetto et al., 1987; Cesario et al., 1987; Billiau, 1981; Taylor-Papadimitriou, and Balkwill, 1982).

Table 2. MAIN CLINICAL APPLICATIONS OF IFN IN VIRAL INFECTIONS

| Disease | Type of IFN | (% CR+PR) |
|---|---|---|
| Chronic hepatitis B | α | 50 |
| Hepatitis NANB | α | 50 |
| Hepatitis delta | α | 50 |
| Papovavirus infections | α | 50 |
| Rhinovirus infections | α | 70 |
| Herpes keratitis" | α,β | 80 |

"In association with Acyclovir

REFERENCES

Ameglio, F., Capobianchi, M.R., Dolei, A., and Tosi, R., 1983, Differential effects of gamma interferon on expression of HLA class II molecules controlled by the DR and DC loci, Infect Immun, 42: 122.

Antonelli, G., and Dianzani, F., 1985, Induction of human interferon-gamma by calcium ionophores: lack of macrophage requirement, IRCS Med Sci, 13: 59.

Antonelli, G., Blalock, J.E., and Dianzani, F., 1985, Generation of a soluble IFN-gamma inducer by oxidation of galactose residues on macrophages, Cell Immunol, 94: 440.

Antonelli, G., and Dianzani, F., 1986, Antigen presentation by specifically sensitized macrophages in interferon-gamma induction, J IFN Res, 6: 535.

Antonelli, G., Amicucci, P., Cefaro, A., Ausiello, C., Malavasi, F., and Dianzani, F., 1988, Mechanism of human interferon-gamma production: involvement of B-2 microglobulin, Cell Immunol, 114: (in press).
Antonelli, G., Dianzani, F., Van Damme, J., Amicucci, P., De Marco, F., and Cefaro, A., 1988, A macrophage-derived factor different from interleukin 1 and able to induce interferon-gamma and lymphoproliferation in resting T lymphocytes, Cell Immunol, 113: 376.

Arenzana-Seisdedos, F., Virelizier, J.L., and Fiers, W., 1985, III. Preferential effects of interferon-gamma on the interleukin 1 secretory potential of fresh or aged human monocytes, J Immunol, 134: 2444.

Beresini, M.H., Lempert, M.J., and Epstein, L.B., 1988, Overlapping polypeptide induction in human fibroblasts in response to treatment with interferon-alpha, interferon-gamma, interleukin 1 alpha, interleukin 1 beta, and tumor necrosis factor, J Immunol, 140: 485.

Baglioni, C., and Maroney, P.A., 1980, Mechanism of action of human interferons: induction of 2',5'-oligo(A) polymerase, J Biol Chem, 255: 8390.

Billiau, A., 1981, Interferon therapy: pharmacokinetic and pharmacological aspects, Arch Virol, 67: 121.

Boraschi, D., Censini, S., and Tagliabue, 1984, Interferon-gamma reduces macrophage-suppressive activity by inhibiting prostaglandin E2 release and inducing interleukin 1 production, J Immunol, 133: 764.

Branca, A.A., and Baglioni, C., 1981, Evidence that type I and II interferons have different receptors, Nature, 294: 768.

Capobianchi, M.R., Facchini, J., Di Marco, P., Antonelli, G., and Dianzani, F., 1985, Induction of alpha interferon by membrane interaction between viral surface and peripheral blood mononuclear cells, Proc Soc Exp Biol Med, 178: 551.

Capobianchi, M.R., De Marco, F., Di Marco, P., and Dianzani, F., 1988, Acid-labile human interferon alpha production by peripheral blood mononuclear cells stimulated by HIV-infected cells, Arch Virol, 99: 9.

Cesario, T, Yousefi, S., Carandang, G., and Tilles, J., 1987, Interferon use during other virus infections, in: "The Interferon System: a current review to 1987", S. Baron, F. Dianzani, G.J. Stanton, W.R. Fleischmann Jr, ed., The University of Texas Medical Branch Series in Biomedical Science, Austin, p. 447.

Chadha, K.C., and Ikossi, M.G., 1987, Acid-labile human interferon alpha, in: "The Interferon System: a current review to 1987", S. Baron, F. Dianzani, G.J. Stanton, W.R. Fleischmann Jr, ed., The University of Texas Medical Branch Series in Biomedical Science, Austin, p. 31.

De Maeyer, E., and De Maeyer-Guignard, J., 1987, Interferon effects on cellular and humoral immunity, in: "The Interferon System: a current review to 1987", S. Baron, F. Dianzani, G.J. Stanton, W.R., Fleischmann Jr, ed., The University of Texas Medical Branch Series in Biomedical Science, Austin, p. 327.

De Maeyer, E., and De Maeyer-Guignard, J., 1981, Interferons as regulatory agents of the immune system, CRC Crit Rev Immunol, 2: 167.

Dianzani, F., and Baron, S., 1975, Unexpectedly rapid action of human interferon in physiological conditions, Nature, 257: 682.

Dianzani, F., Monahan, T.M., Zucca, M., Jordan, C., 1980, Disrupted virus induces interferon in lymphoid cells but not in other cultured cells. In: Kahn A., Hill NO, Dorn G. Eds. Interferon: Properties and Clinical Uses. Dallas, Texas, Wadly Inst. Mol. Med., p. 223.

Dianzani, F., Monahan, T.M., Georgiades, J., and Alperin, J.B., 1980, Human immune interferon: induction in lymphoid cells by a calcium ionophore, Infect Immun, 29: 561.

Dianzani, F., Monahan, T.M., and Santiano, M., 1982, Membrane alteration responsible for the induction of gamma interferon, Infect Immun, 36: 915.

Dianzani, F., and Capobianchi, M.R., 1987, Mechanism of induction of alpha interferon, in: "The Interferon System: a current review to 1987", S. Baron, F. Dianzani, G.J. Stanton, W.R. Fleischmann Jr, ed., The University of Texas Medical Branch Series in Biomedical Science, Austin, p. 21.

Dianzani, F., and Antonelli, G., 1987, Mechanism of induction of gamma interferon, Interferon induction, and its suppression, by viruses. in: "The Interferon System: a current review to 1987", S. Baron, F. Dianzani, G.J. Stanton, W.R. Fleischmann Jr. ed., The University of Texas Medical Branch Series in Biomedical Science, Austin, p. 51.

Dianzani, F., Capobianchi, M.R., Matteucci, D., and Bendinelli, M., 1988, The role of interferon in picornavirus infections, in: "Coxsackieviruses-A general update", M. Bendinelli, H. Friedman, ed., Plenum Press, New York and London.

Friedman, R.M., 1987, Interferon-induced reversion in oncogene-trasformed cells, in: "The Interferon System: a current review to 1987", S. Baron, F. Dianzani, G.J. Stanton, W.R. Fleischmann Jr, ed., The University of Texas Medical Branch Series in Biomedical Science, Austin, p. 409.

Garman, R.D., Kenneth, A.J., Steven, Clarck, C., and Raulet, D.H., 1987, B-cell-stimulatory factor 2 (B2 interferon) functions as a second signal for interleukin 2 production by mature murine T cells, Proc Natl Acad Sci USA, 84: 7629.

Gresser, I., Tovey, M.G., Bandu, M.T., Maury, C., and Brouty-Boy] D., 1976, Role of interferon in the pathogenesis of virus disease in mice as demonstrated by the use of anti-interferon serum, J Exp Med, 144: 1305.

Grossberg, S.E., Kushnaryov, V.M., Macdonald, H.S., and Sedmak J.J., 1986, Nuclear localization of internalized interferons- and , J. IFN Research, 6: 102.

Herberman, R.B., 1984, Interferon and cytotoxic effector cells, in: "Interferon vol. 2: interferon and the immune system", J., Vilcek, and E., De Maeyer, ed., Elsevier Science Publishers BV p. 61.

Horisberger, M.A., Staeheli, P., and Haller, O., 1983, Interferon induces a unique protein in mouse cells bearing a gene for resistance to influenza virus, Proc Natl Acad Sci USA, 80: 1910.

Hunt, J.S., and Wood, G.M., 1986, Interferon-gamma induces class I HLA and B2-microglobulin expression by human amnion cells, J Immunol, 136: 364.

Imanishi, J., Yokota, Y., Kishida, T., Mukainaka, T., Matsud, A., 1975, Phagocytosis-enhancing effect of human leukocyte interferon preparations of human peripheral monocytes in vitro, Acta Virol, 19: 52.

Inglot, A.D., 1983, The hormonal concept of interferon, Arch Virol, 76:1.

Isaacs, A., and Lindenmann, J., 1957, Virus interference: I. The interferon. Proc R Soc Lond (B), 147: 258.

Joklik, W.K., 1985, Interferons, in: "Virology", B.N., Fields, et al, ed., Raven Press, New York, 281.

Kerr, I.M., and Brown, R.E., 1978, pppA2'p5'A2'p5'A: An inhibitor of protein synthesis synthesized with an enzyme fraction from interferon-treated cells. Proc. Natl. Acad. Sci. USA, 75: 256-260.

Klimpel, G.R., Fleischmann, W.R. and Klimpel, K.D., 1982, Gamma interferon (IFNγ) and IFNα/β suppresses murine myeloid colony formation (CFU-C) magnitude of suppression is dependent upon level of colony stimulating factor (CSF). J. Immun., 129: 76.

Klimpel, G.R., and Reyes, V.E., 1987, Interferon as a Regulator of Hematopoiesis, in:"The Interferon System: a current review to 1987", S. Baron, F. Dianzani, G.J. Stanton, W.R. Fleischmann Jr, ed., The University of Texas Medical Branch Series in Biomedical Science, Austin, p. 261.

Lengyel, P., 1982, Biochemistry of interferons and their actions, Ann Rev Biochem, 51: 251.

Marcus, P.I., 1987, Interferon induction, and its suppression, by viruses, in: "The Interferon System: a current review to 1987", S. Baron, F. Dianzani, G.J. Stanton, W.R. Fleischmann Jr, ed., The University of Texas Medical Branch Series in Biomedical Science, Austin, p. 41.

Merigan, T.C., Reed, S.E., Hall, T.S., Tyrrell, D.A.J., 1973, Inhibition of respiratory virus infection by locally applied interferon. Lancet, 1: 563.

Merlin, G., Chebath, J., Benech P., Metz R., and Revel, M., 1983, Molecular cloning and sequence of partial cDNA for interferon-induced (2'-5') oligo (A) synthetase mRNa from human cells. Proc. Natl. Acad. Sci. Usa. 80: 4904.

Perussia, B., Dayton, E., Fanning, T., Thiagarajan, P., Hoxie, J., and Trinchieri, G., 1983, Immune interferon and leukocyte-conditioned medium induce normal and leukemic myeloid cells to differentiate along the monocytic pathway, J Exp Med, 158: 2058.

Petska, S., Langer, J.A., Zoon, K.C., Samuel, C.E., 1987, Interferons and their action. Ann. Rev. Biochem. 56: 727.

Quesada, J.R., 1987, Pharmacokinetics and toxicities of interferons, in: "The Interferon System: a current review to 1987", S. Baron, F. Dianzani, G.J. Stanton, W.R. Fleischmann Jr, ed., The University of Texas Medical Branch Series in Biomedical Science, Austin, p. 415.

Revel, M., 1979, Molecular mechanism involved in the antiviral effects of interferon. In: Interf. 1979, vol.2 edited by I. Gresser, Academic Press- New York, p. 101-163.

Rizzetto, M., Actis, G.C., and Barbara, Luigi, 1987, Interferon use during hepatitis virus infections, in: "The Interferon System: a current review to 1987", S. Baron, F. Dianzani, G.J. Stanton, W.R. Fleischmann Jr, ed., The University of Texas Medical Branch Series in Biomedical Science, Austin, p. 439.

Rossi, G.B., Matarese, G.P., Grappelli, C., Belardelli, F., and Benedetto, A., 1977, Inhibition of Erythroid differentiation of Dimethyl Sulfoxide-induced Friend Leukemia cells by Interferon. Nature, 267: 50.

Samuel, E.C., 1979, Mechanism of interferon action: Posphorilation of protein syntesis inibition factor eIF-2 interferon treated human cells by a rybosome-associated protein kinase possessing site specificity inhibition to heming-regulated rabbit reticulocyte kinase. Proc. Natl. Acad. Sci. USA, 76: 600.

Seghal, P.B., Sagar, A.D., 1980, Heterogeneity of poly(I) poly (C) induced human fibroblast mRNA Species, <u>Nature</u>, 288: 95.

Smith, E.M., and Blalock, E.J., 1987, Interactions of the interferon and endocrine systems, <u>in</u>: "The Interferon System: a current review to 1987", S. Baron, F. Dianzani, G.J. Stanton, W.R. Fleischmann Jr, ed., The University of Texas Medical Branch Series in Biomedical Science, Austin, p. 253.

Stanton, G.J., Weigent, D.A., Fleischmann Jr., W.R., Dianzani, F., ·Baron, S., 1987, State-of-the-Art in Medicine, Interferon Review, <u>Invest Radiol</u>, 22: 259.

Taylor-Papdimitriou, J., and Balkwill, F.R., 1982, Implications for clinical application of new developments in interferon research, <u>Biochim Biphys Acta</u>, 695: 49.

Wabuke-Bunoti, M.A.N., Bennink, J.R., Plotkin, S.A., 1986, Influenza virus-induced encephalopathy in mice: interferon production and natural killer cell activity during acute infection, <u>J Virol</u>, 60(3): 1062.

Zoon, W.C., and Okuno, T., 1987, Interferon receptors, in: "The interferon System: a current review to 1987", S. Baron, F. Dianzani, G.J. Stanton, W.R. Fleischmann Jr, ed. The University of Texas Medical Branch series in Biomedical Science, Austin, p. 221.

# AN INSIGHT INTO POLIOVIRUS BIOLOGY

R. Crainic*, T. Couderc*, A. Martin**, C. Wychowski**,
M. Girard**, and F. Horaud*

*Institut Pasteur, Unité de Virologie Médicale
75724 Paris 15, France
**Institut Pasteur, Unité de Virologie Moléculaire
75724 Paris 15, France

One of the most successful events in the control of infectious diseases has been the development of inactivated and oral poliovaccines in the 1950's. The intensive use of vaccines in some areas of the world, like North America and Europe, has resulted in the disappearance of outbreaks of disease while, in many developing countries, poliomyelitis is still a major public health problem. However, this considerable accomplishment was achieved at a time when we had very little basic knowledge of poliovirus biology. This situation has changed dramatically within the last years, because, thanks to progress in molecular biology and immunochemistry, powerful new tools have become available. This has also created a favourable climate for improving the safety and the efficacy of currently used poliovaccines and for replacing them with a new one which is scientifically better designed. Nonetheless, in the near future the most predictable consequence of contemporary research on poliovirus will be the scientific explanation for poliovaccine efficacy which is still an enigmatic problem. This is the reason why poliovirus antigenicity and virulence has became an important field of investigation in the recent years. Beyond the problem of poliovaccine efficacy, the present studies on poliovirus also have an important cognitive aspect. The simplicity of the structure of poliovirion should allow elucidation of relationships between chemical structure of the genotype and biological functions of the phenotype.

Poliovirus virions are relatively strong immunogens since 0,5 μg of virus is able to induce in an unprimed host a clearcut response in neutralizing antibodies (Svehag and Mandel, 1964). Within the last 5 or 6 years, study of the primary nucleotide structure of the poliovirus genome has been followed by sequence analysis of poliovirus RNA mutants resistant to neutralization by monoclonal antibodies (Mab) (Emini et al, 1983 ; Blondel et al, 1986 ; Minor et al, 1986) and substantiatied by high-resolution X-ray cristallography of poliovirus (Hogle et al, 1985) and rhinovirus (Rossman et al, 1985). These analyses converged and allowed the location and identification of the segments of capsid polypeptides at the surface of the virion which act as immunogens. Table 1 shows the location of immunogenic sites on viral polypeptides of the three poliovirus serotypes.

We focused our attention to the study of the antigenic site 1 (a.a. 89 - 100) of poliovirus type 1 (Horaud et al, 1987). The main reason for concentrating our efforts on this structure was our success in obtaining a Mab with unique properties. Since in our earlier work (Blondel et al, 1982) we showed that isolated structural polypeptide VP1 was able to induce neutralizing antibodies, we decided to generate Mabs using as antigen heat-inactivated poliovirus type 1 virions, generally called C antigen (in contrast to infectious virus, called D antigen). The Mabs obtained in this way were selected for their neutralizing properties. One of the Mabs, called C3, neutralized infectious virus and recognized heat-denatured virions (Blondel et al, 1983). It was assumed that the C3 Mab recognized a continuous epitope on the surface of poliovirus, i.e. a continuous sequence of amino acids at the surface of the native viral protein. This assumption was reinforced by the fact that Mab C3 reacted not only with infectious and heat-denatured virus, but also with isolated structural polypeptide VP1 in both immunoblot and immunoprecipitation tests.

Table 1. Location of antigenic sites in the three poliovirus serotypes.

| Antigenic site | Viral polypeptide location | Position of amino acid changes[1] | Serotype |
|---|---|---|---|
| 1 | VP1 | 89-100 | 1,2,3 |
| 2a | VP1 | 220-223 | 1 |
| 2b | VP2 | 164-172 | 1,3 |
| 2c | VP2 | 270 | 1 |
| 3a | VP1 | 286-290 | 3 |
| 3b | VP3 | 58-60, 70-73, 77-79 | 1,3 |
| 3c | VP2 | 72 | 1 |

[1] The amino acids substituted in the various structural polypeptides in neutralizate escape mutants (Emini et al, 1983 ; Blondel et al, 1986 ; Minor et al, 1986) are clustered on the viral capsid surface as revealed by X-ray analysis of crystallized poliovirus (Hogle et al, 1985).

In order to map topologically the amino acid sequence reacting with Mab C3, a series of plasmids expressing truncated poliovirus VP1 of Mahoney type 1 poliovirus proteins were constructed (Wychowski et al, 1983 ; Van der Werf et al, 1983). The truncated $\beta$-lactamase fusion proteins obtained in E. coli were examined by immunoprecipitation with C3 Mab. This approach allowed the identification of amino acids 92 through 104 of VP1, the VDNPASTTNKDKL sequence, as the C3 neutralization epitope of Mahoney strain.

Another type 1 poliovirus, the attenuated Sabin type 1 strain, is also neutralized by Mab C3. This virus is cleaved by trypsin at a single site, amino acid position 99 [K] of VP1, without loosing its infectivity (Fricks et al, 1985). Trypsin-treated Sabin strain no longer reacts with Mab C3, which indicates that the amino acid sequence around residue 100 of VP1 is implicated in the binding site of this antibody.

A synthetic peptide covering amino acids 92-104 of Mahoney strain VP1 induced antibodies that neutralized type 1 poliovirus when inoculated into rabbits (Horaud et al, 1987). We thus concluded that the epitope recognized by the C3 Mab is a continuous neutralization epitope located in the loop formed by amino acids 92-104.

The pathogenicity of a virus for a particular host depends on several factors, among which the most important are the existence of a specific viral receptor on the surface of susceptible cells and the ability of the virus to replicate efficiently inside the cell. This latter property is controlled by the viral genome, since change in a few nucleotides may dramatically attenuate virulence (Almond, 1987). It might also be supposed that viral proteins situated on the surface of the virion play an important role in virulence of the virus since such structures are necessary for the recognition of the viral-specific cellular receptor. One may therefore ask whether the protein domain in which the neutralization antigenic sites are located on the virion surface also plays a role in virulence. Several approaches developed in recent years have allowed confirmation of the role of structural proteins in virulence/attenuation in all 3 serotypes, (see review by Racaniello, 1987). An important observation was made by La Monica et al. (1987a) who reported that specific amino acid changes within the antigenic site 1 of type 2 poliovirus Lansing strain (the amino acid loop containing residues 89 through 104 of VP1), occurring as a consequence of selection of mutants escaping neutralization with monoclonal antibodies, may result in a noticeable reduction of mouse neurovirulence.

Within the last two years, it has been demonstrated that, by genetic engineering of the poliovirus genome, it was possible to create chimaeric poliovirions, i.e. virions of one serotype carrying on the same capsid its own and a heterotypic antigenic determinant. This avenue has been explored to further our understanding of neurovirulence (Stanway et al, 1986 ; La Monica et al, 1987b ; Murray et al, 1988a ; Kohara et al, 1988). Recently, our group at the Pasteur Institute -Paris (Martin et al, 1988b), simultaneously with the groups of Wimmer and of Racaniello, USA (Murray et al, 1988) were able to demonstrate that, besides the antigenic determinant, the site 1 of poliovirus type 2 also contains a peptide sequence controlling its neurovirulence for mouse. Both groups used basically the same methodology, the site-directed "cartridge mutagenesis" (Kuhn et al, 1987), which facilitates exchange of very small regions of the genome. In both studies the region of the antigenic site 1 of type 1 Mahoney, which is not neurovirulent for mice, was replaced with the equivalent segment derived from the neurovirulent type 2 Lansing strain. It was thus shown that the presence of only 6 amino acids specific for Lansing strain could render the Mahoney virus virulent for mouse (Fig. 1).

This finding indicates that this amino acid sequence 94-102 of VP1 of Lansing type 2 poliovirus is involved in mouse host specificity. Exactly the same sequence is carried by the antigenic site 1 of the attenuated Sabin type 2 virus, which is not neurovirulent for mice. This reinforces the idea that VP1 94-102 amino acid sequence of type 2 poliovirus must be important in virus attachment to murine neurons. It has not yet been established whether the presence of a different residue at position 103 of VP1 of type 2 Sabin virus as compared to the wild Lansing strain affect virus attachement to mouse cells.

It has thus been demonstrated, for the first time, that an amino acid sequence is involved both in an immunogenic site and in animal host range specificity. However, this is not surprising, since these two biological properties depend on segments of the capsid which are expected to be at the surface of the virion.

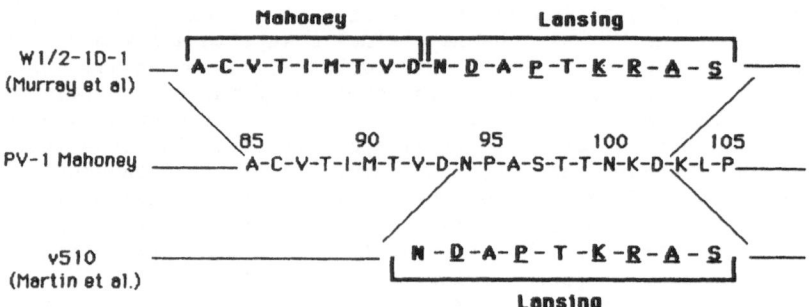

Figure 1. Amino acid sequences mouse neurovirulent chimaeric Mahoney type 1 (PV-1) and Lansing type 2 (PV-2) poliovirus. The amino acid sequence corresponding to the C3 epitope of PV-1 Mahoney strain was exchanged with the sequence of the PV-2 Lansing strain by substituting the corresponding nucleotide sequence in the PV-1 genome. Numbering refers to amino acid positions in VP1 of PV-1. Bold type letters correspond to the one letter code amino acid sequence substituted by Murray et al (1988)or by Martin et al (1988), respectively. The amino acids which are different in PV-2 as compared to PV-1 are underlined.

REFERENCES

Almond, J.W., 1987, The attenuation of poliovirus neurovirulence, Ann. Rev. Microbiol., 41:153.

Blondel, B., Crainic, R., and Horodniceanu, F., 1982, Le polypeptide structural VP1 du poliovirus induit des anticorps neutralisants, C. r. hebd. Séanc. Acad. Sci., Paris, 294:91.

Blondel, B., Akacem, O., Crainic, R., Couillin, P., and Horodniceanu,F., 1983, Detection by monoclonal antibodies of an antigenic determinant critical for poliovirus neutralization present on VP1 and on heat-inactivated virion, Virology, 126:707.

Blondel, B., Crainic, C., Fichot, O., Dufraisse, G., Candrea, A., Diamond, D., Girard, M., and Horaud, F., 1986, Mutations conferring resistance to neutralization with monoclonal antibodies in type 1 poliovirus can be located outside or inside the antibody-binding site, J. of Virol., 57:81.

Emini, A., Kao, S.Y., Lewis, A.J., Wimmer, E., and Crainic, R., 1983, The functional basis of poliovirus neutralization determined with monospecific neutralizing antibodies, J. Virol., 46:466.

Fricks, C.E., Icenogle, J.P., and Hogle, J.M., 1985, Trypsin sensitivity of the Sabin strain of type 1 poliovirus : cleavage sites in virions and related particles, J. Virol, 54:856.

Hogle, M.J., Chow, M., and Filman, J.D., 1985, Three-dimensional structure of poliovirus at 2.9 A resolution, Science, 229:1.

Horaud, F., Crainic, R., Van der Werf, S., Blondel, B., Wychowski, C., Akacem, U., Bruneau, P., Couillin, P., Siffert, U., and Girard, M., 1987, Identification and characterization of a continuous neutralization epitope (C3) present on type 1 poliovirus, Prog. Med. Virol., 34:129.

Kohara, M., Abe, S., Komatsu, T., Tago, K., Arita, M, and Nomoto, A., 1988, A recombinant virus between the Sabin 1 and Sabin 3 vaccine strains of poliovirus as a possible candidate for a new type 3 poliovirus live vaccine strain, J. Virol., 62:2828.

Kuhn, R.J., Tada, H., Ypma-Wong, M.F., Dunn, J.J., Semler, B., and Wimmer, E., 1987, Construction of a "mutagenesis cartridge", for poliovirus genome-linked viral protein : Isolation and characterization of viable and nonviable mutants. Proc. Natl. Acad. Sci. (USA), 85:519.

La Monica, N., Kupsky, W.J., and Racaniello, V.R., 1987a, Reduced mouse neurovirulence of poliovirus Type 2 Lansing antigenic variant selected with monoclonal antibodies, Virology, 161:429.

La Monica, N., Almond, J.W., and Racaniello, V.R., 1987b, A mouse model for poliovirus neurovirulence identifies mutations that attenuate the virus for humans, J. Virol., 61:2917.

Martin, A., Wychowski, C., Couderc, T., Crainic, R., Hogle, J., and Girard, M., 1988, Engineering a poliovirus type 2 antigenic site on a type 1 capsid results in a chimaeric virus which is neurovirulent for mice, Embo J., 7:2839.

Minor, P.D., Ferguson, M., Evans, D.M.A., Almond, J.W., and Icenogle, J.P., 1986, Antigenic structure of polioviruses of serotypes 1, 2 and 3, J. Gen. Virol., 67:1283.

Murray, M.G., Kuhn, R.J., Avita, M., Kawamura, N., Nomoto, A., and Wimmer, E., 1988a, Poliovirus type 1/type 3 antigenic hybrid virus constructed in vitro elicits type 1 and type 3 neutralizing antibodies in rabbits and monkeys, Proc. Natl. Acad. Sci. (USA), 85:3203.

Murray, M.G., Bradley, J., Yang, X.F., Wimmer, E., Moss, E.G., and Racaniello, V.R., 1988b, Poliovirus host range is determined by a short amino acid sequence in neutralization antigenic site 1, _Science_, 241:213.

Racaniello, V.R., 1987, Viral sequences required for neurovirulence of poliovirus, _Bioassays_, 5:266.

Rossman, M.G., Arnold, E., Erickson, J.W., Frankenberger, E.A., Griffith, J.P., Hecht, H.J., Johnson, J.E., Kamer, G., Luo, M., Mosser, A.G., Rueckert, R.R., Sherry, B., and Vriend, G., 1985, The structure of human common cold virus (rhinovirus 14) and its functional relationships to other picornaviruses, _Nature_, Lond., 317:145.

Stanway, G., Hughes, P.J., Westrop, G.D., Evans, D., Dum, G., Minor, P.D., Schild, G.C., and Almond, J.W., 1986, Construction of poliovirus intertypic recombinants by use of cDNA, _J. Virol_, 57:1187.

Svehag, S.E., and Mandel, B., 1964, The formation and properties of poliovirus neutralizing antibody. I. 19s and 7s antibody formation. Differences in kinetics and antigen dose requirement for induction, _J. Exp. Med._, 119:1.

Van der Werf, S., Wychowski, C., Bruneau, P., Blondel, B., Crainic, R., Horodniceanu, F., and Girard, M., 1983, Localization of a poliovirus type 1 neutralization epitope in viral capsid polypeptide VP1, _Proc. Natl. Acad. Sci. (USA)_, 80:5080.

Wychowski, C., Van der Werf, S., Siffert, O., Crainic, R., Bruneau, P., and Girard, M., 1983, A poliovirus type 1 neutralization epitope is located within amino acid residues 93 to 104 of viral capsid polypeptide VP1, _Embo J._, 2:2019.

STRATEGIES FOR THE DEVELOPMENT OF A ROTAVIRUS VACCINE AGAINST INFANTILE

DIARRHEA WITH AN UPDATE ON CLINICAL TRIALS OF ROTAVIRUS VACCINES

Albert Z. Kapikian[1], Jorge Flores[1], Karen Midthun[1], Yasutaka Hoshino[1], Kim Y. Green[1], Mario Gorziglia[1], Kazuo Nishikawa[1], Robert M. Chanock[1], Louis Potash[2], and Irene Perez-Schael[3]

[1]Laboratory of Infectious Diseases, National Institute of Allergy and Infectious Diseases, National Institutes of Health, Bethesda, Maryland, 20892; [2]Flow Laboratories, Inc., McLean, Virginia, 21202; [3]Central University of Venezuela, Caracas, Venezuela

INTRODUCTION

It is an honor and privilege to participate in this Symposium on "The Immune Response to Viral Infections" in this historic, celebrated city of countless renowned splendors where artistic and intellectual giants such as Leonardo da Vinci, Michelangelo, Dante, Machiavelli, Giotto, Botticelli, Galileo, Cellini, The Medici and the navigator Amerigo Vespucci exerted such profound influence, for it was here that Florentine scholars, painters, architects and craftsmen began the Renaissance. Although this pinnacle of human expression was unleashed with such intensity beginning some 500 years ago, thus laying the foundation for striking artistic, cultural and intellectual achievements, it is ironic that this energy did not successfully pollinate concurrently the domain of medicine since progress in discovering the etiologic agents of various scourges of humankind and elucidation of means to eliminate them were not forthcoming for centuries. Indeed, the mortality rate in the 15th century was quite high as documented in the Florentine dowry records of under one year to 19 year old females, with the highest rate occurring, by far, in the less than 5 year-old age group (Morrison et al., 1977).

In this presentation I plan to (1) outline the importance of the scourge of diarrheal diseases, which to this day are major causes of morbidity and mortality in infants and young children, (2) describe progress made in the past 16 years in elucidating the viral agents which cause a major portion of such illnesses, (3) summarize briefly some important properties of one of these viruses, (4) present strategies for the development of vaccines to reduce the enormous toll from viral diarrheal illnesses, and (5) update clinical trials with recently developed vaccines against infantile viral diarrhea.

THE SCOURGE OF DIARRHEAL DISEASES

Diarrheal diseases consistently account for a substantial segment of illnesses in developed countries and a major proportion of morbidity and mortality in the less developed countries with the greatest impact in infants and young children (Kapikian et al., 1986b). For example, infectious gastroenteritis was the second most common

Table 1

ESTIMATED ANNUAL INCIDENCE OF HUMAN ROTAVIRUS GASTROENTERITIS IN THE U.S.A.

| Age Group | Projected 1984 Population | Incidence Rate[a] (percent) | Predicted No. of cases |
|---|---|---|---|
| Under 12 months | 3,733,808 | 11 | 410,718 |
| 12-23 months | 3,684,199 | 40 | 1,473,680 |
| 24-35 months | 3,646,058 | 13 | 473,988 |
| 36-59 months | 7,168,425 | 8 | 573,474 |
| 1-4 years | | | 2,521,142 |
| 5-14 years | 33,514,883 | 5 | 1,675,744 |

[a]From Rodriguez (personal communication, 1984) and Rodriguez et al., 1987.

Adapted from:  Institute of Medicine Report, 1985.

disease experience in the Cleveland Family Study (U.S.A.) accounting for 16% of over 25,000 illnesses during a period of almost 10 years (Dingle et al., 1964). However, the impact of diarrheal diseases is especially striking in the developing countries with estimates of the number of cases ranging from 744 million to 3-5 billion per year (Snyder and Merson, 1982; Walsh and Warren, 1979). The number of deaths from diarrheal diseases in the developing countries was estimated to range from 4.6 million to 5-10 million annually with the major toll in infants and young children. Further dissection of the lower estimate reveals that 12,600 individuals, mostly infants and young children, die of a diarrheal illness each day of the year.

## PROGRESS IN ELUCIDATING THE ETIOLOGIC AGENTS OF DIARRHEA

The quest for etiologic agents of diarrhea was futile until the early 1970's. This was especially frustrating to virologists since it was generally assumed that viruses were responsible for most of these illnesses because bacteria, which accounted for only a small portion of diarrheal episodes, failed to fill this etiologic vacuum (Yow et al., 1970). This frustration was increased with the advent of the tissue culture era during which scores of enteric viruses were discovered but none was found to be the elusive major etiologic agent of viral gastroenteritis.

The discovery in 1972 of the 27 nm Norwalk virus as an etiologic agent of viral gastroenteritis in older children and adults (Kapikian et al., 1972), and the discovery in 1973 of the 70 nm human rotavirus as an etiologic agent of severe diarrhea in infants and young children (Bishop et al., 1973), marked the beginning of a new era in the elucidation of the cause of a substantial proportion of the viral gastroenterides (Figure 1). This lengthy, fallow period preceding the discovery of these diarrhea -causing agents is shown graphically in Figure 2. It is striking that both of these fastidious viruses were discovered without the benefit of tissue culture--rather, they were initially detected by electron microscopy- the Norwalk virus in feces using the technique of immune electron microscopy (Kapikian et al., 1972), and the rotavirus by thin section electron microscopy of duodenal mucosa, and shortly afterwards by electron microscopic examination of feces (Bishop et al., 1973, 1974;  Flewett et al., 1973). The Norwalk virus has yet to be cultivated in tissue culture, and only relatively recently have the rotaviruses been grown efficiently in such cultures (Kapikian and Chanock, 1985b; Sato et al., 1981; Urasawa et al., 1981). The Norwalk virus has been implicated etiologically with 40% of community-wide outbreaks of nonbacterial gastroenteritis in older children and adults (Greenberg et al., 1979; Kaplan et al., 1982b). In addition, approximately 10% of all gastroenteritis outbreaks (bacterial and non-bacterial) are estimated to be associated etiologically with the Norwalk group of viruses (Kaplan et al., 1982a). It is important to note, however, that the Norwalk virus, although capable of inducing mild gastroenteritis in infants and young children, is not the cause of severe diarrhea in this age group (Kapikian and Chanock, 1985b).

Rotaviruses have emerged as the single most important etiologic agents of severe diarrhea in infants and young children in both developed and developing countries. However, the outcome of such illnesses differs markedly in these areas. In developed countries, rotaviruses cause 35-50% of hospitalizations for diarrhea in infants and young children (Kapikian and Chanock, 1985a). Rotavirus illnesses exhibit a marked seasonal distribution, occurring almost exclusively in the cooler months of the year in ' temperate climates (Brandt et al., 1983; Konno et al., 1983). Rotavirus infections are prevalent in all populations regardless of socioeconomic status; thus, by the end of the third year of life over 90% of children have acquired serum antibodies (Kapikian and Chanock, 1985a). An estimate from a family study in the U.S.A. revealed that almost three million children up to 4 years of age develop rotavirus gastroenteritis annually

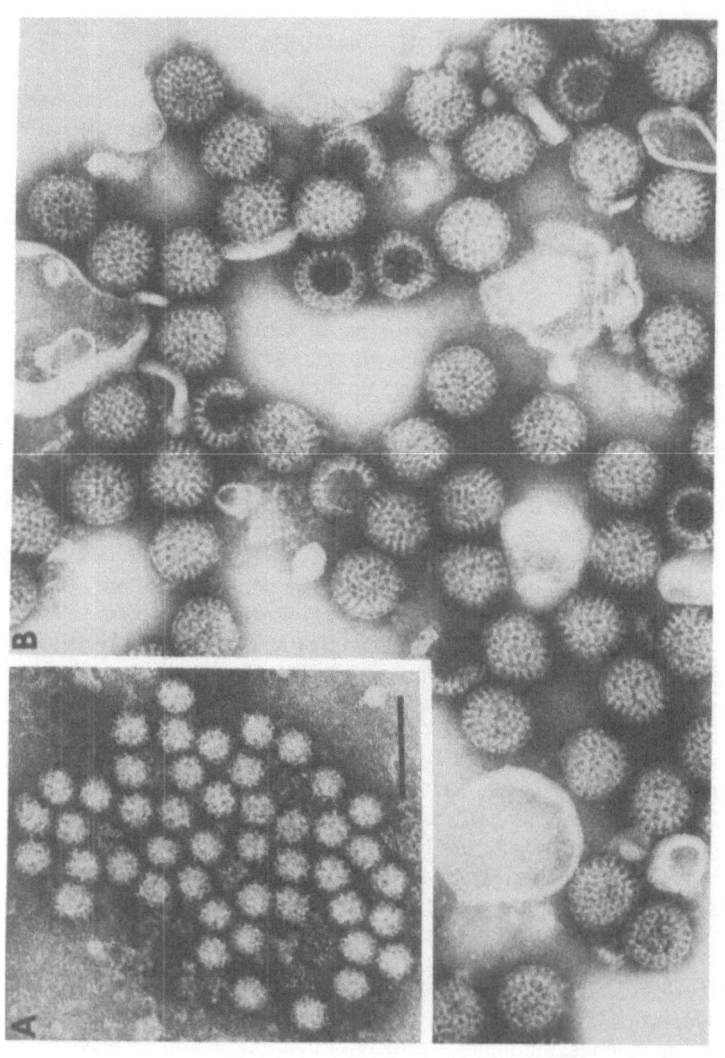

Figure 1A.   A group of Norwalk virus particles observed by immune electron microscopy in
a stool filtrate of a volunteer who developed gastroenteritis after oral administration
of the Norwalk agent (From Kapikian et. al., 1972 [bar added]). The bar=100nm.

B.   Rotavirus particles observed by electron microscopy in a stool filtrate of an infant with
gastroenteritis. The particles appear to have a double capsid. Occasional "empty" particles
are seen. (From: Kapikian et. al., 1973) The bar=100nm.

with the greatest burden in the 12-23 month age group (Table 1) (Institute of Medicine, 1985; Rodriguez et al., 1987). In another U.S.A. estimate based on the annual incidence of rotavirus gastroenteritis involving physician contact, the incidence was greatest in the under 12 month age group (15%) whereas the incidence in the 12-23 month age group was 5%; the total number of such cases up to four years of age was estimated at almost one million per year (Institute of Medicine, 1985; Koopman et al., 1984). Hospitalizations (U.S.A.) due to rotavirus diarrhea are estimated to range from 22,000 individuals annually in a large, pediatric population where medical care was provided by a health-maintenance organization, to over 80,000 annually in a pediatric population where care was provided by private practice (Institute of Medicine, 1985; Koopman et al., 1984; Rodriguez et al., 1980). The disease burden per year in the U.S.A. from rotavirus diarrhea in the 4 year and under age group is estimated to include over one million cases of severe diarrhea and 150 deaths (Table 2).

In developing countries, rotaviruses are characteristically the leading cause of severe diarrhea in infants and young children (Kapikian et al., 1986). For example, in Bangladesh, 46% of children under 2 years of age who visited a treatment center for diarrhea were rotavirus positive (Black et al., 1980). The next most frequently detected diarrheal pathogens were the enterotoxigenic E. coli, detected in 28% in this age group. Moreover, in Egypt, in a study of enteropathogens recovered from 145 infants and young children under 18 months of age with a fatal or potentially fatal diarrheal illness, rotaviruses were the most frequently detected etiologic agent, accounting for 34% of the cases, with the enterotoxigenic E. coli (LT and ST combined) ranking second (27%) (Shukry et al., 1986). Rotaviruses are ubiquitous agents. By the end of the third year of life, over 90% of children in developing countries have already developed serologic evidence of infection, a figure strikingly similar to that in developed countries; no major temporal variations are observed characteristically in the incidence of rotavirus illness in developing areas where seasonal climatic differences are not marked (Kapikian and Chanock, 1985). The burden of rotavirus diarrheal illnesses in children less than 5 years of age in developing countries was estimated to be over 125 million cases of which over 18 million were considered moderately severe or severe (Table 3) (Institute of Medicine, 1986). The number of deaths due to rotavirus diarrhea in this same age group was estimated to be 873,000 annually. This number translates to over 2300 deaths per day or almost 100 per hour from rotavirus diarrhea.

Thus, the need for a rotavirus vaccine is clear since the burden from severe rotavirus gastroenteritis in both developed and developing countries during the first two years of life is so great. Of course, the need is most urgent in the developing countries where the mortality from rotavirus disease is extremely high. Major emphasis is on the development of a live, attenuated orally administered vaccine since animal studies indicated that protection against rotavirus disease was mediated primarily by intestinal immunity (Snodgrass, et al., 1976). In spite of the proven effectiveness of oral rehydration salt solutions in the treatment of diarrheal diseases, the general availability and implementation of this therapy continues to be problematic, and thus the need for a rotavirus vaccine is not reduced.

CHARACTERISTICS OF ROTAVIRUSES IMPORTANT TO VACCINE DEVELOPMENT

A few key points relating to important characteristics of rotaviruses should be made before embarking on a discussion of rotavirus vaccines (Kapikian and Chanock, 1985a): (1) Rotaviruses belong to the family Reoviridae; (2) They are 70 nm in diameter and possess a distinctive double-shelled capsid with the outer margin giving the appearance of the rim of a wheel placed on short spokes radiating from a wide hub ( L. rota=wheel) (Figure 1A) (Flewett et al.,1974); (3) Within the inner capsid is the core which

Table 2

ESTIMATE OF THE BURDEN OF ROTAVIRUS DIARRHEA
IN THE UNITED STATES BY AGE AND SEVERITY

| | Age | | | | | |
|---|---|---|---|---|---|---|
| | Under 1 year | | 1-4 years | | 5-14 years | |
| Severity | No. of cases | Duration (days) | No. of cases | Duration (days) | No. of cases | Duration (days) |
| Mild diarrhea | 164,287 | 3 | 1,636,934 | 3 | 1,675,744 | 3 |
| Severe diarrhea | 246,431 | 5 | 884,208 | 5 | | |
| Dehydration | 13,740 | 4 | 10,126 | 4 | | |
| Death | 75 | | 75 | | | |

Adapted from: Institute of Medicine Report, 1985.

contains the genome consisting of 11 segments of double-stranded RNA ranging from 680-3,300 base pairs in length; (4) Each gene encodes a protein, but only 6 proteins form part of the virus structure as shown in Figure 3; (5) Rotaviruses have 3 important antigenic specificities- group, subgroup, and serotype; group and subgroup are mediated by VP6, the major inner capsid protein and serotype by VP7 and VP4 (the latter formerly designated VP3), both located on the outer capsid. Antigenic properties most relevant for vaccine development are mediated by the VP7 and VP4 proteins, each of which induces neutralizing antibodies (Hoshino et al., 1988). However, differentiation of rotavirus strains into distinct serotypes is specified conventionally by the VP7 protein, since with hyperimmune serum the antigenicity of this protein is predominant; (6) The major group antigen is shared by most human and animal rotaviruses (designated group A rotaviruses) (Bachmann et al., 1984). Some strains do not share this antigen and are thus classified into groups B-F, the "non-group A" rotaviruses, formerly designated "pararotaviruses" (Bridger et al., 1987); (7) Four epidemiologically important human rotavirus serotypes (VP7) have been characterized and are numbered 1-4 (Hoshino et al., 1984; Wyatt et al., 1982). Two new serotypes designated 8 and 9 were recently described, and their importance is still under study (Albert et al., 1985; Clark et al., 1987; Matsuno et al., 1985). Serotypes 5-7 have been detected only in animals; (8) Rotaviruses cause diarrhea in the young of almost all animal species studied.

## STRATEGIES FOR DEVELOPMENT OF A ROTAVIRUS VACCINE

### The Jennerian approach

Strategies for rotavirus vaccine development range from cell culture cultivation of strains derived from humans or animals to the application of newly-developed molecular biologic techniques (Table 4). The most promising and extensively evaluated strategy is based on the concept first used by Edward Jenner in 1798 for human smallpox vaccination in which a related, live attenuated agent derived from a nonhuman host is employed as the immunogen. This Jennerian strategy for development of a rotavirus vaccine was suggested when early studies revealed that human and animal rotaviruses share a common group antigen (Flewett et al.,1974; Kapikian et al. 1974, 1975, 1976; Woode et al., 1976); thus, pediatric patients infected with a human rotavirus developed serologic responses (CF) not only to the infecting human strain but also to certain animal rotavirus strains (Kapikian et al., 1976). The feasibility of this strategy was suggested by animal studies in which calves immunized with a bovine rotavirus (serotype 6) in utero were protected from illness following challenge at birth with a serotypically distinct human rotavirus (Wyatt et al., 1979).

Efficacy trials of a high tissue culture passage level cold-adapted bovine rotavirus NCDV strain vaccine, designated RIT 4237, developed by Smith Kline RIT demonstrated that this orally administered vaccine induced over 80% protection against clinically significant diarrhea in children 8-11 and 6-12 months of age in two separate studies in Finland (Vesikari et al., 1984, 1985). However, later, in several trials in developing countries, the vaccine failed to induce protection against rotavirus diarrhea (De Mol et al., 1986; Hanlon et al., 1987). Thus, this vaccine is no longer being evaluated for clinical efficacy. A related bovine rotavirus strain designated WC3 developed at the Wistar Institute has been evaluated in Phase 1 and early Phase 2 studies with promising results (Clark et al., 1986, 1988).

We have pursued the Jennerian approach with another animal rotavirus, the rhesus rotavirus (RRV) strain MMU-18006, which was recovered from a 3 1/2 month old monkey with diarrhea (Stuker et al., 1980) and which was later found to be

Table 3

THE BURDEN OF ROTAVIRUS DIARRHEA IN DEVELOPING COUNTRIES BY AGE AND SEVERITY

| Severity | Age | | | | | | | |
|---|---|---|---|---|---|---|---|---|
| | Under 5 Years | | 5-14 Years | | 15-59 Years | | 60 Years and Over | |
| | No. of cases | Duration (days) | No. of cases | Duration (days) | No. of cases | Duration (days) | No. of cases | Duration (days) |
| Mild diarrhea | 109,979,000 | 6 | 5,698,000 | 4 | 4,239,000 | 4 | 479,000 | 4 |
| Moderately severe diarrhea | 9,776,000 | 6 | 46,000 | 6 | 34,300 | 5 | 20,200 | 6 |
| Severe diarrhea | 8,729,000 | 7 | | 7 | | 7 | | 7 |
| Death | 873,000 | | | | | | | |

Adapted from: Institute of Medicine Report, 1986.

antigenically similar (via VP7) to human rotavirus serotype 3 by neutralization (Hoshino et al., 1984; Kapikian et al., 1985, 1986, 1988; Wyatt et al., 1982). This vaccine virus has been passaged 9 times in primary or secondary monkey kidney cell cultures and 7 times in DBS FRhL-2 cells, a semi-continuous diploid cell strain from rhesus monkey lung which was developed by the Office of Biologics (FDA) as a potential cell substrate for vaccine manufacture (Wallace et al., 1973).

Extensive, step-wise phase 1 studies with this orally administered vaccine in adults, and later in older children, infants and neonates, revealed a variable pattern of reactogenicity and a consistent pattern of satisfactory antigenicity with different doses in different locations (Anderson et al., 1986; Christy et al., 1986; Flores et al., 1988, Gothefors et al., 1989, Kapikian et al., 1985, 1986, 1988; Losonsky et al., 1986; Perez-Schael et al., 1987; Rennels et al., 1987b; Vesikari et al., 1986). However, a $10^4$ PFU oral dose was found to be safe and acceptably reactogenic, inducing a transient febrile response in about one-third of the vaccinees in certain locations, and a serum antibody response in over 50% of vaccinees. The absence of reactions in the neonatal period following vaccination in Venezuela was especially encouraging since vaccination soon after birth is a desirable goal (Table 5) (Flores et al., 1988). Perhaps high levels of passively-acquired maternal rotavirus antibody and/or host factors prevented vaccine reactions in this age group.

As shown in Table 6, twelve field trials involving over 1000 infants are completed, in progress, or planned with the RRV vaccine. The results from the Caracas trial (study #2) in 1-10 month old children in which 151 children received the vaccine and the same number the placebo were very encouraging since: (1) an overall efficacy rate of 64% against any rotavirus diarrhea was observed; (2) among the 1-4 month age group the vaccine efficacy was 82% against any rotavirus diarrhea; and (3) vaccine efficacy reached 90% for the more severely affected patients when the illnesses were graded by a scoring system ($\geq$ 8) (Table 7) (Flores et al., 1987; Kapikian et al., 1988, Perez-Schael et al., 1989). RRV vaccine trials were also encouraging in Maryland (study#1), Sweden (study#5) and Finland (study#8), but the vaccine failed to protect in a study of 2-4 month old infants in Rochester (study#3) and 2-5 month old infants in Arizona (study#4) (Gothefors et al., 1989; Christy et al., 1989; Kapikian et al., 1988; Rennels et al., 1987a; Vesikari et al., 1989).

These inconsistent results were difficult to understand until practical methods for identification of rotavirus serotypes became available (Taniguchi et al., 1987). The predominant infecting rotavirus in Venezuela was serotype 3 (13 of 23) the same serotype as the vaccine strain, whereas in the Rochester study it was serotype 1 (29 of 30) and in the Arizona study only 1 of 15 typeable strains belonged to serotype 3 (Christy et al., 1989; Kapikian et al., 1988; Perez-Schael et al., 1989). Thus, it appears that serotype-specific immunity is necessary for protection against rotavirus diarrhea in infants undergoing initial infection. The importance of serotype-specific serum neutralizing antibodies in protection against homotypic rotavirus infection or illness was shown in a study of gastroenteritis outbreaks in an orphanage in Japan (Chiba et al., 1986).

Modified Jennerian Approach

Since the Jennerian approach failed to protect against heterotypic rotavirus strains in the youngest age groups, we have modified this strategy with the goal of developing an effective multivalent vaccine containing viruses with the VP7 specificity of each of the 4 epidemiologically important serotypes and the attenuation phenotype of RRV. In this modified strategy, reassortant rotaviruses have been prepared by coinfecting cell cultures with both the RRV strain MMU18006 and a serotypically distinct human rotavirus under selective pressure of antibody against the rhesus rotavirus (Midthun et

Table 4

SELECTED APPROACHES TO DEVELOPMENT OF A HUMAN
ROTOVIRUS (RV) VACCINE

A. LIVE ATTENUATED ANIMAL RV STRAIN (THE JENNERIAN APPROACH)
    1. Bovine RV strain NCDV (RIT 4237), UK, WC3
    2. Simian RV strain MMU 18006 (RRV)

B. LIVE ATTENUATED HUMAN RV- ANIMAL RV REASSORTANT (MODIFIED
   JENNERIAN APPROACH)
    1. Bovine RV strain UK with single gene (encoding VP7)
       from human RV serotypes 1,2,3, or 4
    2. Simian RV strain MMU18006 (RRV) with single gene
       (encoding VP7) from human RV serotypes 1,2, or 4

C. LIVE ARTIFICIALLY ATTENUATED HUMAN RV STRAINS
    1. Cold adapted strains
    2. Multiply passaged tissue culture strains
    3. Strains passaged in heterologous hosts

D. LIVE NATURALLY ATTENUATED HUMAN NEONATAL RV STRAINS
    1. Serotypes 1 (M37), 2 (1076), 3 (McN13), 4 (ST3)

E. LIVE REASSORTANTS BETWEEN HUMAN RV STRAINS
    1. Ten genes from a neonatal strain and a single
       gene (encoding VP7) from serotypes 1,2,3, or 4
    2. Ten genes from attenuated RV from serotype 1,3,4
       family and a single gene (encoding VP7) from
       serotype 2

F. INACTIVATED HUMAN RV STRAINS
    1. Serotypes 1,2,3, or 4

G. RECOMBINANT DNA TECHNOLOGY
    1. Expression of VP7 and/or VP4 in viral or bacterial
       vectors
    2. Synthetic peptides


Table 5

CLINICAL REACTIONS RESULTING FROM ORAL ADMINISTRATION
OF $10^4$ PFU OF RHESUS ROTAVIRUS VACCINE TO NEWBORN INFANTS

| Clinical Reactions | Inoculum Administered | |
| --- | --- | --- |
| | Placebo | Vaccine |
| No. of infants studied | 20 | 20 |
| No. of infants with | | |
| Fever (≥38.1°C) | 0 | 1 |
| Diarrhea[a] | 1 (1)[b] | 2 (1) |
| Vomiting | 1 | 0 |
| Physiologic jaundice | 1 | 2 |

[a] Three or more liquid or semiliquid stools within a 24-hour
   period.
[b] Numbers in parentheses are the duration in days of indicated
   reaction.

From: Flores et al., 1988.

al., 1985, 1986). "Shuffling" (reassortment) of the rotavirus genes in cells infected by the two strains produces reassortant progeny with various combinations of genes. With this procedure, single gene substitution reassortants have been derived for human rotavirus serotypes 1,2, or 4 with each reassortant having 10 RRV genes and a single human rotavirus gene, that which encodes the major neutralization antigen VP7. This process is shown schematically for a human rotavirus serotype 2-RRV reassortant in figure 4.

Phase 1 studies of individual reassortants in adults and in progressively younger age groups revealed that these vaccine candidates were safe, that is they did not exhibit significant reactogenicity but were similar in this regard to the $10^4$ PFU dose of RRV vaccine. Their antigenicity was also similar to that of the RRV parent virus (Flores et al., 1989; Halsey et al., 1988, Kapikian et al., 1988). In addition, a bivalent formulation was also found to be comparable to RRV vaccine. We therefore extended our studies to evaluate a quadrivalent rotavirus vaccine formulation consisting of $0.25 \times 10^4$ PFU of D (human rotavirus serotype 1) x RRV, $0.25 \times 10^4$ PFU of DS-1 (human rotavirus serotype 2) x RRV, $0.25 \times 10^4$ PFU of RRV (serotype 3), and $0.25 \times 10^4$ PFU of ST-3 (human rotavirus serotype 4) x RRV in 2-5 month old infants (Perez-Schael et al., 1989). Since the quadrivalent vaccine formulation was safe, being similar to the $10^4$ PFU dose of the RRV vaccine in reactogenicity, additional infants in the same age group were given the same four components orally except that the dose of each component was increased to $0.5 \times 10^4$ PFU. Significant reactions were not observed. However, the number of seroresponses to each serotype by neutralization was disappointing as shown in tables 8 and 9. Before commencing the large single dose field trial shown in table 6 (study#11), a quadrivalent vaccine consisting of a $10^4$ PFU dose of each component will be evaluated for reactogenicity and antigenicity to determine if such a formulation would induce a higher rate of seroresponses to each component.

## Non-Jennerian Approaches to Rotavirus Vaccination

Other approaches to rotavirus vaccination are also being considered. A most promising strategy entails the use of a neonatal rotavirus strain(s) which appear to be naturally attenuated (Christie et al., 1975). The use of such a strain is based on the observation that neonates who developed a subclinical rotavirus infection during the first 14 days of life with one such strain were protected against clinically significant rotavirus diarrhea during a 3-year follow-up (Bishop et al., 1983). Another approach involves preparing reassortants between two human rotavirus strains with desirable VP4 and VP7 components (Hoshino et al., 1987).

Approaches to rotavirus vaccination based on the application of recently described genetic engineering principles include (1) the expression of VP7 and/or VP4 with bacterial or viral expression vectors and (2) the production of synthetic peptides.

## CONCLUSION

The past 16 years has seen a burst of activity in the elucidation of etiologic agents of viral gastroenteritis. Hopefully, this form of medical "renaissance" will lead to the development of a successful vaccine and begin to fulfill the goal of eliminating a major scourge of humanity-diarrheal diseases of infants and young children.

## Acknowledgments

We thank H.D. James, Jr., A.L. Pittman, M. Finbloom, J. Sears, J. Valdesuso, M.M. Sereno, R. Jones, J. Jackson, C. Banks, E. Williams, and S. Chang for assistance in these studies.

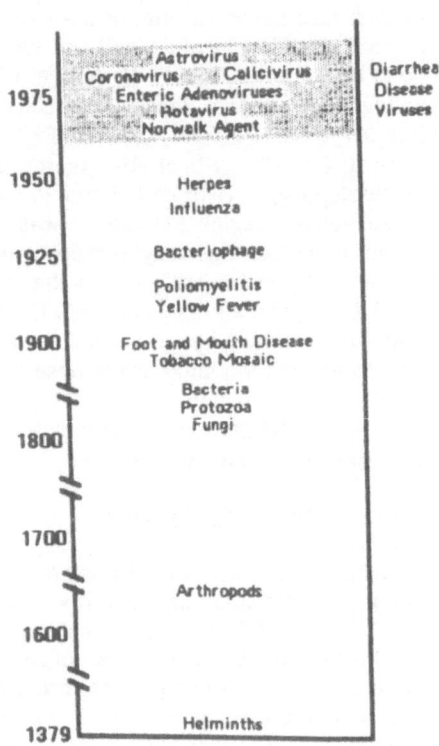

Figure 2.  Cumulative discovery of infectious disease agents.
(From: Sidwell, 1986)

## ROTAVIRUS GENE CODING ASSIGNMENTS

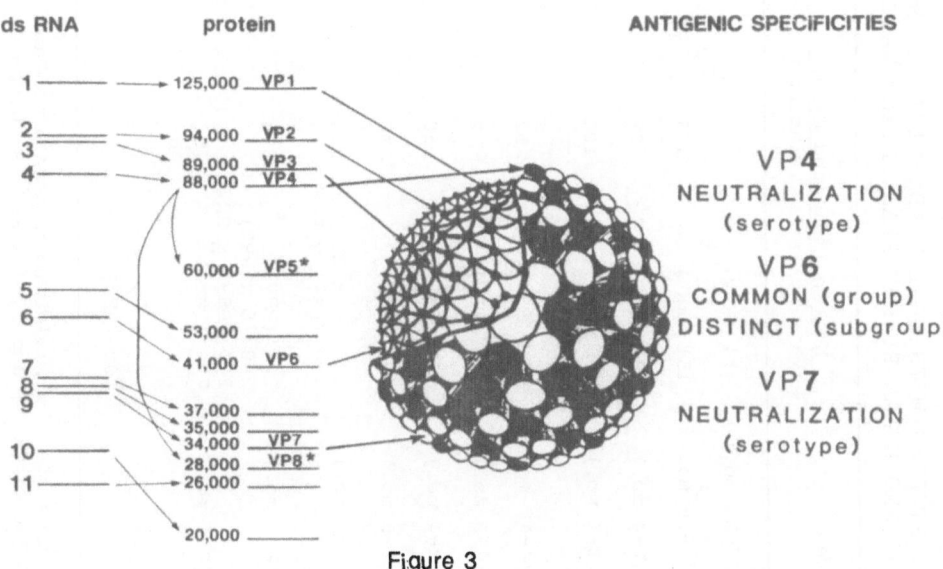

Figure 3

(From: Flores and Kapikian, 1989)

Table 6

STATUS OF PHASE 2 STUDIES WITH ORALLY ADMINISTERED RHESUS ROTAVIRUS (RRV) VACCINE AND HUMAN RV-RRV REASSORTANT VACCINES

| Institution (Investigator) | No. Children (Age) | No. in Indicated Group Vaccine | Placebo | Status |
|---|---|---|---|---|
| (1) Univ. of Maryland (Rennels, Levine) | 27 (5-20 mo.) | 14 (RRV)* | 13 | Completed |
| (2) Central Univ. Venezuela (Perez-Schael, Flores) | 302 (1-10 mo.) | 151 (RRV) | 151 | Completed |
| (3) Univ. of Rochester (Christy, Dolin) | 103 (2-4 mo.) | 85 (RRV) | 88 | Completed |
| (4) Johns Hopkins Univ., Ariz. (Santosham, Sack) | 301 (2-5 mo.) | 100 (RRV)<br>99 (RIT) | 102 | Completed |
| (5) Univ. of Umea, Sweden (Gothefors, Wadell) | 104 (4-12 mo.) | 54 (RRV) | 52 | Completed |
| (6) Univ. of Umea, Sweden (Gothefors, Wadell) | 35 (2-5 mo.) | 23 (RRV) | 12 | In progress |
| (7) Univ. of Maryland (Rennels, Levine) | 100 (2-4 mo.) | 50 (RRV) | 50 | In progress |
| (8) Univ. of Tampere, Finland (Vesikari) | 200 (2-5 mo.) | 100 (RRV) | 100 | In progress |
| (9) Univ. of Tampere, Finland (Vesikari) | 360 (2-4 mo.) | 120 (D x RRV)<br>120 (DS-1 x RRV) | 120 | In progress |
| (10) Institute of Nutrition, Lima, Peru (Lanata, Black) | 800 (2-4 mo.) | 200 (RRV)<br>200 (D x RRV)<br>200 (DS-1 x RRV) | 200 | To begin |
| (11) Central Univ. Venezuela (Perez-Schael, Flores) | 3,500 (2-4 mo.) | 1,750 (Quad)** | 1,750 | To begin |
| (12) Univ. of Rochester (Christy, Dolin) | 300 (2-4 mo.) | 100 (D x RRV)<br>100 (RRV) | 100 | To begin |

* = Vaccine composition shown in parentheses.  ** = Quadrivalent Rotavirus Vaccine consisting of RRV (serotype 3), D x RRV (serotype 1), DS-1 x RRV (serotype 2) and ST-3 x RRV (serotype 4).
Adapted from Kapikian et al., 1988.

# PRODUCTION OF REASSORTANT ROTAVIRUS VACCINE

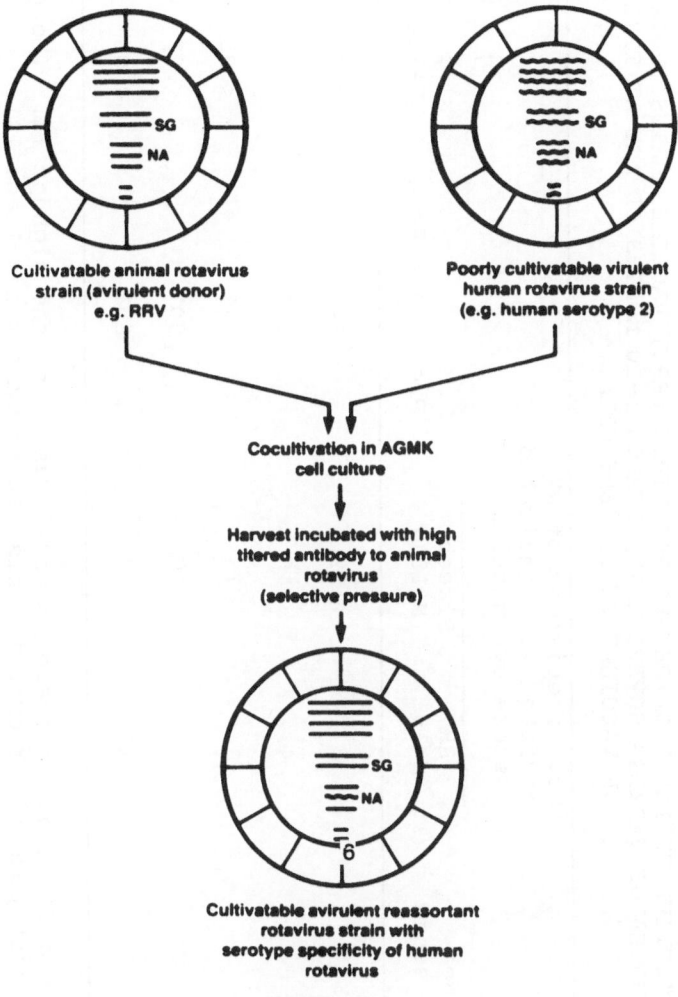

Figure 4

(Adapted from: Kapikian et al., 1986
(From: Kapikian et al., 1989)

Table 7

PROTECTIVE EFFICACY OF RHESUS ROTAVIRUS (RRV) VACCINE AGAINST ROTAVIRUS DIARRHEA OF VARYING SEVERITY IN INFANTS 1-10 MONTHS OF AGE AT TIME OF VACCINATION IN VENEZUELA.

| RV Diarrhea severity score * | No. of episodes of RV diarrhea with indicated score by inoculum | | Protection Rate | P Value |
|---|---|---|---|---|
| | RRV (N=151) | Placebo (N=151) | | |
| Any, 3-13 | 8 | 22 | 64% | <.01 |
| $\geq 6$ | 4 | 15 | 73% | <.01 |
| $\geq 7$ | 3 | 13 | 77% | <.05 |
| $\geq 8$ | 1 | 10 | 90% | <.01 |
| $\geq 9$ | 1 | 7 | 85% | <.05 |

* The lowest RV diarrhea score observed was 3 and the highest 13. (Lowest and highest possible scores in this rating system are 1 and 14)

Adapted from: Perez-Schael et al., 1989.

82

Table 8

SEROLOGICAL RESPONSES TO A QUADRIVALENT ROTAVIRUS VACCINE
PREPARATION CONSISTING OF 0.25 x $10^4$ PFU OF EACH COMPONENT:

D x RRV, (SEROTYPE 1), DS1 x RRV (SEROTYPE 2),

RRV (SEROTYPE 3), AND ST3 x RRV (SEROTYPE 4)

| ASSAY | RESPONSE TO INDICATED ROTAVIRUS SEROTYPE | | | |
| --- | --- | --- | --- | --- |
| | Serotype 1 | Serotype 2 | Serotype 3 | Serotype 4 |
| Plaque reduction neutralization | 3/22*(Wa) 14% | 4/26 (DS1) 15% | 1/19 (P) 5% 12/19 (RRV) 63% | 1/11 (ST3) 9% |
| ELISA IgA | – | – | 23/25 (RRV) 92% | |

* number of infants exhibiting $\geq$4-fold antibody response/number of infants tested. Virus employed as antigen in indicated assay shown in parentheses.

Adapted from: Perez-Schael et al., 1989.

Table 9

SEROLOGICAL RESPONSES TO A QUADRIVALENT ROTAVIRUS VACCINE
PREPARATION CONSISTING OF 0.5 x 10$^4$ PFU OF EACH COMPONENT:

D x RRV (SEROTYPE 1), DS1 x RRV (SEROTYPE 2),

RRV (SEROTYPE 3), AND ST3 x RRV (SEROTYPE 4)

| ASSAY | RESPONSES TO INDICATED ROTAVIRUS SEROTYPE | | |
| --- | --- | --- | --- |
| | Serotype 1 | Serotype 2 | Serotype 3 |
| Plaque reduction neutralization | 3/12* (Wa) 25% | 3/11 (DS1) 27% | 4/12 (P) 33% 9/11 (RRV) 81% |
| ELISA IgA | | | 10/14 (RRV) 71% |

* = number of infants exhibiting antibody response/number of infants tested. Virus used as antigen in each assay shown in parentheses.

Adapted from: Perez-Schael et al., 1989.

REFERENCES

Albert, M.J., Unicomb, L.E., Bishop, R.F., 1987, Cultivation and characterization of human rotaviruses with "supershort" RNA patterns, J. Clin. Microbiol., 25:183-185.

Anderson, E.L., Belshe, R.B., Bartram, J. Crookshanks, Newman, F., Chanock, R.M., Kapikian, A.Z., 1986, Evaluation of rhesus rotavirus vaccine (MMU 18006) in infants and young children, J. Infect. Dis., 153:823-831.

Bachmann, P., Bishop, R.F., Flewett, T.H., Kapikian, A.Z., Mathan, M.M., Zissis, G., 1984, Nomenclature of human rotaviruses: designation of subgroup and serotypes, Bull. WHO., 62:501-503.

Bishop, R.F., Barnes, G.L., Cipriani, E., Liund, J.S., 1983, Clinical immunity after neonatal rotavirus infection: A prospective longitudinal study in young children, New Engl. J. Med., 309:72-76.

Bishop, R.F., Davidson, G.P., Holmes, I.H., Ruck, B.J., 1973, Virus particles in epithelial cells of duodenal mucosa from children with viral gastroenteritis, Lancet, 2:1281-1283.

Bishop, R.F., Davidson, G.P., Holmes, I.H., Ruck, B.J., 1974, Detection of a new virus by electron microscopy of fecal extracts from children with acute gastroenteritis, Lancet, 1:149-151.

Black, R.E., Merson, M.H., Mizanur Rahman, A.S.M., Yunus, M., Alim, A.R.M.A., Huq, I., Yolken. R.H., Curlin, G.T., 1980,. A two-year study of bacterial, viral and parasitic agents associated with diarrhea in rural Bangladesh, J. Infect. Dis., 142:660-664.

Brandt, C.D., Kim, H.W., Rodriguez, W.J., Arrobio, J.O., Jeffries, B.C., Stallings, E.P., Lewis, C., Miles, A.J., Chanock, R.M., Kapikian, A.Z., Parrott, R.H., 1983, Pediatric viral gastroenteritis during eight years of study, J. Clin. Microbiol., 18:71-78.

Bridger, J.C.,1987, Novel rotaviruses in animals and man. In Ciba Foundation Symposium 128 "Novel Diarrhoea Viruses". New York. J. Wiley and Sons, pp. 5—15.

Chiba, S., Yokoyama, T. Nakata, S., Morita, Y., Urasawa, T., Taniguchi, K., Urasawa, S., Nakao, T., 1986, Prospective effect of naturally acquired homotypic and heterotypic rotavirus antibodies, Lancet, 2:417-421.

Christy, C., Madore, H.P., Gala C., Pincus, P., Vosefski, D., Hoshino, Y., Kapikian, A., Dolin, R., 1989, Field trial of rhesus rotavirus vaccine in infants, Ped. Inf. Dis. J., in press.

Christy, C., Madore, H.P., Treanor, J.J., Pray, K., Kapikian, A.Z., Chanock, R.M., Dolin, R., 1986, Safety and immunogenicity of live, attenuated monkey rotavirus vaccine, J. Infect. Dis., 154:1045-1047.

Chrystie, I.L., Totterdell, B.M., Baker, M.J., Banatvala, J.E., 1975, Rotavirus infections in a maternity unit, Lancet, 2:79 .

Clark, H.F., Borian F.E., Bell, L.M., Modesto, K., Gouvea, V., Plotkin, S.A., 1988, Protective effect of WC3 vaccine against rotavirus diarrhea in infants during a predominantly serotype 1 rotavirus season, J. Infect. Dis., 158:570-587.

Clark, H.F., Furukawa, T., Bell. L.M., Offit, P.A., Perrella, P.A., Plotkin, S.A., 1986, Immune response of infants and children to low-passage bovine rotavirus (strain WC3), Am. J. Dis. Child.,140:350-356.

Clark, H.F., Hoshino, Y., Bell, L.M., Grabb, J., Hess, G., Bachmann, P., Offit, P.A., 1987, Rotavirus isolate WI61 representing a presumptive new human serotype. J. Clin. Microbiol., 25:1757-1762.

De Mol, P., Zissis, G., Butzler, J.P., Mutwewingabo, A., André, F.E., 1986, Failure of live, attenuated oral rotavirus vaccine, Lancet, 2:108.

Dingle, J.H., Badger, G.F. Jordan, W.J., 1964, Illness in the home. A study of 25,000 illnesses in a group of Cleveland families. Cleveland, Ohio: Western Reserve University Press, pp. 19-32.

Flewett, T.H., Bryden, A.S., Davies, H., 1973, Virus particles in gastroenteritis, Lancet, 2:1497.

Flewett, T.H., Bryden, A.S., Davies, H., Woode, G.N., Bridger, J.C., Derrick, J.M. ,1974, Relationship between virus from acute gastroenteritis of children and newborn calves, Lancet, 2:61-63.

Flores, J., Kapikian, A.Z., 1989, Rotavirus, in:Tropical and Geographical Medicine, 2nd edition, Warren, K.S., Mahmoud, A.A.F. (eds.) McGraw—Hill Book Company, in press.

Flores, J., Daoud, G., Daoud, N., Puig, M., Martinez, M., Perez—Schael, I., Shaw, R., Greenberg, H.B., Midthun, K., Kapikian, A.Z., 1988, Reactogenicity and antigenicity of rhesus rotavirus vaccine (MMU-18006) in newborn infants in Venezuela, Pediatr. Infect. Dis. J., 7:776-780.

Flores, J., Perez—Schael, I., Blanco, M., Vilar, M., Daoud, N., Midthun, K., Kapikian, A.Z., 1989, Reactogenicity and antigenicity of two human-rhesus rotavirus reassortant vaccine candidates serotypes 1 and 2 in Venezuelan infants, J. Clin. Microbiol., 27: (March) 1989, (in press).

Flores, J., Perez—Schael, I., Gonzalez, M., Garcia, D., Perez, M., Daoud, N., Cunto, W., Kapikian, A.Z., 1987, Protection against rotavirus diarrhoea by rhesus rotavirus vaccine in Venezuelan children, Lancet, 1:882-884.

Gothefors, L., Wadell, G., Juto, P., Taniguchi, K., Kapikian, A.Z., Glass, R.I., 1989, Prolonged efficacy of rhesus rotavirus vaccine in Swedish children, J. Infect. Dis. (in press).

Greenberg, H.B., Valdesuso, J., Yolken, R.H., Gangarosa, E., Gary, W., Wyatt, R., Konno, T., Suzuki, H., Chanock, R.M., Kapikian, A.Z., 1979, Role of Norwalk virus in outbreaks of nonbacterial gastroenteritis, J. Infect. Dis., 139:564-568.

Halsey, N.A., Anderson, E.L., Sears, S.D., Steinhoff, M., Wilson, M., Belshe, R.B., Midthun, K., Kapikian, A.Z., Chanock, R.M., Samorodin R., Burns, B., Clements, M.L., 1988, Human-rhesus reassortant vaccines: safety and immunogenicity in adults, infants and children, J. Infect. Dis., 158: 1261-1267.

Hanlon, P., Hanlon, L., Marsh, W., Byass, P., Shenton, F., Hassan—King, M., Hobe, O., Sillah, H., Hayes, R., Boge, B.H.M., Whittle, H.C., and Greenwood, B.M., 1987, Trial of an attenuated bovine rotavirus vaccine (RIT 4237) in Gambian infants, Lancet, 1:1342-1345.

Hoshino, Y., Saif, L.J., Sereno, M.M., Chanock, R.M., Kapikian, A.Z., 1988, Infection immunity of piglets to either VP3 or VP7 outer capsid proteins confers resistance to challenge with a virulent rotavirus bearing the corresponding antigen, J. Virol., 62:744-748.

Hoshino, Y., Wyatt, R.G., Greenberg, H.B., Flores, J., Kapikian, A.Z., 1984, Serotypic similarity and diversity of rotaviruses of mammalian and avian origin as studied by plaque reduction neutralization, J. Infect. Dis., 149:694-702.

Institute of Medicine, 1985, Prospects for immunizing against rotavirus, In: New vaccine development. Establishing priorities. Vol. I. Diseases of importance in the United States, National Academy Press, Washington, D.C., pp. 410-423.

Institute of Medicine, 1986, The prospects for immunizing against rotavirus. In: New vaccine development. Establishing priorities. Vol 2. Diseases of importance in developing countries, National Academy Press, Washington, D.C. pp. 308-318.

Kapikian A.Z., Chanock, R.M., 1985a, Rotaviruses. In: Virology. Fields, B.N., Knipe, D.N., Chanock, R.M., Melnick, J.L., Roizman, B., Shope, R.E. (eds). Raven Press. N.Y. pp. 863-906.

Kapikian, A.Z., Chanock, R.M. ,1985b, Norwalk group of viruses. In: Virology. Fields, B.N., Knipe, D.N., Chanock, R.M., Melnick, J.L., Roizman, B., Shope, R.E. (eds). Raven Press. N.Y. pp. 1495-1517.

Kapikian, A.Z., Cline, W.L., Kim, H.W., Kalica, A.R., Wyatt, R.G., Van Kirk, D.H., Chanock, R.M., James, H.D., Jr., Vaughn, A.L.,1976, Antigenic relationships among five reovirus—like (RVL) agents by complement fixation (CF) and development of a new CF antigen for the human RVL agent of infantile gastroenteritis, Proc. Soc. Biol. Med., 152:535-539.

Kapikian, A.Z., Cline, W.L., Mebus, C.A., Wyatt, R.G., Kalica, A.R., James, H.D., Jr., VanKirk, D., Chanock, R.M., Kim. H.W., 1975, New complement-fixation test for the human reovirus—like agent of infantile gastroenteritis. Nebraska calf diarrhea virus used as antigen, Lancet, 1:1056-1061.

Kapikian, A.Z., Flores, J., Hoshino, Y., Glass, R.I., Midthun, K., Gorziglia, M., Chanock, R.M., 1986b, Rotavirus: the major etiologic agent of severe infantile diarrhea may be controllable by a "Jennerian" approach to vaccination, J. Infect. Dis., 153:815-822.

Kapikian, A.Z., Flores, J., Hoshino, Y., Midthun, K., Green, K., Gorziglia, M., Chanock, R.M., Potash, L., Perez-Schael, I., Gonzalez, M., Vesikari, T., Gothefors, L., Wadell, G., Glass,R.I., Levine, M.M., Rennels, M.B., Losonsky, G., Christy, C., Dolin, R., Anderson, E.L., Belshe, R.B., Wright, P., Santosham M., Halsey, N.A., Clements, M.L., Sears, S.D., Steinhoff, M.C., Black, R.E., Rationale for the development of a rotavirus vaccine for infants and young children. In: Progress in Vaccinology (G. Talwar, ed.) Springer-Verlag, 1989, (in press).

Kapikian, A.Z., Flores, J., Midthun, K., Hoshino, Y., Green, K.Y., Gorziglia, M., Taniguchi, K., Nishikawa, K., Chanock, R.M., Potash, L., Perez—Schael, I., Dolin, R., Christy, C., Santosham, M., Halsey, N.A., Clements, M.L., Sears, S.D., Black, R.E., Levine, M.M., Losonsky, G.A., Rennels, M.B., Gothefors, L., Wadell, G., Glass, R.I., Vesikari, T., Anderson, E.L., Belshe, R.B., Wright, P.F., Urasawa, S. , 1988, Development of a rotavirus vaccine by a "Jennerian" and "modified Jennerian" approach. In: Vaccines 88. (Ginsberg, H., Brown, F., Lerner, R.A., Chanock, R.M., eds.) New York: Cold Spring Harbor Laboratory, pp. 151-158.

Kapikian, A.Z., Hoshino, Y., Flores, J., Midthun, K., Glass, R.I., Nakagomi, O., Nakagomi, T., Chanock, R.M., Potash, L., Levine, M.M., Dolin, R., Wright, P.F., Belshe, R.E., Anderson, E.L., Gothefors, L., Wadell, G., Perez-Schael, I., 1986a, Alternative approaches to the development of a rotavirus vaccine. In: Development of vaccines and drugs against diarrhea. 11th Nobel conference, Stockholm, Sweden, 1985 (Holmgren, J., Lindberg, A., Mollby, R., eds.) Studentlitteratur, Lund, Sweden. pp. 192-214.

Kapikian, A.Z., Kim, H.W., Wyatt, R.G., Rodriguez, W.J., Cline, W.L., Parrott, R.H., Chanock, R.M., 1974, Reovirus-like agent in stools: association with infantile diarrhea and development of serologic tests, Science, 185:1049-1053.

Kapikian, A.Z., Midthun, K., Hoshino, Y., Flores, J., Wyatt, R.G., Glass, R.I., Askaa, J., Nakagomi, O., Nakagomi, T., Chanock, R.M., Levine, M.M., Clements, M.L., Dolin, R., Wright, P.F., Belshe, R.B., Anderson, E.L., Potash, L., 1985, Rhesus rotavirus: a candidate vaccine for prevention of human rotavirus disease. In: Vaccines 85. (Lerner, R.A., Chanock, R.M., Brown, F., eds.) Molecular and chemical basis of resistance to parasitic, bacterial, and viral diseases. New York: Cold Spring Harbor Laboratory, pp. 357-367.

Kapikian, A.Z., Wyatt, R.G., Dolin, R., Thornhill, T.S., Kalica, A.R., Chanock, R.M., 1972, Visualization by immune electron microscopy of a 27 nm particle associated with acute infectious nonbacterial gastroenteritis, J. Virol., 10: 1075-1081.

Kaplan, J.E., Feldman, R., Campbell, D.S., Lookabaugh, C., Gary, G.W., 1982a, The frequency of the Norwalk-like pattern illness in outbreaks of acute gastroenteritis, Am. J. Pub. Health, 12:1329-1332.

Kaplan, J.E., Gary, G.W., Baron, R.C., Singh, N., Schonberger, L.B., Feldman, R., Greenberg, H.B., 1982b, Epidemiology of Norwalk gastroenteritis and the role of Norwalk virus in outbreaks of acute nonbacterial gastroenteritis, Ann. Int. Med., 96:756-761.

Konno, T., Suzuki, H., Katsushima, N., Imai, A., Tazawa, F., Kutsuzawa, T., Kitaoka, S., Sakamoto, M., Yazaki, N., Ishida, N., 1983, Influence of temperature and relative humidity on human rotavirus infection in Japan, J. Infect. Dis., 147:125 -128.

Koopman, J.S., Turkish, V.J., Monto, S., Gouvea, V., Srivastava, S., Isaacson, R.E., 1984, Patterns and etiology of diarrhea in three clinical settings, Am. J. Epidemiol., 119:114-123.

Losonsky, G.A., Rennels, M.B., Kapikian, A.Z., Midthun, K., Ferra, P.J., Fortier, D.N., Hoffman, K.M., Baig, A., Levine, M.M., 1986, Safety, infectivity transmissibility and immunogenicity of rhesus rotavirus vaccine (MMU18006) in infants, Pediatr. Infect. Dis., 5:25-29.

Matsuno, S., Hasegawa, A., Mukoyama, A., Inouye, S., 1985, A candidate for a new serotype of human rotavirus, J. Virol., 54:623-624.

Midthun, K., Greenberg, H.B., Hoshino, Y., Kapikian, A.Z., Wyatt, R.G., Chanock, R.M., 1985, Reassortant rotaviruses as potential live rotavirus vaccine candidates, J. Virol., 53:949-954.

Midthun, K., Hoshino, Y., Kapikian, A.Z., Chanock, R.M., 1986, Single gene substitution rotavirus reassortants containing the major neutralization protein (VP7) of human rotavirus serotype 4, J. Clin. Microbiol., 24:822-826.

Morrison, A.S., Kirshner, J., Molho, A., 1977, Life cycle events in 15th century Florence: records of the Monte Delle Doti, Am. J. Epidemiol., 106:487-492.

Perez—Schael, I., Blanco, M., Vilar, M., Garcia, D., Gonzalez, R., Kapikian, A.Z., Flores, J., 1989, Clinical studies of a quadrivalent rotavirus vaccine in Venezuelan infants. To be submitted to J. Clin. Microbiol.

Perez—Schael, I., Gonzalez, M., Daoud, N., Perez, M., Soto, I., García, D., Daoud, G., Kapikian, A.Z., Flores, J., 1987, Reactogenicity and antigenicity of the rhesus rotavirus vaccine in Venezuelan children, J. Infect. Dis., 155:334-338.

Rennels, M.B., Losonsky, G.A., Levine, M.M., Kapikian, A.Z., and the Clinical Study Group (Fortier, D.N., Sutton, J.M., Ferra, P.J., Hoffman, K.M.), 1987a, Preliminary evaluation of the efficacy of rhesus rotavirus vaccine strain MMU—18006 in young children, Pediatr. Infect. Dis. J., 5:587-588.

Rennels, M.B, Losonsky, G.A., Shindledecker, F.N.P., Hughes, T.P., Kapikian, A.Z., Levine, M.M., and the Clinical Study Group (Ferra, P.J., Fortier, D.N., Sutton, J.M.), 1987b, Immunogenicity and reactogenicity of lower doses of rhesus rotavirus vaccine strain MMU—18006 in young children, Pediatr. Infect. Dis. J., 6:260-264.

Rodriguez, W.J., Kim, H.W., Brandt, C.D., Bise, B., Kapikian, A.Z., Chanock, R.M., Curlin, G., Parrott, R.H., 1980, Rotavirus gastroenteritis in the Washington, D.C., area: incidence of cases resulting in admission to the hospital. Am. J. Dis. Child., 134:777-779.

Rodriguez, W.J., Kim, H.W., Brandt, C.D., Schwartz, R.H., Gardner, M.K., Jeffries, B., Parrott, R.H., Kaslow, R.A., Smith, J.A., Kapikian, A.Z., 1987, Longitudinal study of rotavirus infection and gastroenteritis in families served by a pediatric medical practice: clinical and epidemiologic observations, Pediatr. Infect. Dis. J., 6:170-176.

Sato, K., Inaba, Y., Shinozaki, T., Fujii, R., Matumoto, M., 1981, Isolation of human rotavirus in cell culture, Arch. Virol., 69:155-160.

Shukry, S., Zaki, A.M., Shoukry, I., Tagi, M.E., Hamed, Z., 1986, Detection of enteropathogens in fatal and potentially fatal diarrheas in Cairo, Egypt, J. Clin. Microbiol., 24:959-962.

Sidwell, R.W., 1986. Overview of viral agents in pediatric enteric infection, Pediatric Infect. J., 5(S):44-45.

Snodgrass, D.R., Smith, W., Gray, E.W. Herring, J.A.,1976, A rotavirus in lambs with diarrhea, Res. Vet. Sci., 20:113-114, 1976.

Stuker, G., Oshiro, L., Schmidt, N.J., 1980, Antigenic composition of two new rotaviruses from rhesus monkeys, J. Clin. Microbiol., 11:202-203.

Snyder, J.D., Merson, M.H., 1982, The magnitude of the global problem of acute diarrhoeal disease: a review of surveillance data, Bull. WHO., 60:605-613.

Taniguchi, K., Urasawa, T., Morita, Y., Greenberg, H.B., Urasawa, S., 1987, Direct serotyping of human rotavirus in stools by an enzyme-linked immunosorbent assay using serotype 1-,2-, 3-, and 4- specific monoclonal antibodies to VP7, J. Infect. Dis., 155:1159-1166.

Urasawa, T., Urasawa, S., Taniguchi, K., 1981, Sequential passages of human rotavirus in MA—104 cells, Microbiol. Immun., 25:1025-1035.

Vesikari, T., Isolauri, E., Delem, A., D'Hondt, E., André, F.E., Beards, G.M., Flewett, T.H., 1985, Clinical efficacy of the RIT 4237 live attenuated bovine rotavirus vaccine in infants vaccinated before rotavirus epidemic, J. Pediatr., 107:189-194.

Vesikari, T., Isolauri, E., D'Hondt, E., Delem, A., André, F.E., 1984, Protection of infants against rotavirus diarrhoea by RIT 4237 attenuated bovine rotavirus strain vaccine, Lancet, 1:977-981.

Vesikari, T., Kapikian, A.Z., Delem, A., Zissis, G., 1986, A comparative trial of rhesus monkey (RRV—1) and bovine (RIT4237) oral rotavirus vaccines in young children, J. Infect. Dis., 153:832-839.

Vesikari, T., Rautanen, T., Varis, T., Beards, G.M., Kapikian, A.Z., 1989, Clinical trial of rhesus rotavirus candidate vaccine (strain MMU 18006) in children vaccinated between 2 and 5 months of age, To be submitted to Am. J. Dis. Child.

Wallace, R.E., Vasinton, P.J., Petricciani, J.C., Hopps, H.E., Lorenz, D.E, Kadanka, Z., 1973, Development of a diploid cell line from fetal rhesus monkey lung for virus vaccine production, In vitro, 8:323-332.

Walsh, J.A., Warren, K.S. 1979. Selective primary health care. An interim strategy for disease control in developing countries, New Engl. J. Med., 301:967-974.

Woode, G.N., Bridger, J.C., Jones, J.M., Flewett, T.H., Bryden, A.S., Davies, H.A., White, G.B., 1976, Morphological and antigenic relationships between viruses (rotaviruses) from acute gastroenteritis of children, calves, piglets, mice and foals, Infect. Immun., 14:804-810.

Wright, P.F., Tajima, T., Thompson, J., Kokubun, K., Kapikian, A.Z., Carson, D.T., 1987, Candidate rotavirus vaccine (rhesus rotavirus strain) in children: an evaluation, Pediatr., 80:473-480.

Wyatt, R.G., Greenberg, H.B., James, W.D., Pittman, A.L., Kalica, A.R., Flores, J., Chanock, R.M., Kapikian, A.Z., 1982, Definition of human rotavirus serotypes by plaque reduction assay, Infect. Immun., 37:110-115.

Wyatt, R.G., Kapikian, A.Z., Mebus, C.A., 1983, Induction of cross—reactive serum neutralizing antibody to human rotavirus in calves after in utero administration of bovine rotavirus, J. Clin. Microbiol., 18:505-508.

Wyatt, R.G., Mebus, C.A., Yolken, R.H., Kalica, A.R., James, H.D., Jr., Kapikian, A.Z., Chanock, R.M., 1979, Rotaviral immunity in gnotobiotic calves: heterologous resistance to human virus induced by bovine virus, Science, 203:548-550.

Yow, M.D., Melnick, J.L., Blattner, R.J., Stephenson, W.B., Robinson, N.M., Burkhardt, M.A., 1970, The association of viruses and bacteria with infantile diarrhea, Am. J. Epidemiol., 92:33-39.

# LYMPHOCYTIC CHORIOMENINGITIS VIRUS-SPECIFIC DELAYED-TYPE HYPERSENSITIVITY REACTION IN MICE

Fritz Lehmann-Grube and Demetrius Moskophidis

Heinrich-Pette-Institut für Experimentelle Virologie und Immunologie an der Universität Hamburg, Martinistrasse 52, 2000 Hamburg 20 Federal Republic of Germany

## INTRODUCTION

The first description of a local delayed-type hypersensitivity (DTH) reaction dates as far back as 1798, when E. Jenner commented on the response following the cutaneous application of poxvirus in a woman, who, 31 years previously, had had cowpox: "It is remarkable that variolous matter, when the system is disposed to reject it, should excite inflammation on the part to which it is applied more speedily than when it produces the Small Pox. Indeed it becomes almost a criterion by which we can determine whether the infection will be received or not. It seems as if a change, which endures through life, had been produced in the action, or disposition to action, in the vessels of the skin; and it is remarkable too, that whether this change has been effected by the Small Pox, or the Cow Pox, that the disposition to sudden cuticular inflammation is the same on the application of variolous matter".

About 100 years later, C. Janson vaccinated children on successive days and observed the accelerated appearance of lesions in sites of later application of the virus, and R. Koch described a phenomenon that was to become the tuberculin reaction. First quantitative studies were performed around the turn of the century by C. von Pirquet, who also recognized the immunologic nature. Since this time, the literature about DTH has steadily grown, and today it is close to impossible to be aware of everything that has been done.

Despite the long and intensive preoccupation with this form of immunity little is known of the basic mechanisms, and the same must be said of the biological meaning. Does DTH protect against viruses and certain bacteria and parasites or does it represent an immunological overreaction with pathologic consequences? Is the latter the price that has to be paid for the defense against sometimes life-threatening infections?

We became confronted with these questions when trying to understand the mechanism by which a higher organism controls a virus infection. Our model is the lymphocytic choriomeningitis (LCM) virus-infected mouse, which, under appropriate experimental conditions, allows studying the local DTH reaction observed as swelling of the foot after intraplantar inoculation of the virus (Hotchin, 1962a; Tosolini and Mims, 1971; Lehmann-Grube and Löhler, 1981; Lehmann-Grube, 1988) and also a probably related phenomenon, the severe, often fatal immunopathologic illness following intracerebral infection (Hotchin, 1962b; Cole and Nathanson, 1974; Lehmann-Grube and Löhler, 1981). Since alterations of the murine cells in which the virus multiplies are all but absent (Lehmann-Grube et al., 1983), the findings are not likely to be confounded by direct viral cytopathology.

METHODS

Inbred mice of the strains CBA/J and C57BL/6J were used and a variety of C57BL/10J mice congenic for defined regions of the major histocompatibility gene complex (MHC). All animals were purchased as "specific-pathogen free" and thus maintained in our animal quarters. Strain "WE" LCM virus (Rivers and Scott, 1935) was inoculated intravenously (i.v.) or subcutaneously (s.c.) into the pad of one hind foot. To determine the infectious titer, the foot was weighed and homogenized with a defined volume of physiological saline. The homogenate was cleared by centrifugation, snap frozen to -70°C, and titrated as soon as possible in L cell cultures or mice, and values thus obtained are expressed as mouse infectious units (IU) (Lehmann-Grube et al., 1985).

To achieve adoptive immunization of virus-infected recipients, spleen cells were taken from donor mice that had been infected 7 or 8 days previously by i.v. inoculation of $10^3$ IU (Lehmann-Grube et al., 1985).

For depletion of defined cells in vivo ("serologic surgery"), "negative selection" in vitro, and immunocytochemistry, monoclonal antibodies (MAb) with similar specificities but different properties were employed; these were obtained either as ascitic fluids from rats or mice or as cell culture media. For serologic surgery MAb was injected once i.v. into infected mice (Moskophidis et al., 1987) and for negative selection in vitro cells were incubated first with suitably diluted MAb and subsequently with suitably diluted rabbit serum as source of complement. Recipient mice received volume equivalents of cells, meaning that cells were counted before incubation with immune reagents and subsequently not adjusted for losses due to treatment.

To eliminate monocytes from the circulation, the mice were γ-irradiated with 900 rad. This procedure is based on the well-founded assumption that the precursor cells reside in the bone marrow, where they rapidly multiply (van Furth, 1986), which renders them exquisitely susceptible to ionizing radiation. For at least 8 days following treatment, no monocytes were detected in the blood.

LCM virus-specific DTH was measured as swelling of a

hind foot following the intraplantar inoculation of LCM virus and is expressed as ratio of dorso-ventral thickness of the infected versus thickness of the uninfected foot.

Recombinant human interleukin-2 (IL-2) was the gift of Sandoz Forschungsinstitut, GmbH, Wien.

RESULTS

After s.c. infection into a rear footpad, the LCM virus multiplies locally to concentrations around $10^7$ IU per gram of tissue, the initial rate depending on the infectious inoculum. The virus is also dispersed throughout the body and multiplies rapidly in the tissues, especially in spleen and lymph nodes. There, immunologically relevant cells are sensitized that migrate to the foot, where they cause inflammation which can be measured as swelling (Fig. 1). Beginning on day 8, the virus is eliminated to a certain level which may vary among mice of different strains but usually lies between $10^4$ and $10^5$ IU per gram (Fig. 2), a concentration that is maintained for weeks and even months. While this latter phenomenon is interesting, here we are concerned with the events during the period of rapid clearance between days 8 and 12 after infection.

Local swelling and virus elimination can be greatly accelerated by adoptive immunization, i.e. infusion of spleen cells from syngeneic donor mice undergoing an immunizing infection (Table 1). (For further details, see Lehmann-Grube, 1988.)

To find out which cells cause the swelling, the method of "serologic surgery" was employed, in which mice are infected and subsequently depleted of defined cells by treatment with MAb specifically directed against the cell that is to be removed. Fig. 3 demonstrates what happens when Thy-1 MAb, which is known to remove essentially all T lymphocytes from the peripheral lymphatic system (Opitz et al., 1982), is injected once on days 5, 6, or 7 after infection; the swelling is abruptly terminated, even when the antibody is injected late. We conclude that for the local LCM virus-specific DTH reaction T lymphocytes are required.

T cells belong to two subclasses whose interactions with other cells are restricted by MHC-encoded molecules of either class I or class II. They are functionally (predominantly, not exclusively) categorized as cytotoxic (CTL) and helper/inducer, respectively, and are marked by differentiation antigens CD8 (Lyt-2) and CD4 (L3T4) (Cantor and Boyse, 1976; Swain, 1981; Dialynas et al., 1983). For obtaining information as to which of these subsets participate in the local DTH reaction, we again employed serologic surgery. The findings (Fig. 4) revealed that the LCM virus-specific murine DTH reaction consists of two phases which develop largely independently of each other and are mediated by (first) CD8+ T lymphocytes and (subsequently) CD4+ T lymphocytes. Further experiments, in which virus-infected mice were adoptively

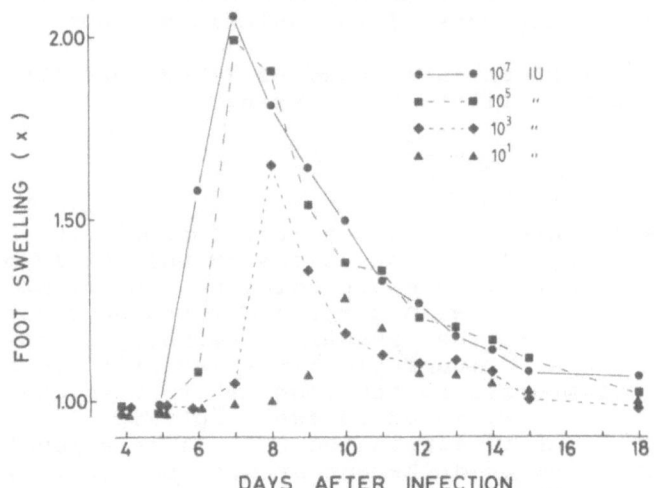

Fig. 1. Swelling of feet of CBA/J mice after intraplantar
inoculation of different quantities of LCM virus.
Swelling is expressed as the factor by which the
dorso-ventral thickness of the inoculated hind foot
exceeded the dorso-ventral thickness of the con-
tralateral control foot. Data points denote means of
10 mice.

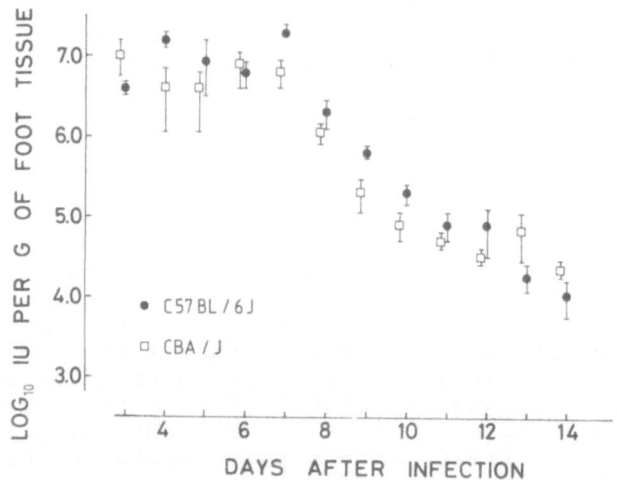

Fig. 2. Elimination of LCM virus from feet of CBA/J and
C57BL/6J mice. At intervals after s.c. inoculation of
$10^5$ IU into right hind feet, these were homogenized
and the virus was titrated. Data points and vertical
bars denote means of virus concentrations and standard errors, respectively, in four mice.

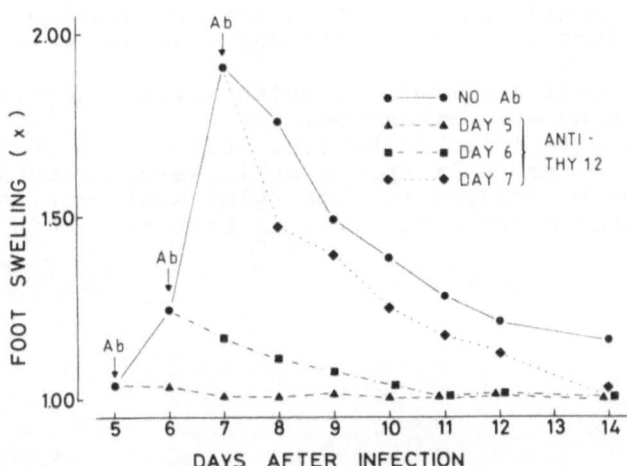

Fig. 3. Effect of depletion of CBA/J mice of T lymphocytes on
LCM virus-induced foot swelling. Five, 6, or 7 days
after s.c. inoculation of $10^5$ IU into right hind
feet, mice were injected intraperitoneally with 0.3
ml mouse ascitic fluid containing approximately 2 mg
of Thy-1.2 MAb (Opitz et al., 1982). Beginning 5 days
after infection and at 24-hour intervals thereafter,
dorso-ventral thicknesses of both feet were measured
and ratios calculated. Data points signify means of
five mice.

Table 1.  Effect of Adoptive Immunization of C57BL/6J Mice with Syngeneic Immune Spleen T Lymphocytes on Virus Concentration in Feet[a]

| Day after Transfer[b] | Cells Transferred[c] | | % Residual Infectivity |
|---|---|---|---|
| | 0 | $4\times10^7$ | |
| 3 | $1.2\pm0.2\times10^7$ [d] | $4.7\pm2.0\times10^6$ | 39 |
| 4 | $9.0\pm3.1\times10^6$ | $1.7\pm1.2\times10^6$ | 19 |
| 5 | $7.6\pm3.4\times10^6$ | $1.7\pm1.3\times10^5$ | 2 |

[a]Twenty-two hours after intraplantar inoculation of $10^5$ IU mice were inoculated i.v. with day-7-immune spleen T lymphocytes.
[b]Time after cell transfer, at which virus concentration in recipient mice was determined.
[c]Donor mice were infected by i.v. inoculation of $10^3$ IU, and 7 days later their spleen cells were enriched for T lymphocytes by passage through nylon wool columns.
[d]Mean $\pm$ standard error IU per g of foot tissue in five mice.

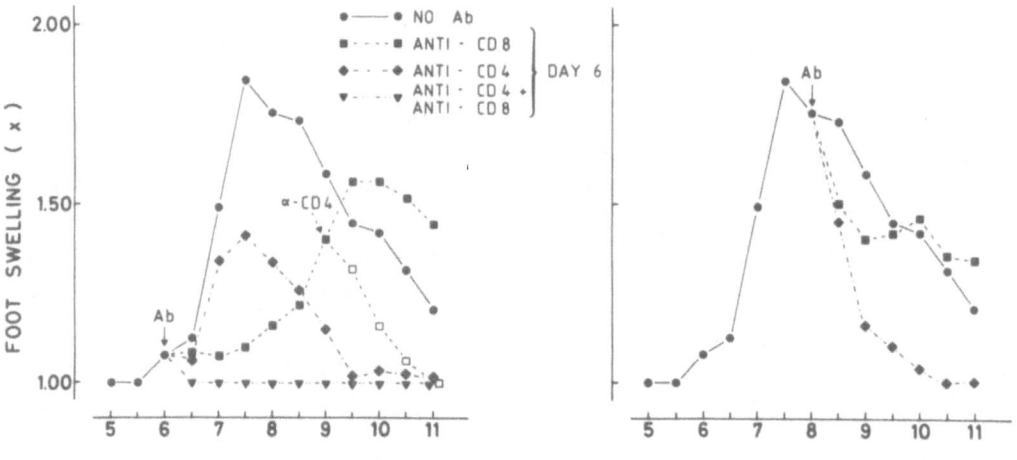

DAYS AFTER INFECTION

Fig. 4.  Effect of depletion of CBA/J mice of T lymphocyte subsets on LCM virus-induced foot swelling. Six (left) or 8 (right) days after intraplantar inoculation of $10^5$ IU, mice were injected once i.v. with ca. 80 μg per g weight of specific CD4 or the same quantity of specific CD8 MAb (Cobbold et al., 1984) or (only day 6 after infection) both. Half the mice that had received CD8 MAb on day 6 were injected 3 days later (9 days after infection) with CD4 MAb. Beginning 5 days after infection and at 24-hour intervals thereafter, dorso-ventral thicknesses of both feet were measured and ratios calculated. Data points signify means of five mice.

immunized with immune splenocytes negatively selected for either subclass of T lymphocytes, led to the same results. Finally, we adoptively immunized infected mice with immune cells from donors differing with regard to defined regions of the MHC. All combinations of interest were thus analyzed, which led again to the conclusion that the local reaction consists of two portions, in which T lymphocytes participate that are restricted by K and D(L) region or I region gene products, respectively (Moskophidis and Lehmann-Grube, 1989).

To investigate the relevance of T lymphocyte subtypes for controlling the infection, similar methods were utilized. Mice were infected by intraplantar inoculation and subsequently depleted of T lymphocytes or their subtypes by treatment with MAb. Removal of all T lymphocytes resulted in inability to eliminate the virus. The same was found when the mice were depleted of CD8+ cells, whereas removal of CD4+ cells led to retardation of virus elimination (Moskophidis and Lehmann-Grube, 1989; Table 2); this latter effect was slight but seen in every experiment of this kind. Our conclusion that CD8+ class I-restricted CTL are mainly responsible for controlling the infection was confirmed in a great number of experiments, in which virus-infected mice were adoptively immunized with negatively selected immune spleen cells (by treatment in vitro with CD8 or CD4 MAb plus complement) from syngeneic donors or with immune splenocytes from C57BL/10 mice congenic with regard to defined regions of the MHC.

A closer look at Fig. 4 reveals that during the second CD4+ cell-mediated part of the reaction the feet were more swollen than in the infected but otherwise untreated control mice, which probably reflects the higher local antigen (virus) concentration in CD8+ cell-depleted animals (Table 2). In contrast, swelling during the first CD8+ cell-mediated phase was less than in the controls, which was not as easily explained. As already mentioned, virus elimination, too, was always suboptimal in CD4+ cell-depleted mice, which was also the case when adoptive immunization was with immune cells that had been selected for CD8+ (freed of CD4+) cells or whose MHC compatibility was limited to K or D region. On the other hand, the effect of adoptive immunization could be improved by adding to the inoculum the corresponding immune cells selected for CD4+ cells or extending MHC K/D compatibility to the I region.

CD4+ helper cells are regarded as the principal producers of IL-2, which led to the assumption that lack of this lymphokine might be the main reason for the CD8+ cells' impaired ability to mediate a DTH reaction and to eliminate the virus. Mice were infected by footpad inoculation of the virus and 24 hours later transfused with day-8-immune splenocytes negatively selected for CD8+ T lymphocytes. In comparison with the animals that had received unselected immune cells, the swelling was reduced but was fully restored by treatment of each animal at 8-hour intervals with 10,000 units of human IL-2. In parallel, these mice's ability to control the infection was determined. Again, transfer of CD8+ cells alone resulted in virus elimination, which was less than in mice that had been transfused with unselected immune cells, and the antiviral effect could be restored by systemic

Table 2. Effect of Depletion of CBA/J Mice of T Lymphocyte Subsets by Treatment with MAb on Ability to Eliminate LCM Virus from Foot Tissue

| Day after Infection[a] | Antibody Treatment[b] | | | |
|---|---|---|---|---|
| | No Antibody | Anti-CD8 | Anti-CD4 | Rat Control[c] |
| 6.5 | $1.0 \pm 0.2 \times 10^6$ [d] | – | – | – |
| 7.5 | $1.1 \pm 0.4 \times 10^7$ | – | – | – |
| 8.5 | $6.0 \pm 1.0 \times 10^5$ | $2.0 \pm 1.2 \times 10^7$ | $5.1 \pm 2.4 \times 10^5$ | $1.4 \pm 0.5 \times 10^6$ |
| 10.5 | $1.5 \pm 0.1 \times 10^5$ | $1.2 \pm 0.2 \times 10^7$ | $1.8 \pm 0.7 \times 10^5$ | $1.4 \pm 0.3 \times 10^5$ |
| 11.5 | $8.1 \pm 1.7 \times 10^4$ | $1.9 \pm 0.7 \times 10^7$ | $2.3 \pm 0.5 \times 10^5$ | $9.2 \pm 5.2 \times 10^4$ |

[a] Intraplantar inoculation of $10^5$ IU.

[b] Seven days after infection each mouse was inoculated with 2.5 mg of specific rat MAb (Cobbold et al., 1984).

[c] Irrelevant human glycophorin A MAb (Cobbold et al., 1984).

[d] Mean ± standard error IU per g of foot tissue in four mice.

treatment with IL-2. It appears that the functioning of the
CD8+ cells is improved by IL-2. Assuming that this lymphokine
is normally set free by activated CD4+ T lymphocytes and
knowing that these cells are not by themselves antivirally
effective, we assign them an accessory role.

To answer the question whether and to what extent
circulating monocytes participate in the DTH reaction, mice
were depleted of monocytes, which turned out to be a difficult task, but γ-irradiation with 900 rad and the rigorous
removal of adherent elements from the cell inoculum proved to
be effective; in these animals, a thorough search failed to
locate any cell which morphologically, with regard to enzyme
activity, or immunocytochemically resembled a monocyte. When
these mice were footpad-infected and adoptively immunized,
the swelling was marked, although, in comparison with unirradiated recipients, delayed in appearance and somewhat reduced
in magnitude. The swollen tissue proved to be essentially
free of cells with mononuclear phagocyte characteristics.
Depleting these mice of CD4+ cells by serologic surgery did
not affect the swelling, whereas depletion of CD8+ cells
rapidly reduced it to almost zero level. The conclusion that
the CD4+ cell-mediated portion of the local response requires
monocytes, whereas the CD8+ cell-mediated portion does not,
was confirmed by experiments in which monocyte-deprived mice
were adoptively immunized with splenocytes from MHC-disparate
donors. Hardly any swelling was seen in donor/recipient
combinations in which the I region was identical, whereas
under otherwise the same conditions the swelling in D region-
compatible combination was as high in irradiated mice as it
was in unirradiated controls.

The role monocytes play in virus elimination from the
foot was also ascertained. Mice were γ-irradiated with 900
rad, infected 24 hours later by intraplantar inoculation of
virus, and transfused 48 hours later (24 hours after infection) with day-8-immune splenocytes which had been freed of
adhering elements. Sixty hours after cell transfer the mice
were depleted of either CD4+ or CD8+ cells (by treatment
with MAb); one group remained untreated. The subsequent
determination of virus concentration in the feet disclosed
that transfer of immune cells into unirradiated and irradiated recipients had reduced the virus by 99 and 90%,
respectively. Subsequent removal of CD4+ cells did not alter
the result, whereas removal of CD8+ cells completely abolished the antiviral effect of immune cell transfer. Obviously
CD8+ CTL are antivirally active even in the absence of monocytes, albeit with reduced efficiency.


DISCUSSION

Immunity may be defined as the altered state of an
organism, which is induced by antigen ("non-self") and
revealed as accelerated and heightened response to the same
antigen. Immunity is customarily subdivided into a humoral
and a cellular branch, in which B lymphocytes and T lymphocytes, respectively, are the main elements. DTH is generally seen as a special case of cell-mediated immunity, which
is revealed as local, sometimes systemic, inflammatory
reaction developing within 24 or 48 hours after application

of the antigen. DTH reactions vary considerably in appearance, as do the underlying mechanisms, but the central event is generally assumed to be interaction between antigen, which is presented in context with class II MHC-encoded molecules on special cells, and antigen-specific CD4[+] T$_{DTH}$ cells. As a consequence, lymphokines are released which cause the influx of inflammatory cells, especially monocytes.

The widely held view that DTH is principally mediated by class II-restricted T$_{DTH}$ lymphocytes is at variance with the long-known observation that there are DTH reactions which are mediated also by class I-restricted T lymphocytes (Vadas et al., 1977; Sunday et al., 1980), and where the antigen is infectious virus, such as influenza virus (Leung and Ada, 1980), reovirus (Weiner et al., 1980), Sendai virus (Ertl, 1981) and dengue virus (Pang et al., 1984), this seems to be the rule. For some time the LCM virus has been thought to be an exception, because the virus-specific DTH reaction had been described as being associated only with T lymphocytes restricted by K and D region-encoded class I molecules (Zinkernagel, 1976), but, as our findings show, this virus follows the general pattern.

One may ask why some antigens are recognized by T lymphocytes in the context of both classes of MHC-encoded molecules, whereas others are seen only together with class II molecules; the answer is mode of antigen presentation. For both influenza virus and Sendai virus it has been shown that infectivity is associated with either form of DTH reaction, whereas inactivated virus elicits only a class II molecule-restricted response (Leung and Ada, 1980; Ertl, 1981). Why antigens (or antigenic fragments) synthesized during intracellular replication of the virus on one hand or resulting from its catabolic break-down on the other combine with different MHC-encoded molecules and thus activate T lymphocytes belonging to different subsets is not well understood, but in view of current efforts in a number of laboratories a satisfactory explanation is likely to be given soon (Morrison et al., 1986; Hosaka et al., 1988; Townsend et al., 1988).

Probably the main function of the immune system is to protect against infectious agents, for which humoral and cellular mechanisms are available. For terminating a primary virus infection, cell-mediated immunity is of supreme importance with T lymphocytes as the central elements (Mims, 1987). Of these, CD8[+] CTL appear to be conditio sine qua non for the control of such diverse viruses as influenza virus (Yap et al., 1978; Lukacher et al., 1984), murine cytomegalovirus (Ho, 1980; Reddehase et al., 1987), ectromelia virus (Kees and Blanden, 1976), and LCM virus (Zinkernagel and Welsh, 1976; Byrne and Oldstone, 1984; Moskophidis et al., 1987; data presented here). Even for the termination of infection of mice with Listeria monocytogenes, which for a long time had been assumed to depend principally on CD4[+] cells, it emerges that CD8[+] T lymphocytes are essential (Bishop and Hinrichs, 1987; Mielke et al., 1988). Thus, CTL may be regarded as the main effector cells in the defense against viruses (and presumably other intracellular parasites), although there is at least one example where CTL are of uncertain relevance, namely the herpes simplex virus in the mouse (Nash et al., 1985; Martin and Rouse, 1987). The question as to how CTL function in vivo cannot yet be answered to everybody's satisfaction. We believe that they release or induce other cells to release lymphokines that interfere with virus replication in the surrounding tissue (Lehmann-Grube et al., 1988). The evidence is indirect, but the same is true with regard to the assumption that virus replication is blocked by CTL-induced destruction of infected cells (Zinkernagel and Doherty, 1979).

Whether CD4[+] T lymphocytes are also required for terminating a primary virus infection is an open question. For several viruses it has been shown that antiviral CTL are generated in vivo without the help of CD4[+] cells (Buller et al., 1987; Nash et al., 1987; Ahmed et al., 1988). In contrast, the production of antibodies against numerous antigens (Coulie et al., 1985; Wofsy et al., 1985; Qin et al., 1987) including viruses (Cobbold et al., 1984; Buller et al., 1987; Leist et al., 1987; Moskophidis et al., 1987; Ahmed et al., 1988) depends on CD4[+] helper cells, but the value of antiviral antibodies for the termination of a primary virus infection is uncertain (although they probably contribute to protection against reinfection).

Our observations show that the LCM virus-specific local DTH reaction consists of two separable phases, of which the first is mediated by CD8[+] T lymphocytes and develops without participation of monocytes, whereas the second is mediated by CD4[+] T lymphocytes and requires monocytes. This dichotomy has to be kept in mind when trying to understand the biological relevance of DTH during LCM virus infection of the mouse.

Of the two phases, only the CD8[+] cell-mediated first one is associated with antiviral activity. True, for optimal efficiency cooperation of CD4[+] cells is required, but these cells can be replaced by IL-2, and we assume that they fulfill some accessory function, for instance by improving the microenvironment for the CTL; there was never even a hint that in the LCM virus-infected mouse CD4[+] cells by themselves might be antivirally active. The antiviral capacity of LCM-specific CTL is also less than optimal in the absence of monocytes. Whether this indicates that the latter cells are directly involved in virus elimination or that they function as accessory cells, or whether this simply reflects shortage of cells presenting antigen to CD4[+] cells (which then are insufficiently activated for their accessory function) cannot be decided from the present data. We have previously shown that mononuclear phagocytes (to which the monocytes belong) are not essential for terminating the LCM virus infection in the murine spleen (Lehmann-Grube et al., 1987).

If the CD8[+] CTL are primarily important for controlling the infection, what then are the tasks of the CD4[+] helper/inducer cells? Two have already been mentioned, namely to help in antibody production (where they are indispensable) and to improve the performance of the CD8[+] CTL (where their indispensability is not as certain); they are also causally involved in the local inflammation, which, however, accompanies both phases and whose relevance is unknown (see below). Otherwise, we are not aware of further functions of the CD4[+] helper/inducer cells, at least not in connection with protection against virus infections.

Inflammation, the hallmark of DTH, is probably caused by lymphokines, whose identification we currently pursue. Knowing that different cells of the immune system participate in the two parts, we assume that different lymphokines are involved. One candidate that may be active early is IL-2, which is known to be released by CD8+ T lymphocytes (Mizuochi et al., 1986; Heeg et al., 1987) and to increase the leakiness of vessels (Rosenstein et al., 1986); this latter property would tally with the observation that the local reaction is initially characterized by edema. Later, elements of the immune system are present that are qualitatively not different from the ones seen in cellular infiltrates in other tissues of the LCM virus-infected mouse (J. Löhler, pers. comm.). Our attempts to elucidate whether the local inflammation by itself plays some antiviral role have failed so far, because we have not succeeded in preventing its development without compromising the immune system in general. The long-known propensity of activated T lymphocytes to migrate (immunologically non-specifically) into sites of inflammation (Prendergast, 1964; Koster et al., 1971) could result in accumulation of virus-specific CTL, but it should be pointed out that the virus is not eliminated more efficiently from the foot than it is from other murine tissues. As to the swelling, the visible and measurable outward sign of the DTH reaction, it is probably a consequence of anatomical and/or physiological peculiarities of the skin of the footpad and perhaps other areas of the mouse's body (Traub and Kesting, 1963).

There remains the possibility that the inflammation accompanying both phases of the local response causes, if it develops in other tissues, the severe immunopathologic disease often seen in LCM virus-infected mice (Hotchin, 1962b; Cole and Nathanson, 1974; Lehmann-Grube and Löhler, 1981); Thomson et al. (1983) have postulated that the characteristic illness of mice following intracerebral inoculation of the LCM virus reflects DTH rather than cytotoxicity. Inasmuch as LCM virus-specific DTH seems to be mediated by T lymphocytes of both subclasses, the first part of this assumption may be right, but not the second; our colleague J. Löhler has shown that murine lymphocytic choriomeningitis does not develop in mice depleted by serologic surgery of CD8+ cells either early (before intracerebral inoculation of the virus) or late (just before clinical signs appear) in infection. By way of contrast, depletion of CD4+ cells spared a few mice and somewhat prolonged survival times but did not prevent the development of clinical signs (pers. commun.). It appears that CTL are involved in the pathogenesis, whereas CD4+ cells, again, play some supportive role. Other findings have led to similar conclusions (Doherty and Allan, 1985). Whether the immunopathologic illness due to CTL is causally related with control of the infection or whether it reflects parallel events is not yet known.

We are aware of the danger of drawing general conclusions on the basis of observations with one model. On the other hand, it is not likely that animals and man protect themselves against different viruses by principally different mechanisms. We, therefore, venture to conclude with the following remarks. Provided the LCM virus-induced foot swelling is the outward sign of a local DTH reaction and provided local DTH reactions represent focal magnifications of general phenomena, then any DTH reaction is a tissue-specific manifestation of immunity as defined above and does not reflect a special form. CD8⁺ lymphocytes control primary virus (as well as certain bacterial and parasitic) infections, whereas CD4⁺ lymphocytes are essential for antibody production, thereby shielding against reinfection. Probably both, when activated, release lymphokines with the capacity to cause inflammation, the biologic function of which is uncertain, except that it may be the basis for immunopathologic phenomena that seem to accompany many virus diseases.

## ACKNOWLEDGMENT

The Heinrich-Pette-Institut is financially supported by Freie und Hansestadt Hamburg and Bundesministerium für Jugend, Familie, Frauen und Gesundheit. This work was aided by a research grant from the Deutsche Forschungsgemeinschaft.

## REFERENCES

Ahmed, R., Butler, L.D., and Bhatti, L., 1988, T4⁺ T helper cell function in vivo: differential requirement for induction of antiviral cytotoxic T-cell and antibody responses, J. Virol., 62:2102.

Bishop, D.K., and Hinrichs, D.J., 1987, Adoptive transfer of immunity to Listeria monocytogenes. The influence of in vitro stimulation on lymphocyte subset requirements, J. Immunol., 139:2005.

Buller, R.M.L., Holmes, K.L., Hügin, A., Frederickson, T.N., and Morse, H.C., 1987, Induction of cytotoxic T-cell responses in vivo in the absence of CD4 helper cells, Nature, 328:77.

Byrne, J.A., and Oldstone, M.B.A., 1984, Biology of cloned cytotoxic T lymphocytes specific for lymphocytic choriomeningitis virus: clearance of virus in vivo, J. Virol., 51:682.

Cantor, H., and Boyse, E.A., 1976, Regulation of cellular and humoral immune responses by T-cell subclasses, Cold Spring Harbor Symp. Quant. Biol., 41:23.

Cobbold, S.P., Jayasuriya, A., Nash, A., Prospero, T.D., and Waldmann, H., 1984, Therapy with monoclonal antibodies by elimination of T-cell subsets in vivo, Nature, 312:548.

Cole, G.A., and Nathanson, N., 1974, Lymphocytic choriomeningitis. Pathogenesis, Progr. Med. Virol., 18:94.

Coulie, P.G., Coutelier, J.-P., Uyttenhove, C., Lambotte, P., and van Snick, J., 1985, In vivo suppression of T-dependent antibody responses by treatment with a monoclonal anti-L3T4 antibody, Eur. J. Immunol., 15:638.

Dialynas, D.P., Wilde, D.B., Marrack, P., Pierres, A., Wall, K.A., Havran, W., Otten, G., Loken, M.R., Pierres, M., Kappler, J., and Fitch, F.W., 1983, Characterization of the murine antigenic determinant, designated L3T4a, recognized by monoclonal antibody GK1.5: expression of L3T4a by functional T cell clones appears to correlate primarily with class II MHC antigen-reactivity, Immunol. Rev., 74:29.

Doherty, P.C., and Allan, J.E., 1985, Participation of cyclophosphamide-resistant T cells in murine lymphocytic choriomeningitis, Scand. J. Immunol., 21:127.

Ertl, H.C.J., 1981, Adoptive transfer of delayed-type hypersensitivity to Sendai virus. I. Induction of two different subsets of T lymphocytes which differ in H-2 restriction as well as in the Lyt phenotype, Cell. Immunol., 62:38.

Heeg, K., Steeg, C., Hardt, C., and Wagner, H., 1987, Identification of interleukin 2-producing T helper cells within murine Lyt-2⁺ T lymphocytes: frequency, specificity and clonal segregation from Lyt-2⁺ precursors of cytotoxic T lymphocytes, Eur. J. Immunol., 17:229.

Ho, M., 1980, Role of specific cytotoxic lymphocytes in cellular immunity against murine cytomegalovirus, Infect. Immun., 27:767.

Hosaka, Y., Sasao, F., Yamanaka, K., Bennink, J.R., and Yewdell, J.W., 1988, Recognition of noninfectious influenza virus by class I-restricted murine cytotoxic T lymphocytes, J. Immunol., 140:606.

Hotchin, J., 1962a, The foot pad reaction of mice to lymphocytic choriomeningitis virus, Virology, 17:214.

Hotchin, J., 1962b, The biology of lymphocytic choriomeningitis infection: virus-induced immune disease, Cold Spring Harbor Symp. Quant. Biol., 27:479.

Janson, C., 1891, Versuche zur Erlangung künstlicher Immunität bei Variola vaccina, Cbl. Bakt. Parasitenk. (Jena), 10:40.

Jenner, E., 1798, "An Inquiry Into the Causes and Effects of the Variolae Vaccinae, a Disease Discovered in Some of the Western Counties of England, Particularly Gloucestershire, and Known by the Name of the Cow Pox," Sampson Low, London.

Kees, U., and Blanden, R.V., 1976, A single genetic element in H-2K affects mouse T-cell antiviral function in poxvirus infection, J. Exp. Med., 143:450.

Koch, R., 1891, Fortsetzung der Mittheilungen über ein Heilmittel gegen Tuberculose, Dtsch. med. Wschr., 17:101.

Koster, F.T., McGregor, D.D., and Mackaness, G.B., 1971, The mediator of cellular immunity. II. Migration of immunologically committed lymphocytes into inflammatory exudates, J. Exp. Med., 133:400.

Lehmann-Grube, F., 1988, Mechanism of recovery from acute virus infection. VI. Replication of lymphocytic choriomeningitis virus in and clearance from the foot of the mouse, J. Gen. Virol., 69:1883.

Lehmann-Grube, F., Assmann, U., Löliger, C., Moskophidis, D., and Löhler, J., 1985, Mechanism of recovery from acute virus infection. I. Role of T lymphocytes in the clearance of lymphocytic choriomeningitis virus from spleens of mice, J. Immunol., 134:608.

Lehmann-Grube, F., Krenz, I., Krahnert, T., Schwachenwald, R., Moskophidis, D., Löhler, J., and Villeda Posada, C.J., 1987, Mechanism of recovery from acute virus infection. IV. Questionable role of mononuclear phagocytes in the clearance of lymphocytic choriomeningitis virus from spleens of mice, J. Immunol., 138:2282.

Lehmann-Grube, F., and Löhler, J., 1981, Immunopathologic alterations of lymphatic tissues of mice infected with lymphocytic choriomeningitis virus. II. Pathogenetic mechanism, Labor. Invest., 44:205.

Lehmann-Grube, F., Martínez Peralta, L., Bruns, M., and Löhler, J., 1983, Persistent infection of mice with the lymphocytic choriomeningitis virus, <u>Compr. Virol.</u>, 18:43.

Lehmann-Grube, F., Moskophidis, D., and Löhler, J., 1988, Recovery from acute virus infection. Role of cytotoxic T lymphocytes in the elimination of lymphocytic choriomeningitis virus from spleens of mice, <u>Ann. N.Y. Acad. Sci.</u>, 532:238.

Leist, T.P., Cobbold, S.P., Waldmann, H., Aguet, M., and Zinkernagel, R.M., 1987, Functional analysis of T lymphocyte subsets in antiviral host defense. <u>J. Immunol.</u>, 138:2278.

Leung, K.N., and Ada, G.L., 1980, Two T-cell populations mediate delayed-type hypersensitivity to murine influenza virus infection, <u>Scand. J. Immunol.</u>, 12:481.

Lukacher, A.E., Braciale, V.L., and Braciale, T.J., 1984, In vivo effector function of influenza virus-specific cytotoxic T lymphocyte clones is highly specific. <u>J. Exp. Med.</u>, 160:814.

Martin, S., and Rouse, B.T., 1987, The mechanisms of anti-viral immunity induced by a vaccinia virus recombinant expressing herpes simplex virus type 1 glycoprotein D: clearance of local infection, <u>J. Immunol.</u>, 138:3431.

Mielke, M.E.A., Ehlers, S., and Hahn, H., 1988, T-cell subsets in delayed-type hypersensitivity, protection, and granuloma formation in primary and secondary <u>Listeria</u> infection in mice: superior role of Lyt-2[+] cells in acquired immunity, <u>Infect. Immun.</u>, 56:1920.

Mims, C.A., 1987, "The Pathogenesis of Infectious Disease," third ed., Academic Press, London.

Mizuochi, T., Ono, S., Malek, T.R., and Singer, A., 1986, Characterization of two distinct primary T cell populations that secrete interleukin 2 upon recogni-tion of class I or class II major histocompatibility antigens, <u>J. Exp. Med.</u>, 163:603.

Morrison, L.A., Lukacher, A.E., Braciale, V.L., Fan, D.P., and Braciale, T.J., 1986, Differences in antigen presentation to MHC class I- and class II-restricted influenza virus-specific cytolytic T lymphocyte clones, <u>J. Exp. Med.</u>, 163:903.

Moskophidis, D., Cobbold, S.P., Waldmann, H., and Lehmann-Grube, F., 1987, Mechanism of recovery from acute virus infection: treatment of lymphocytic choriomenin-gitis virus-infected mice with monoclonal antibodies reveals that Lyt-2[+] T lymphocytes mediate clearance of virus and regulate the antiviral antibody response, <u>J. Virol.</u>, 61:1867.

Moskophidis, D., and Lehmann-Grube, F., 1989, Virus-induced delayed-type hypersensitivity reaction is sequentially mediated by CD8[+] and CD4[+] T lymphocytes, <u>Proc. Natl. Acad. Sci. USA</u>, in press.

Nash, A.A., Jayasuriya, A., Phelan, J., Cobbold, S.P., Waldmann, H., and Prospero, T., 1987, Different roles for L3T4[+] and Lyt2[+] T cell subsets in the control of an acute herpes simplex virus infection of the skin and nervous system, <u>J. Gen. Virol.</u>, 68:825.

Nash, A.A., Leung, K.-N., and Wildy, P., 1985, The T-cell-mediated immune response of mice to herpes simplex virus, <u>in</u>: "The Herpes Viruses," Vol. 4, B. Roizman and C. Lopez, eds, Plenum Publishing Co., New York.

Opitz, H.G., Opitz, U., Hewlett, G., and Schlumberger, H.D., 1982, A new model for investigations of T-cell functions in mice: differential immunosuppressive effects of two monoclonal anti-Thy-1.2 antibodies, Immunobiology, 160:438.

Pang, T., Devi, S., Yeen, W.P., McKenzie, I.F.C., and Leong, Y.K., 1984, Lyt phenotype and H-2 compatibility requirements of effector cells in the delayed-type hypersensitivity response to dengue virus infection, Infect. Immun., 43:429.

Prendergast, R.A., 1964, Cellular specificity in the homograft reaction, J. Exp. Med., 119:377.

Qin, S., Cobbold, S., Tighe, H., Benjamin, R., and Waldmann, H., 1987, CD4 monoclonal antibody pairs for immunosuppression and tolerance induction, Eur. J. Immunol., 17:1159.

Reddehase, M.J., Mutter, W., Münch, K., Bühring, H.-J., and Koszinowski, U.H., 1987, CD8-positive T lymphocytes specific for murine cytomegalovirus immediate-early antigens mediate protective immunity, J. Virol., 61:3102.

Rivers, T.M., and Scott, T.F.M., 1935, Meningitis in man caused by a filterable virus, Science, 81:439.

Rosenstein, M., Ettinghausen, S.E., and Rosenberg, S.A., 1986, Extravasation of intravascular fluid mediated by the systemic administration of recombinant interleukin 2, J. Immunol., 137:1735.

Sunday, M.E., Benacerraf, B., and Dorf, M.E., 1980, Hapten-specific T cell responses to 4-hydroxy-3-nitrophenyl acetyl. VI. Evidence for different T cell receptors in cells that mediate H-2I-restricted and H-2D-restricted cutaneous sensitivity responses, J. Exp. Med., 152:1554.

Swain, S.L., 1981, Significance of Lyt phenotypes: Lyt2 antibodies block activities of T cells that recognize class 1 major histocompatibility complex antigens regardless of their function, Proc. Natl. Acad. Sci. USA, 78:7101.

Thomsen, A.R., Bro-Jørgensen, K., and Volkert, M., 1983, Fatal meningitis following lymphocytic choriomeningitis virus infection reflects delayed-type hypersensitivity rather than cytotoxicity, Scand. J. Immunol., 17:139.

Tosolini, F.A., and Mims, C.A., 1971, Effect of murine strain and viral strain on the pathogenesis of lymphocytic choriomeningitis infection and a study of footpad responses, J. Infect. Dis., 123:134.

Townsend, A., Bastin, J., Gould, K., Brownlee, G., Andres, M., Coupar, B., Boyle, D., Chan, S., and Smith, G., 1988, Defective presentation to class I-restricted cytotoxic T lymphocytes in vaccinia-infected cells is overcome by enhanced degradation of antigen, J. Exp. Med., 168:1211.

Traub, E., and Kesting, F., 1963, Further observations on the behavior of the cells in murine LCM, Arch. ges. Virusforsch., 13:452.

Vadas, M.A., Miller, J.F.A.P., Whitelaw, A.M., and Gamble, J.R., 1977, Regulation by the H-2 gene complex of delayed type hypersensitivity, Immunogenetics, 4:137.

van Furth, R., 1986, Overview: the mononuclear phagocyte system, in: "Handbook of Experimental Immunology," Vol. 2: "Cellular Immunology," fourth ed., D.M. Weir, L.A. Herzenberg, C. Blackwell, and L.A. Herzenberg, eds, Blackwell Scientific Publications, Oxford, London, Edinburgh, Boston, Palo Alto, Melbourne.

von Pirquet, C., 1907, "Klinische Studien über Vakzination und Vakzinale Allergie," Franz Deuticke, Leipzig.

Weiner, H.L., Greene, M.I., and Fields, B.N., 1980, Delayed hypersensitivity in mice infected with reovirus. I. Identification of host and viral gene products responsible for the immune response, J. Immunol., 125:278.

Wofsy, D., Mayes, D.C., Woodcock, J., and Seaman, W.E., 1985, Inhibition of humoral immunity in vivo by monoclonal antibody to L3T4: studies with soluble antigens in intact mice, J. Immunol., 135:1698.

Yap, K.L., Ada, G.L., and McKenzie, I.F.C., 1978, Transfer of specific cytotoxic T lymphocytes protects mice inoculated with influenza virus, Nature, 273:238.

Zinkernagel, R.M., 1976, H-2 restriction of virus-specific T-cell-mediated effector functions in vivo. II. Adoptive transfer of delayed-type hypersensitivity to murine lymphocytic choriomeningitis virus is restricted by the K and D region of H-2, J. Exp. Med., 144:776.

Zinkernagel, R.M., and Doherty, P.C., 1979, MHC-restricted cytotoxic T cells: studies on the biological role of polymorphic major transplantation antigens determining T-cell restriction-specificity, function, and responsiveness, Adv. Immunol., 27:51.

Zinkernagel, R.M., and Welsh, R.M., 1976, H-2 compatibility requirement for virus-specific T cell-mediated effector functions in vivo. I. Specificity of T cells conferring antiviral protection against lymphocytic choriomeningitis virus is associated with H-2K and H-2D, J. Immunol., 117:1495.

# RECOGNITION OF INFLUENZA A VIRUS BY HUMAN CYTOTOXIC T LYMPHOCYTES

Andrew J. McMichael and Frances M. Gotch

University of Oxford
Institute of Molecular Medicine
John Radcliffe Hospital, Oxford. U.K.

Cytotoxic T lymphocytes (CTL) are an important component of the immune system carrying the CD8 surface glycoprotein and forming a significant part of the total lymphoid population. The role of CTL appears to be to kill cells of the body that differ from normal and this occurs during infection with intracellular parasites, particularly viruses. It is likely that CTL have evolved as a major defence mechanism against virus infections. CTL are detected in vitro by their ability to kill infected target cells that are labelled with chromium-51. There is evidence that CTL lyse infected cells in vivo (1) and they may also act by releasing lymphokines such as Interferon-Gamma (2). When infected cells are killed they can no longer replicate new virus particles, and if the cell is already doomed this is an effective way of controlling a virus infection. However the constraints on CTL antigen receptors preclude them from recognising and neutralising virus particules directly. The natural role of CTL therefore is likely to be quite distinct from that of the antibody response.

In the last ten years a considerable body of data has been accumulated defining the role of CTL in influenza virus infection in mice and in humans It has been shown that transfer of cultured cytotoxic T lymphocytes to mice infected with influenza A virus results in rapid virus clearance (3,4). Since this has been achieved with cloned CTL, the possibility that the effect is mediated by passenger cells may be ruled out (4). In vivo the CTL response peaks before the IgM response (5) and this is therefore a very early immune reaction. In humans the level of memory cells for the CTL response correlated with protection against and clearance of influenza A virus in volunteers (6). In nude, athymic mice susceptibility to death through influenza virus infection is greater than in normal mice and survivors secrete virus chronically (7). Nude mice will recover, if given cytotoxic T lymphocytes (8). All of these data support the view that the CTL immune response is a major component of the natural defence against influenza virus infection indicating that CTL are able to clear the virus during the recovery phase.

The specificity of CTL for virus proved puzzling for many years but is now becoming more clearly understood. Following the work of Zinkernagel and Doherty (9) it was found that influenza virus specific CTL were restricted by antigens of the major histocompatibility complex (MHC)

(10,11,12). Thus CTL specific for influenza virus lysed only histo-compatible target cells. This was initially shown in mice and then in humans (10,11,12). These results implied that there was some interaction between virus antigen and MHC antigen on the surface of infected cells and that the CTL receptor would react with this complex. The alternative view that all CTL had a receptor for self MHC antigen in addition to a receptor for virus antigen has been disproved by the finding of a single receptor on cloned T cells that show MHC restricted recognition (13).

The nature of the virus antigen recognised by CTL has been well studied. An early observation was that CTL induced with one strain of influenza A virus, such as A/X31(H3N2) could recognise targets infected with any other strain of influenza A virus, e.g. A/USSR(H1N1) (10-12). Initially, specificity was erroneously thought to be entirely directed at the surface proteins of the virus, which are present in abundance on infected cells but which are highly polymorphic. The exact nature of the virus antigen recognised by CTL was resolved when it became possible to insert single influenza genes into target cells. This was first done by transfection of viral cDNA into mouse L cells, when it was found that a sub-population of influenza virus specific CTL did recognise haemagglutinin (14). However, such CTL were not cross-reactive between the haemagglutinin subtypes and subsequent transfection experiments showed that a major component of the murine CTL response was directed to virus nucleoprotein (15). This followed the identification of a mouse CTL clone that showed specificity for nucleoprotein (16).

An alternative strategy for inserting a single influenza virus protein into target cells was to use recombinant vaccinia virus (17). In this way it was found by Yewdell et al (18) that mouse CTL recognised nucleoprotein inserted into target cells. It was further shown by Gotch et al (19) that human CTL recognised not only nucleoprotein but the matrix protein and polymerase protein PB2 inserted into MHC matched target cells. All of these experiments confirmed that CTL were recognising internal proteins of the virus, which are conserved between different virus strains. This explained the apparent crossreactivity of CTL recognition of different influenza strains. However, the results posed a new question: How do CTL recognise internal virus proteins which are difficult to detect with antibodies on the surface of virus infected cells. One possibility was that CTL might recognise antigens which had been processed to be presented as peptide fragments on the surface of infected cells. The idea was attractive because it was known that the helper T cells recognise foreign antigens that have been degraded to peptides and then presented at the surface in association with MHC class II molecules (20). The first test of this proposal was the insertion of truncated genes for the nucleoprotein into target cells and the demonstration that they were recognised by CTL (21). These experiments also demonstrated that different mouse strains responded to different regions of the nucleoprotein molecule. Townsend et al (22) also showed that recognition of the haemagglutinin was not dependent on that molecule being inserted into the cell membrane; the leader sequence was deleted from the haemagglutinin gene and the protein that was made was shown to be shortlived and cytoplasmic, but cells transfected with this gene were efficiently recognised by haemagglutinin specific CTL.

The proposal that CTL recognise fragments of virus proteins was directly tested by incubating target cells with short synthetic peptides based on nucleoprotein sequence (23). It was found that such target cells were recognised by nucleoprotein specific CTL in mice and in humans and that different MHC class I types interacted with different peptide sequences (23-25). The selective effect on the epitope imposed by the MHC type was dramatic. So far in the nucleoprotein four distinct regions have been

110

defined, each of which is presented to CTL by a different MHC molecule. In humans the epitope that is presented by HLA37 appears to be selected by every individual with B37 (25). So far this epitope has not been presented by any other HLA class I antigen. This result therefore explains the phenomenon of MHC restriction where the CTL response is determined in its fine epitope specificity by the MHC class I type of the responding individual. It is also clear that some individuals with some HLA types may not respond to particular virus proteins. Thus individuals with HLA A2 respond to the influenza matrix protein (19) and a particular epitope (residues 57-68) (26) but do not respond with this MHC antigen to other virus proteins. Conversely no mouse strain has yet been described which responds to influenza A matrix protein, they all respond to nucleoprotein however. Therefore this finding at once implies that HLA type (MHC type) plays a fundamental role in determining the nature of the CTL response to a virus. This would be most simply explained if the peptide epitopes actively bind to the presenting MHC molecule. The high degree of polymorphism of MHC antigens may in part be determined by the variety of viruses against which the population must be capable of defending itself.

The idea that peptide epitopes bind to MHC class I molecules received very strong support when the crystal structure of HLA A2 was reported by Bjorkman et al (27,28). They found that there was unidentified electron density, possibly peptide, located in a groove between two alpha helices on the top of the molecule. The residues of the helices which point into the groove include most of the polymorphic residues in the class I MHC molecule. Many of the remainder were found in the floor of the groove, an eight stranded beta sheet. Thus the class I MHC molecule appears to be a specialised peptide binding structure with a variable region located around its peptide binding site. Thus the previously defined inter-relationship between MHC type and peptide epitope was explained.

There are two major issues that remain to be resolved. The first is how does the peptide bind to the MHC molecule. Is it possible to predict from an HLA sequence the nature of the peptide that will interact? The second question is how does the virus protein become degraded and where does it interact with the MHC molecule?

Current work in our laboratory addresses the first issue. We have chosen to work on the peptide associated with HLA A2, which was defined from the influenza A virus matrix sequence (26). We have tested this peptide for recognition by CTL clones with mutant HLA A2 molecules (29). These have been cloned and subjected to site-directed mutagenesis by J. Santos-Aguado and J.L. Strominger and transfected into a suitable target cell. Our experiments have shown that mutations of amino acids located around the binding groove (at positions 9, 66, 70, 152 and 156) abrogate recognition of the peptide by some or all CTL clones specific for HLA-A2 plus the matrix peptide. Mutations at other sites (residues 43, 74, 107) had no effect. Such results give some indication as to the minimum size for the peptide. It should make contact with residues 66 and 152 of HLA A2 which are approximately 5 amino acids apart, if the peptide fits as an extended chain, or ten amino acids apart if the peptide fits as an alpha helix. Some mutations only affect recognition of the peptide of some CTL clones. This implies that the peptide binds to these mutants but that its fine positioning may vary slightly, affecting some T cells and not others.

In collaboration with J. Rothbard we have also explored the peptide, by changing amino acids at single positions (30). By doing this we have identified the shortest peptide that functions, 60-68. We have shown that there is a core region in the middle of this peptide, from residues 61-65 which plays a major role in specificity. Some mutations affect recognition by some T cell clones and not others, others abrogate recognition but compete for binding HLA A2. From this information we find that positions 61, 62 and 63 most likely interact with HLA A2. Residues 60, 64 and 65 probably face the T cell receptor. These orientations could be achieved with the peptide in an alpha helix and we have suggested that the peptide fits as an alpha helix with residues 60, 64 and 65 facing the T cell receptor and residues 62 and 63 facing HLA A2.

These experiments have been aided by the recent finding of a second HLA A2 restricted peptide epitope derived from influenza B virus (P. Robbins, J. Rothbard, A. McMichael, unpublished). We have started to exchange residues from one peptide to the other which should enable us to identify which residues interact with HLA A2 and should therefore be exchangeable. Conversely, substitution of residues that interact with the T cell receptor should give analogue peptides that compete for binding to A2 even though they are not recognised by the T cell receptor. From these experiments it should be possible to build up a model of how the peptide fits into the groove of the MHC class I molecule. In parallel with these experiments attempts are being made to physically bind peptide to purified HLA A2 and then define its orientation in a crystal.

The other major unresolved issue is how peptide is processed and where it interacts with the MHC class I molecule. While it has been relatively straightforward to exchange peptides that bind to cell free class II molecules (31), this has not yet been the case for class I. One of the differences between the class I and class II MHC presenting systems may be that class II molecules are designed to recirculate in lysosomes and exchange peptide in an acid environment whereas class I MHC molecules may not naturally exchange peptide. We and others, have found, for instance that for competition to occur between two related peptides, it is necessary to pre-incubate with a large excess of the competitor (30,32). Evan a short pre-incubation with the index peptide before adding the competitor abrogates competition. This implies that once the index peptide is bound it is extremely difficult to compete it off the MHC class I molecule. This competition is only found in the pool of molecules competing to bind the class I molecule. Furthermore, once bound, the class I molecule may retain peptide for a long time. We have found that target cells may remain as targets and be recognised by CTL for as long as 72 hours after a short incubation period with peptide (Gotch, McMichael, unpublished observations).

The peptide and HLA class I molecule may associate as the class I molecule folds. An alternative possibility is that peptide binds to HLA molecules in a membrane vesicle. In either event the protein has to be degraded and the peptide has to cross a membrane. There are a number of ways in which this might occur, for instance the membrane associated ubiquitin system might be involved in this process (33). The fine details of this process, when resolved, may provide clues as to how the system could be manipulated. One point that has emerged is that a particular MHC type may be able to present an artificially synthesised peptide but that this peptide may not be produced by the natural processes of degrading a particular intracellular protein. Two examples has been described, one where primary responses have been induced with peptide generated CTL that recognised peptide but not virus infected cells (34). The other was the demonstration that influenza A matrix specific CTL, which were HLA A2 restricted, cross-reacted with cells bearing HLA A69 (which differ by only 6 amino acids) in the presence of peptide but not when the cells are infected by influenza virus. This implies that the natural product cannot interact with HLA A69 although the short peptide can (35).

The importance of CTL immunity to influenza virus and other virus is becoming increasingly realised. The fine details of the molecular associations involved are being actively investigated. It is particularly fascinating that the polymorphic HLA system imposes constraints on specificity and implies that HLA type will be important in the immune response to important virus infections. Conversely some of the variability in clinical response to acute virus infections may therefore be attributable to HLA type.

Acknowledgement. We are grateful to the Medical Research Council for support and to Mrs. Clare Crew for help in preparing the manuscript.

REFERENCES

1. R.M. Zinkernagel, E. Haenseler, T. Leist, A. Cerny, H. Hengartner and A.Althage. T cell mediated hepatitis in mice infected with lymphocytic choriomeningitis virus. Liver cell destruction by H-2 class I-restricted virus-specific cytotoxic T cells as a physiological correlate of the 51Cr release assay? J.Exp.Med. 164:1075.(1986).
2. A.G. Morris, Y.L. Lin and B.A. Askonas. Immune interferon release when a cloned cytotoxic T cell line meets its correct influenza infected target cells. Nature 295:150. (1982).
3. K.L. Yap, G.L. Ada, I.F.C. McKenzie. Transfer of specific cytotoxic T lymphocytes protect mice inoculated with influenza virus. Nature 273:238. (1978).
4. Y.L. Lin and B.A. Askonas. Biological properties of an influenza A virus specific killer T cell clone. J.Exp.Med. 154:225. (1981).
5. K.L. Yap and G.L. Ada. Cytotoxic T cells in the lungs of mice infected with influenza A virus. Scand.J.Immunol. 7:73 (1978).
6. A.J. McMichael,F.M. Gotch, G.R. Noble and P.A.S. Beare. Cytotoxic T-cell immunity to influenza. N.Engl.J.Med. 309:13. (1983).
7. M.A. Wells, P. Albrecht and F.A. Ennis. Recovery from a viral respiratory tract infection.I.Influenza pneumonia in normal and T deficient mice. J.Immunol. 126:1036. (1981).
8. M.A. Wells, F.A. Ennis and P. Albrecht. Recovery from a viral respiratory tract infection.II.Passive transfer of immune spleen cells to mice with influenza pneumonia. J.Immunol. 126:1042. (1981).
9. R.M. Zinkernagel and P.C. Doherty. H-2 compatibility requirement for T-cell mediated lysis of target cells infected with lymphocytic choriomeningitis virus. J.Exp.Med. 141:1427. (1975).
10. H.J. Zweerink, S.A. Courtneidge, J.J. Skehel, M.J. Crumpton and B.A. Askonas. Cytotoxic T cells kill influenza virus-infected cells but do not distinguish between serologically distinct type A viruses. Nature 267:354. (1977).
11. T.J. Braciale. Immunologic recognition of influenza-virus infected cells. Generation of virus strain specific and cross reactive subpopulations of cytotoxic T-cells in the response to type A influenza viruese of different subtypes. Cell.Immunol. 33:423. (1977).
12. A.J. McMichael, A.Ting, H.J. Zweerink and B.A. Askonas. HLA restriction of cell mediated lysis of influenza virus infected human cells. Nature 270:524. (1977=.
13. Z. Dembic, W.Haas, S. Weiss et al. Transfer of specificity of murine alpha and beta T-cell receptor genes. Nature 320:232. (1986).
14. T.J. Braciale, V.L. Braciale, T.J. Henkel, J. Sambrook and M-J. Gething. Cytotoxic T lymphocyte recognition of the influenza haemagglutinin gene product expressed by DNA-mediated gene transfer. J.Exp.Med. 159:341. (1984).
15. A.R.M. Townsend, A.J. McMichael, N.P. Carter, J.A. Huddleston and G.G. Brownlee. Cytotoxic T cell recognition of the influenza nucleoprotein and haemagglutinin expressed in transfected mouse L cells. Cell. 39:13.(1984).
16. A.R.M. Townsend and J.J. Skehel. Influenza A specific cytotoxic T-cell clones that do not recognise viral glycoproteins. Nature 300:655.(1982)

17. J.R. Bennink, J.W. Yewdell, G.L. Smith, C. Moller and B. Moss.
Recombinant vaccinia virus primes and stimulates influenza haemagglutinin-specific cytotoxic T cells. Nature 311:578. (1984).

18. J.W. Yewdell, J.R. Bennink, G.L. Smith and B. Moss. Influenza A virus
nucleoprotein is a major target antigen for cross reactive anti-
influenza A virus cytotoxic T lymphocytes. P.N.A.S. (USA) 82:1785.
(1985)

19. F.M. Gotch, A.J. McMichael, G.Smith and B. Moss. Identification of
viral molecules recognized by influenza-specific human cytotoxic T
lymphocytes. J.Exp.Med. 165:408. (1987).

20. R.H. Schwartz. T lymphocyte recognition of antigen in association
with gene products of the major histocompatibility complex.
Ann.Rev.Immunol. 3:237. (1985).

21. A.R.M. Townsend, F.M. Gotch and J. Davey. Cytotoxic T-cells recognise
fragments of the influenza nucleoprotein. Cell 42:457. (1985).

22. A.R.M. Townsend, J. Bastin, K. Gould and G.G. Brownlee. Cytotoxic T
lymphocytes recognise influenza haemagglutinin that lacks a signal
sequence. Nature 234:575. (1986).

23. A.R.M. Townsend, J. Rothbard, F.M. Gotch, G. Bahadur, D. Wraith and
A.J. McMichael. The epitopes of influenza nucleoprotein recognized
by cytotoxic T lymphocytes can be defined with short synthetic
peptides. Cell 44:959. (1986).

24. J. Bastin, J. Rothbard, J. Davey, I. Jones and A. Townsend. Use of
synthetic peptides of influenza nucleoprotein to define epitopes
recognized by class I restricted cytotoxic T lymphocytes. J.Exp.Med.
165:1508. (1987).

25. A.J. McMichael, F.M. Gotch, and J. Rothbard. HLA B37 determines an
influenza A virus nucleoprotein epitope recognized by cytotoxic T
lymphocytes. J.Exp.Med. 164:1397. (1986).

26. F. Gotch, J. Rothbard, J. Howland, A. Townsend and A. McMichael.
Cytotoxic T lymphocytes recognise a fragment of influenza virus
matrix protein in association with HLA-A2. Nature 326:881. (1987).

27. P.J. Bjorkman, M.A. Saper, B. Samraoui, W.S. Bennett, J.L. Strominger
and D.C. Wiley. Structure of the human class I histocompatibility
antigen, HLA-A2. Nature. 329:511. (1987).

28. P.J. Bjorkman, M.A. Saper, B. Samraoui, W.S. Bennett, J.L. Strominger
and D.C. Wiley. The foreign antigen binding site and T cell
recognition regions of class I histocompatibility antigens.
Nature. 329:512. (1987).

29. A.J. McMichael, F. Gotch, J. Santos-Aguado and J. Strominger. The
effect of mutations and variations of HLA-A2 on recognition of a
virus peptide epitope by cytotoxic T lymphocytes. P.N.A.S. (USA).
In press. (1988).

30. F. Gotch, A. McMichael and J. Rothbard. Recognition of influenza A
matrix protein by HLA A2 restricted cytotoxic T lymphocytes. Use of
analogues to orientate the matrix peptide in the HLA A2 binding site.
J.Exp.Med. In press (1988).

31. S. Buus, A. Sette, S.M. Colon, D.M. Jenis and H.M. Grey. Isolation and
characterization of antigen-Ia complexes involved in T cell
recognition. Cell 47:1071. (1986).

32. H. Bodmer, J. Bastin, B. Askonas and A. Townsend. Influenza specific
cytotoxic T cell recognition is inhibited by peptides unrelated in
both sequence and MHC restriction. Immunology. In press (1988).

33. A. Hershko and A. Ciechanover. The ubiquitin pathway for the
degradation of intracellular proteins. Prog.Nucleic Acid.Res.Mol.
Biol. 33:19 (1986).

34. F. Carbone, M. Moore, J. Sheil and M. Bevan. Induction of cytotoxic T
lymphocytes by primary in vitro stimulation with peptides.
J. Exp.Med. 167:1767. (1988).

35. H. Bodmer, F. Gotch and A.McMichael. Cross-restricted T cells reveal
low responder allele due to processing of viral antigens. Submitted.
(1988).

114

# MOLECULAR AND GENETIC ASPECTS OF THE IMMUNE RESPONSES TO HEPATITIS B VIRAL ANTIGENS

David R. Milich

Department of Molecular Biology
Research Institute of Scripps Clinic
La Jolla, California USA 92037

SUMMARY

In the absence of an inbred animal model of hepatitis B virus (HBV) infection, several laboratories have chosen to study the murine immune response to HBV-encoded proteins as immunogens as opposed to an infectious agent. This article reviews the immunogenicity, the fine specificity of T and B cell recognition of HBV antigens, and the genetic influences that regulate these responses. It is anticipated that this approach will increase our understanding of immune-mediated viral clearance mechanisms during HBV infection, and may provide the framework for the design of second and third generation HBV vaccines.

## DISCUSSION

The clinical consequences of HBV infection are extremely variable. It is likely that nonviral, host factors are involved in the pathogenesis of hepatocellular injury since the hepatitis B virus is not directly cytopathic for hepatocytes.[1] It has been suggested that variation in immune responsiveness to HBV infection may, at least partially, account for the diversity of clinical syndromes including fulminant, acute, chronic active and chronic persistent hepatitis as well as the asymptomatic chronic carrier state.[2]

The specific serological marker of HBV infection is the hepatitis B surface antigen (HBsAg) which is present as the envelope in the intact virion and as free circulating filamentous and spherical 22-nm subviral particles. The HBsAg is composed of a major polypeptide, P25, and its glycosylated form, GP28. Additional polypeptides of higher molecular weight (P39/GP42 and GP33/GP36) have recently been identified.[3,4] The larger polypeptides share the 226 amino acids of P25 (S region) at the C terminus and possess additional residues at the N terminus. The pre-S(2) region consists of 55 residues N-terminal to the S region[5], and the pre-S(1) region consists of 119 residues N-terminal to the pre-S(2) region[4] (Fig. 1). Herein, HBsAg particle preparations are designated by virtue of the highest molecular weight polypeptide present (i.e. HBsAg/P39, HBsAg/GP33 and HBsAg/P25).

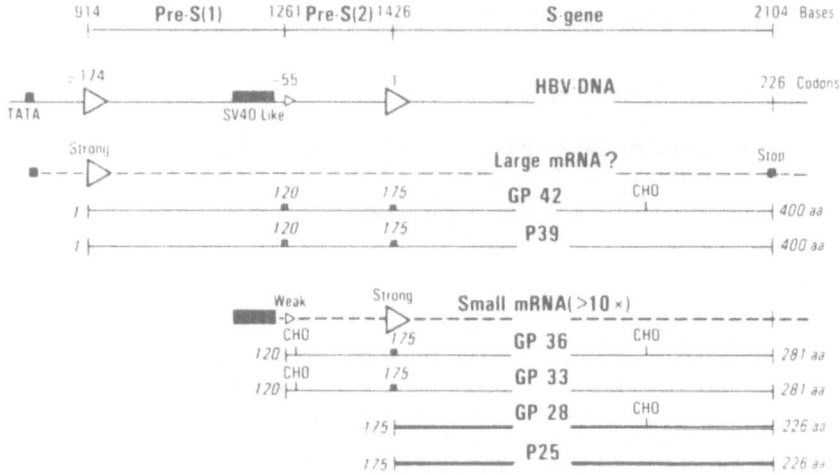

FIGURE 1.  Representation of the translation of the 3 coterminal envelope polypeptides of HBsAg (P39, GP33, and P25). The P39 polypeptide is translated from a putative large mRNA, and GP33 and P25 are derived from shorter, more abundant mRNAs. Amino acid positions are depicted from the N terminus (1) to the C terminus (400). Adapted from Heermann et al. (4)

The nucleocapsid of the HBV is a 27-nm particle composed of multiple copies of a single polypeptide (P21), and the intact structure exhibits hepatitis B core antigenicity (HBcAg). A nonparticulate form of HBcAg designated HBeAg may be present in the serum during HBV infection.

## Immunogenetics of the humoral response to HBV envelope antigens

Early studies indicated marked strain variation in antibody production after immunization with HBsAg/P25.[6,7] Studies in H-2 congenic and intra-H-2 recombinant strains confirmed a hierarchy of immune responsiveness to HBsAg/P25 determined by the H-2 haplotype of the responding strain, and identified high ($H-2^{d,q}$), intermediate ($H-2^a > H-2^b > H-2^k$); and nonresponder ($H-2^{f,s}$) phenotypes.[8]

When recombinant HBsAg particles containing the larger molecular weight polypeptides became available, studies were extended to examine genetic influences on the immune responses to pre-S sequences. Antibody production to the pre-S(2) region after immunization with HBsAg/GP33 is also H-2 restricted, however, the hierarchy of response status differs from that of the S region[9] (Table 1). It was also notable that the pre-S(2) region was significantly more immunogenic than the S region in terms of primary antibody production in vivo. Additionally, immunization with HBsAg/GP33 was capable of bypassing nonresponse to the S region in HBsAg/P25 nonresponder B10.S mice.[9] The immune response to the pre-S(1) region of HBsAg/P39 is also

TABLE I.  Influence of H-2 Genotype on the Humoral Response to HBsAg Particles of Varied Composition.

| Immunogen | Strain | H-2 | Specific Antibody Titer (1/dilution) | | |
|---|---|---|---|---|---|
| | | | S | pre-S(2) | pre-S(1) |
| HBsAg/P25 | B10.D2 | d | 81,920 | 0 | 0 |
| | B10 | b | 20,480 | 0 | 0 |
| | B10.BR | k | 5,120 | 0 | 0 |
| | B10.S | s | 0 | 0 | 0 |
| | B10.M | f | 0 | 0 | 0 |
| HBsAg/P33 | B10.D2 | | 40,960 | 10,240 | 0 |
| | B10 | | 10,240 | 40,960 | 0 |
| | B10.BR | | 1,280 | 2,560 | 0 |
| | B10.S | | 640[a] | 10,240 | 0 |
| | B10.M | | 0 | 0 | 0 |
| HBsAg/P39 | B10.D2 | | 81,920 | 5,120 | 640 |
| | B10 | | 20,480 | 40,960 | 10,240 |
| | B10.BR | | 5,120 | 1,280 | 2,560 |
| | B10.S | | 5,120[a] | 10,240 | 1,280 |
| | B10.M | | 10,240[a] | 1,280[a] | 10,240 |

[a]Represents an antibody response to a specific region of HBsAg which is not observed when the strain is immunized with that same antigen (i.e. B10.S is nonresponsive to the S region when immunized with HBsAg/p25). From Ref. 10.

influenced by H-2-linked genes, but again the hierarchy of response status differs from the responses to the S and pre-S(2) regions[10] (Table 1). HBsAg/P39 immunization elicited anti-pre-S(1)-specific antibody in all strains, and further-more elicited anti-S and anti-pre-S(2) responses in all strains including an S-specific response in the "nonresponder" B10.S strain, and S and pre-S(2)-specific responses in the "nonresponder" B10.M strain. These data indicated that distinct H-2-linked genes influence S, pre-S(2), and pre-S(1)-specific antibody production in vivo.

## T cell recognition of pre-S regions of HBsAg can circumvent nonresponse to the S and pre-S(2) regions

The ability of HBsAg/GP33 immunization to bypass S region nonresponder status in the B10.S strain and HBsAg/P39 immunization to bypass S and pre-S(2) region nonresponder status in the B10.M strain, suggests that although these strains lack an S-specific T cell response they must possess pre-S-specific T cells that can help B cell clones specific for S as well as pre-S region determinants. The T cell responses after HBsAg/GP33 immunization were regulated by H-2-linked genes, and the H-2-linked variation in T cell IL-2 production correlated with _in vivo_ anti-pre-S(2) antibody production such that B10 > B10.D2 > B10.S > B10.BR > B10.M.[11] This hierarchy of T cell response status differed from that

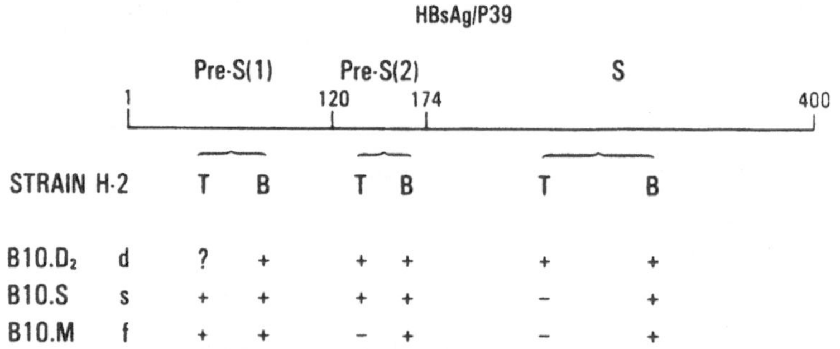

HBsAg/P39

| | | Pre-S(1) | | Pre-S(2) | | S | |
|---|---|---|---|---|---|---|---|
| | | 1 | | 120 | 174 | | 400 |
| STRAIN | H-2 | T | B | T | B | T | B |
| B10.D₂ | d | ? | + | + | + | + | + |
| B10.S | s | + | + | + | + | − | + |
| B10.M | f | + | + | − | + | − | + |

FIGURE 2. Summary of T cell proliferative responses (T) and _in vivo_ antibody production (B) of H-2 congenic, murine strains immunized with HBsAg/P39. A plus in the T column represents significant, dose-dependent, T cell proliferation and IL-2 production. A plus in the B column represents significant antibody production after secondary immunization.

observed after HBsAg/P25 immunization indicating that distinct H-2-linked genes can influence S and pre-S(2)-specific T cell responses. However, since S and pre-S(2) region determinants exists on the same polypeptide (GP33), it is likely that T helper (Th) cells specific for a determinant on one region are capable of providing functional help to B cell clones recognizing a determinant on the other region. The responses of the B10.S strain seem to confirm this possibility, and suggest that a pre-S(2) determinant(s) is acting as a "carrier" for S-region "hapten" in this strain.

After HBsAg/P39 immunization in the B10.M strain, only HBsAg/P39 induced a proliferative T cell response _in vivo_ and not HBsAg/P25 or HBsAg/GP33, indicating an exclusive pre-S(1)-specific T cell response.[10] A summary of T and B cell responses after HBsAg/P39 immunization of three representative H-2-congenic strains is shown in Fig. 2.

The B10.D2 strain is responsive at the T cell level to the S and pre-S(2) regions, and possibly the pre-S(1) region (not determined). The B10.S strain is only responsive to the pre-S regions at the T cell level. The B10.M strain only recognizes the pre-S(1) region at the T cell level. However, all strains produce antibody specific for each of the regions due to the fact that T cell recognition of a single region is sufficient to provide help for multiple B cell specificities present on the three regions of HBsAg.

## Fine specificity of antibody and T cell recognition of the pre-S regions of HBsAg

Additional studies have been aimed at identifying T cell and B cell (antibody) recognition sites within the pre-S regions of HBsAg. These studies have been greatly facilitated by the fact that the pre-S regions of HBsAg possess continuous[9,12] as opposed to discontinuous or conformational antibody determinants unlike the S region, which requires intact disulfide bonds for full antigenicity.[13] Nevertheless, numerous groups utilizing synthetic peptides have identified putative antibody binding sites mostly within the major hydrophilic domain (residues 284-324) of the S region[14-19] (reviewed in [20]). Two synthetic peptides derived from the pre-S(2) region sequence elicit antibodies cross-reactive with the native pre-S(2) region, and bind antibodies raised to the native protein. Denoting the amino terminus of the pre-S(2) region as residue 120, synthetic peptides p120-145[12] and p133-151[21] were shown to bind human antibodies elicited by HBV infection. Further analysis using a combination of truncated synthetic peptides and monoclonal antibodies revealed that the murine antibody response to the pre-S(2) region is predominantly focused on residues 133 through 143, and two distinct but overlapping epitopes were identified as p133-139 and p137-143[22]. An independent report of a pre-S(2)-specific monoclonal antibody, which binds the p132-145 sequence, confirms the dominance of this region of pre-S(2) as an antibody recognition site.[23] Sera from HBV-infected patients also bind p133-145 and the constituent overlapping epitopes within this sequence.[22] A number of pre-S(1)-specific antibody binding sites recognized by murine and human sera have also been recently elucidated including p1-21[24], p32-53, p41-53, p94-105, and p106-117[25] (see Fig. 3). Although these data indicate that the pre-S(1) region possesses continuous B cell determinants, the presence of conformational epitopes within the pre-S(1) region has also been suggested.[26]

Whereas, the antibody response after immunization with the peptide 120-145 is directed to the C terminus (residues 133-143), examination of the T cell response localized a predominant T cell recognition site to the N-terminal 120-132 residues in the H-2$^{q,s}$ haplotypes.[27] The presence of this T cell site explains the immunogenicity of the unconjugated peptide 120-145 (i.e. a primary antibody titer of 1:512,000). In contrast to the total crossreactivity of anti-p133-143 antibodies with native HBsAg/GP33 particles, p120-132-primed T cells do not recognize this sequence on HBsAg/GP33. Therefore, identification of a peptide sequence which elicits significant T cell activation does not guarantee that the resulting T cell population(s) will recognize the same sequence in the native protein. In support of this, it was shown that the anti-p120-145 response is regulated by distinct H-2-linked genes as compared to the anti-native pre-S(2) response.[28] In contrast to p120-132, pre-S(1)-specific T cell recognition sites have been defined which are relevant to both the peptide sequence and native HBsAg/P39. The pre-S(1) sequences p12-21 and p94-117 can induce and elicit HBsAg/P39-specific T cell activation.[25]

## A single synthetic T cell determinant can prime Th cell function for antibody production to multiple B cell epitopes on HBsAg: T cell fine specificity can influence B cell fine specificity

Identification of a number of T cell and B cell recognition sites within the pre-S(1) and pre-S(2) regions of HBsAg permitted examination of the influence of T cell fine specificity on antibody fine specificity. This was accomplished by examining the ability of distinct peptide-primed T cell populations to provide functional T cell help for a series of B cell specificities on HBsAg. Two pre-S(1)-specific T cell determinants were chosen (i.e. p12-21, p94-117), which induced only minimal antibody responses. This allowed priming of T helper (Th) cells with peptides and subsequent determination of in vivo antibody production after challenge with a suboptimal dose of HBsAg/P39 in the same animal, as opposed to performing transfer experiments. This approach requires the memory T cells primed by immunization with peptide to be recalled by challenge with native particles, indicating the relevance of the synthetic T cell site to the native molecule. Using this protocol it was demonstrated that priming with a single synthetic peptide, p12-21, elicited Th function resulting in in vivo antibody production to p16-27, p133-140, p135-145 in the pre-S region and group- and subtype-specific determinants in the S region[25] (Fig. 3). Similarly, priming with p94-117 elicited Th function resulting in in vivo antibody production specific for p32-53, p94-105, p106-117, p133-140, p135-145 in the pre-S region, but did not prime antibody production to S region determinants (Fig. 3). These results indicate that T cells primed to a single determinant are sufficient to provide functional help to multiple B cell clones, which recognize distinct epitopes on a complex, particulate antigen.

Note that the pre-S(1)-specific, T cell recognition sites, p12-21 and p94-117, primed antibody production specific for unique as well as common B cell determinants. For example, p94-117 primed an anti-p32-53 response, whereas, p12-21 did not. This data provides strong evidence that the fine specificity of the Th cell can influence the fine specificity of the antibody produced. Berzofsky[29] has proposed a T cell-B cell reciprocity circuit in which B cell immunoglobulin receptor-antigen-Ia interactions may limit T cell specificity, which in turn limits B cell specificity. In the context of this hypothesis, the B cell clone specific for the p32-53 epitope may present the P39 polypeptide in the context of Ia in such a way as to be recognized by the T cell clone(s) specific for p94-117, but not by the p12-21-specific

T cell clone(s), and therefore will not receive the necessary
Th cell signals from p12-21-primed T cells.

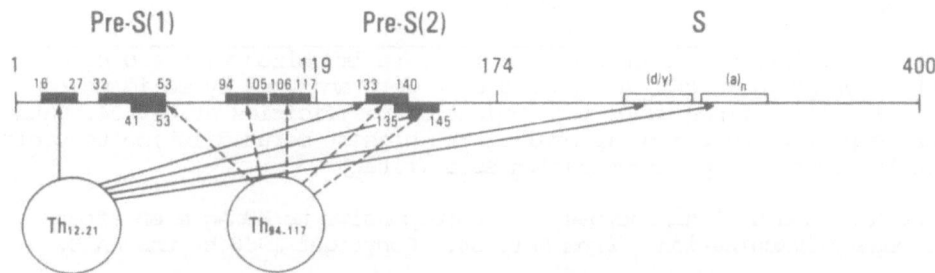

FIGURE 3.  Summary of T and B cell recognition of HBsAg/P39.
Defined antibody binding sites are represented by
solid boxes, and the $NH_2$ and COOH-terminal coordinates are depicted.  Serologically determined
antibody binding sites [i.e., S region group (a)
and subtype (d/y specific)] for which there are no
consensus sequences are represented by open boxes.
T cell helper (Th) activity and the antibody
specificities elicited are represented by arrows.
The specificities of the Th cells are depicted.
Modified from Ref. 25.

## The nucleocapsid of HBV is significantly more immunogenic than the envelope

Envelope and nucleocapsid-specific cellular immune
responses  have been suggested to be important in virus
elimination and the attendant hepatocellular injury[30,31], and
vaccination with both antigens has been reported to protect
against HBV infection.[32,33]  Comparative studies of murine
antibody production revealed that anti-HBc responses were
significantly greater (at least 80-fold) than anti-HBs
responses in all strains tested (Table 2).  The influence of
H-2-linked genes on the anti-HBc response is apparent, and no
nonresponder strains have been identified.  The comparative
magnitudes of the anti-HBc and anti-HBs responses, and the
lack of nonresponsiveness to HBcAg are consistent with the
human immune responses to these HBV antigens.

TABLE II. Comparison of Primary Antibody Responses After Immunization
With HBsAg and HBcAg.

| Strain[a] | H-2 | Anti-HBs (titer) | Anti-HBc (titer) |
|-----------|-----|------------------|------------------|
| B10       | b   | 256              | 40,960           |
| B10.D2    | d   | 1024             | 81,920           |
| B10.S     | s   | 0[b]             | 163,840          |
| B10.BR    | k   | 32               | 163,840          |
| B10.M     | f   | 0[b]             | 20,480           |
| C$_3$H.Q  | q   | 2048             | 327,680          |
| Balb/c    | d   | 1024             | 327,680          |

[a]Groups of five mice from each strain were immunized with 4.0 μg of
HBsAg or HBcAg in CFA, and pooled sera were analyzed by solid phase
RIA for IgG antibodies of the indicated specificities at day 24. Data
are expressed as the reciprocal of the highest serum dilution to yield
4X the counts of preimmunization sera (titer).

[b]The H-2$^s$ and H-2$^f$ haplotypes are nonresponsive to HBsAg even after
secondary immunization. From Ref. 34. Copyright 1986 by the AAAS.

    Studies of the murine cellular response to HBcAg
revealed that HBcAg was an extremely efficient immunogen at
the T cell level as measured by the in vivo dose required to
induce T cell sensitization (1.0 μg), and the minimal in
vitro concentration required to elicit IL-2 production from
primed-T cells (0.03 ng/ml).[35]  This degree of T cell
immunogenicity is approximately 100-fold that observed for
HBsAg subviral particles.[36]  Similarly, studies of the human
T cell response to HBcAg also suggest that HBcAg is an
efficient T cell antigen.[37]

    Examination of the fine specificity of T cell recogni-
tion of HBcAg revealed that HBcAg-specific T cells from a
variety of strains recognized multiple but distinct sites
within the HBcAg/HBeAg sequence[35] (HBeAg lacks the C-terminal
34 residues of HBcAg).  T cell recognition sites have been
defined by 12-21 residue synthetic peptides.  Each strain
recognized a predominant T cell determinant, and the fine
specificity of this recognition process was dependent on the
H-2 haplotype of the responding strain:  For example, H-2$^s$
strains recognized p120-131, H-2$^b$ strains recognized p129-
140, H-2$^{f,q}$ strains recognized p100-120, and H-2$^d$ mice
recognized p85-100 predominantly[35,38] (see Fig. 4).  This
murine model predicts that a human outbred population would
exhibit similar complexity, and individuals may recognize
distinct T cell sites in the context of their HLA genotype.

    To determine the functional ability of synthetic T cell
sites to prime Th cells and induce antibody production in
vivo, the HBcAg-specific peptide 120-140 and N-terminal
(p120-131) and C-terminal (p129-140) fragments were chosen
for study in the B10.S and B10 strains.[38]  Mice were primed
with peptide and challenged with a suboptimal dose of HBcAg,

and serum anti-HBc measured. In the B10.S strain, p120-140-primed and p120-131-primed mice produced IgG, anti-HBc efficiently 7 days after the challenge, whereas, p129-140-primed mice did not (Fig. 5). Similarly, priming with p120-140 elicited anti-HBc in the B10 strain. In contrast to the B10.S strain, the C-terminal p129-140 contained the active T cell site. These results were consistent with the T cell proliferation (Tp) results and indicate a concordance between Tp and Th cell fine specificities.[38]

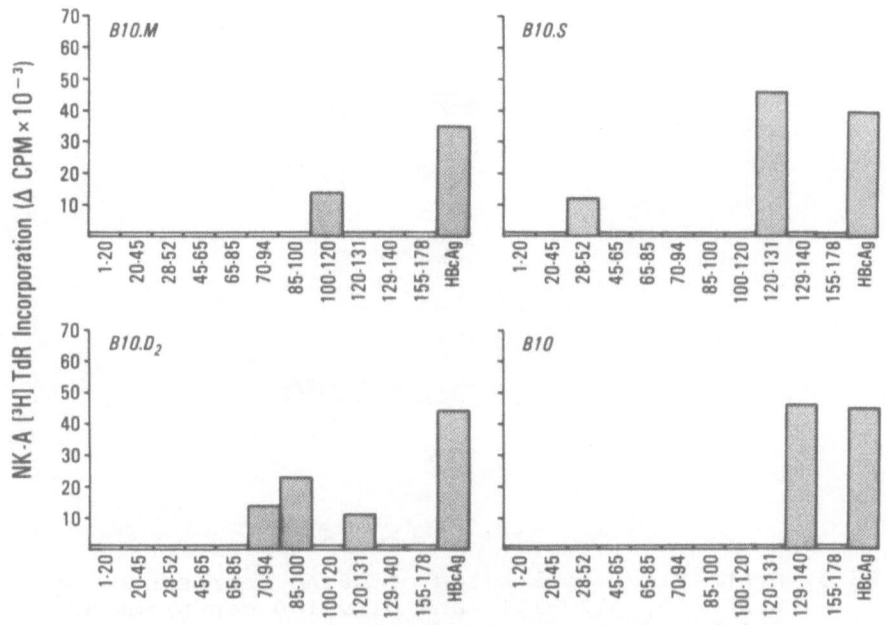

FIGURE 4. Localization of T cell sites within the HBcAg/HBeAg sequence using synthetic peptide analogs. Mice were immunized with 4 μg of HBcAg and PLN cells were harvested 8 d post-immunization and cultured _in vitro_ with the synthetic peptide fragments shown or HBcAg as the positive control. T cell activation was measured by IL-2 production elicited by the optimal concentration of peptide (16-64 μg/ml) or HBcAg (0.5 μg/ml). Modified from Ref. 35.

## HBcAg can induce antibody responses via both a T cell dependent and a T cell independent pathway.

Although HBcAg is an efficient T cell immunogen, this characteristic alone cannot explain the extremely high levels of antibody produced after primary immunization with HBcAg. The nonparticulate form of HBcAg, namely HBeAg is cross-reactive with HBcAg at the T cell level[38], but is a relatively poor immunogen in mice in terms of _in vivo_

antibody production (unpublished).  Therefore, the ability
of HBcAg to activate B cells directly was examined by
immunizing Balb/c euthymic and Balb/c athymic (nude) mice
with a mixture of HBcAg and HBsAg.  HBcAg was able to induce
antibody production in athymic mice, whereas the comparably
sized HBsAg particle was not immunogenic in athymic Balb/c

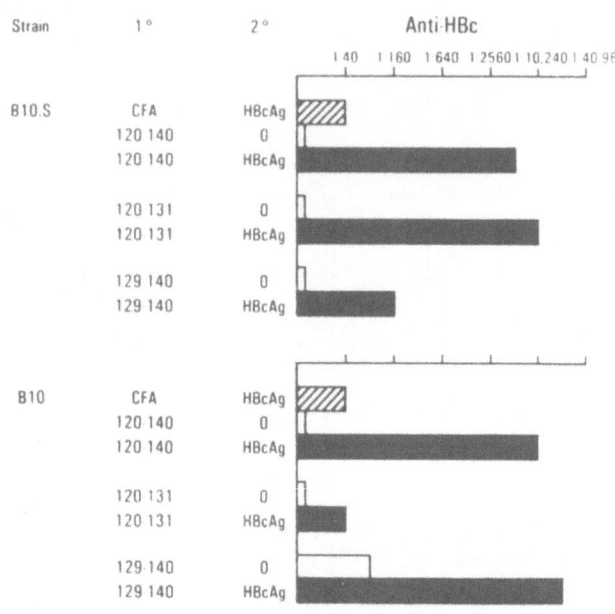

FIGURE 5.  Synthetic T cell sites of HBcAg represented by
          p120-140, p120-131, and p129-140 can prime Th cells
          which induce anti-HBc production in vivo.  B10.S
          (upper panel) or B10 (lower panel) mice were primed
          (1°) by immunization with 100 μg of either p120-
          140, p120-131, p129-140 in CFA or CFA alone.  After
          three weeks, the primed mice were challenged (2°)
          with either a suboptimal dose of HBcAg (0.1 μg) in
          incomplete adjuvant or with adjuvant alone (0).
          Seven days after the challenge dose, sera were
          collected, pooled and analyzed for IgG class, anti-
          HBc antibody by solid-phase RIA.  The anti-HBc
          titer is expressed as the highest serum dilution
          required to yield four times the counts of sera
          before immunization.  From Ref. 38.


mice[34] (Fig. 6).  The T cell independence required that the
HBcAg be particulate because denatured HBcAg and nonparticulate HBeAg were not immunogenic in athymic mice.  The
immunogenic effects of an antigen that possesses both T cell-
independent and T cell-dependent characteristics may be
synergistic in the presence of competent T cells.  This
property of HBcAg may explain its enhanced immunogenicity in
the mouse model and during HBV infection.

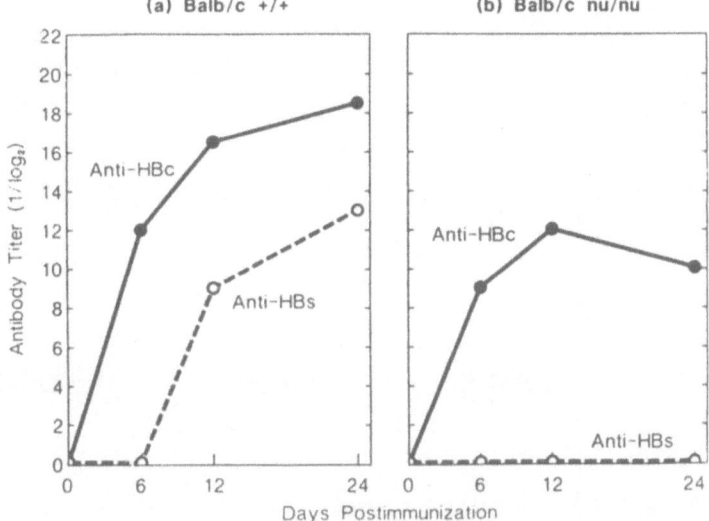

(a) Balb/c +/+          (b) Balb/c nu/nu

FIGURE 6. The HBcAg can function as a T cell-independent antigen. Groups of five BALB/c euthymic (+/+) (A) or BALB/c athymic (nu/nu) (B) mice were immunized intraperitoneally with a mixture of rHBcAg (8 μg) and HBsAg (8 μg) in CFA. Serum samples obtained before and 6, 12, and 24 days after immunization were pooled and analyzed for anti-HBc and anti-HBs activity. Data are expressed as the reciprocal of the $\log_2$ of the highest serum dilution to yield four times the counts of sera before immunization.

## HBcAg-specific T cells can prime antibody production to HBsAg: Intermolecular/intrastructural T cell help

Due to the marked immunogenicity of HBcAg and the observation in the influenza system that nucleocapsid-specific T cells could elicit hemagglutinin-specific antibody production[39], the ability of HBcAg-primed T cell to function as Th cells for antibody production to envelope (HBsAg) epitopes was examined.[40] B10.S mice primed with HBcAg and challenged with a mixture of HBcAg and HBsAg/P39 produced no anti-HBs, however, mice challenged with virions produced anti-S, anti-pre-S(2), and anti-pre-S(1)-specific antibodies (Fig. 7). To confirm the T cell nature of this effect, the identical experiment was performed using the synthetic T cell recognition site, p120-140, as the priming antigen. The results obtained were similar to those using native HBcAg as the priming antigen. This result indicated that HBcAg-primed T cells could function to help anti-envelope antibody production to multiple epitopes, and the Th cell activity did not require that HBcAg and HBsAg be present on the same molecule (intermolecular), but did require that they be within the same particle (intrastructural).[40]

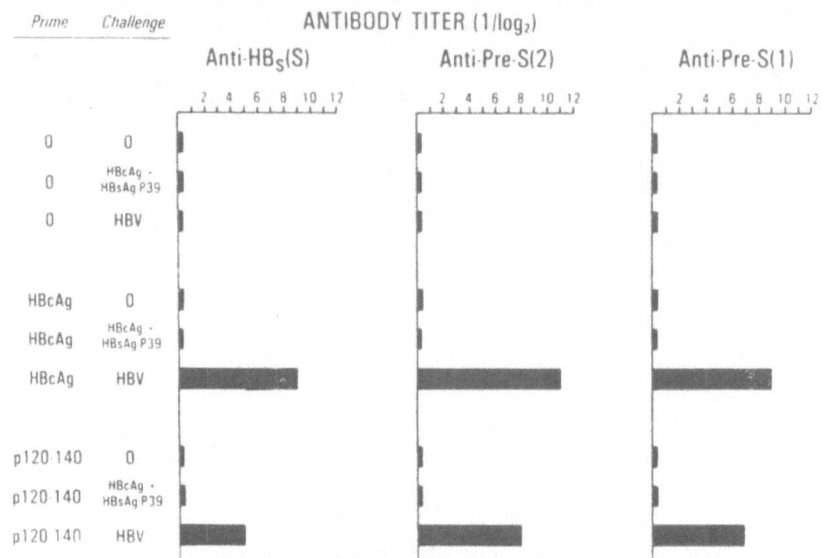

FIGURE 7. HBcAg-primed and p120-140-primed Th cells of B10.S mice can induce antibody production specific for the envelope of HBV. B10.S mice were primed by immunization with either CFA alone (0, upper panel), 4.0 $\mu$g of HBcAg in CFA (middle panel), or 100 $\mu$g of the synthetic peptide p120-140 in CFA (lower panel). Three weeks after priming, mice were challenged either with incomplete adjuvant alone (0), or a suboptimal dose of a mixture of HBcAg (0.1 $\mu$g) and HBsAg/P39 (0.6 $\mu$g), or with HBV (0.4 $\mu$g) in incomplete adjuvant. Seven days after the challenge dose, sera were collected, pooled, and analyzed for IgG antibody specific for the S, pre-S(2), and pre-S(1) regions of HBsAg by solid-phase RIA. The antibody titers are expressed as the reciprocal of the $\log_2$ of the highest serum dilution required to yield four times the counts of sera before immunization. From Ref. 40.

Production of antibodies specific for the S, and pre-S regions occurs during resolving acute HBV infection, but not during the acute phase or subsequently in chronic asymptomatic infection (reviewed in [41]). However, a subset of chronically infected HBV patients has been reported to produce only anti-pre-S(1) in the absence of anti-pre-S(2) or anti-S antibodies, which correlated with seroconversion from HBeAg to anti-HBe status and viral clearance.[42] It has been proposed that progression to chronic infection requires, or is a consequence of nonresponsiveness to the envelope antigens at the T cell level, and furthermore that anti-pre-S(1) production in the absence of anti-pre-S(2) and anti-S production is mediated by Th cells specific for HBcAg/HBeAg determinants.[40,43]

Although antibodies to HBcAg do not prevent infection, the fact that HBcAg/HBeAg-specific Th cells can elicit anti-envelope antibodies, which are virus neutralizing[32,44], may explain the reported ability of HBcAg vaccination to protect against HBV liver disease.[32,33]  Furthermore, since HBcAg/HBeAg-specific Th cells were shown to induce anti-S antibody production in S region nonresponder mice, this represents another mechanism of circumventing HBsAg non-responsiveness.[40]  This observation can be applied to vaccine development.  The HBcAg could be used as a T cell carrier for HBsAg by coupling the two antigens either chemically or by recombinant DNA technology.  Alternatively, synthetic HBcAg/HBeAg Th cell epitopes may be coupled or genetically engineered into HBsAg particles or polypeptides as discussed in the next section.

A totally synthetic HBV immunogen comprised of nucleocapsid T-cell sites and an envelope B cell epitope.

Because the p120-140 sequence of HBcAg was shown to encompass distinct Th cell recognition sites for B10.S and B10 mice, the ability of p120-140, coupled directly to a synthetic B cell epitope, to act as a T cell carrier moiety was examined.[38]  The B cell epitope chosen was the pre-S(2) region peptide p133-140, which was previously shown to represent a dominant antibody binding site within the pre-S(2) region.[22]  The unconjugated p133-140 sequence of the pre-S(2) region is nonimmunogenic.  B10.S, B10 and B10.BR mice were immunized with a composite peptide composed of residues 120-140 from the HBcAg sequence and residues 133-140 from the pre-S(2) region of the envelope designated c120-140-(133-140).  The B10.S and B10 strains produced antibody to the envelope B cell epitope (p133-140), which was highly crossreactive with native HBsAg/GP33 (Table III).

TABLE III.  The HBcAg-specific p120-140 sequence can function as a T cell carrier for a synthetic pre-S(2) region B cell epitope (133-140).

| Strain | Immunogen[a] | Time | Antibody Titer(1/dilution)[b] | | | |
|---|---|---|---|---|---|---|
| | | | c120-140 | HBcAg | (133-140) | HBsAg/GP33 |
| B10.S | c120-140- | Pre | 0 | 0 | 0 | 0 |
| | (133-140) | 1° | 10,240 | 0 | 640 | 1,280 |
| | | 2° | 40,960 | 1,280 | 10,240 | 10,240 |
| B10 | c120-140- | Pre | 0 | 0 | 0 | 0 |
| | (133-140) | 1° | 1,280 | 0 | 0 | 0 |
| | | 2° | 81,920 | 5,120 | 2,560 | 1,280 |
| B10.BR | c120-140- | Pre | 0 | 0 | 0 | 0 |
| | (133-140) | 1° | 0 | 0 | 0 | 0 |
| | | 2° | 0 | 0 | 0 | 0 |

[a]The indicated strains were immunized i.p. with 100 μg of c120-140-(133-140) in CFA, and boosted with 50 μg i.p. in incomplete adjuvant. Sera were collected preimmunization (Pre), 3 weeks after the primary immunization (1°), and 2 weeks after the secondary immunization (2°).

[b]Antibody (IgG) specific for the indicated antigens was measured by solid-phase RIA, and expressed as the reciprocal of the dilution to yield 4X the counts of preimmunization sera.  From Ref. 38.

The B10.BR strain does not recognize the p120-140 HBcAg sequence at the T cell level, and predictably this strain was a nonresponder to immunization with c120-140-(133-140).

In order to confirm that the predicted sites within the composite immunogen were functioning as T cell recognition sites, c120-140-(133-140)-immunized mice were evaluated at the T cell level as well. B10.S, c120-140-(133-140)-primed T cells responded to c120-140, the N-terminal fragment, p120-131, and to native HBcAg. B10, c120-140-(133-140)-primed T cells were activated by c120-140, the C-terminal fragment, p129-140, and native HBcAg. The B cell epitope (133-140) and native HBsAg/GP33 were non-stimulatory in both strains. B10.BR, c120-140-(133-140)-primed T cells were nonresponsive to the entire antigen panel.[38] These results demonstrate the feasibility of constructing complex synthetic immunogens representing multiple proteins of a pathogen, and capable of engaging both T and B cells relevant to the native antigens.

CONCLUSION

This experimental model offers a unique opportunity to study the genetic, cellular, and molecular basis for variable immune responsiveness to HBV-encoded antigens. The linkage between the MHC and the regulation of immune responsiveness to HBsAg in mice has been extended to the human immune response by reports of an association between HLA phenotype and low to nonresponsiveness to recent HBsAg vaccines.[45,46] The murine model has provided a means of studying the immunogenicity of the pre-S regions of HBsAg and has elucidated the independent H-2-linked genes regulating antibody production to pre-S and S region determinants. The ability to circumvent genetic nonresponsiveness has implications for the design of future HBV vaccines. The murine model afforded the opportunity to examine the ability of HBcAg to activate B cells directly, and to prime Th cells capable of eliciting anti-envelope antibody production. These observations have potential clinical relevance, and may explain the ability of HBcAg vaccination to protect against HBV infection. The murine system has also facilitated the mapping of T cell and B cell recognition sites within HBV proteins, which at least conceptually enhances the prospects for development of a synthetic HBV vaccine. Antibody binding sites appear to be similar in the human and murine systems. The extent to which murine and human T cell repertoires overlap is not known. However, the probable chemical and structural constraints imposed on T cell antigenicity and the fact that MHC-encoded class II molecules and the cellular mechanisms mediating T cell recognition are conserved across species make such overlaps appear likely. Several S region T cell sites have been identified which activate both murine and human T cells[47], and an S region T cell site recently identified as immunodominant for a number of human T cell clones[48] has not been shown to be active in mice. At the very least, investigation at the level of an animal model will provide insight, and focus the search for T cell recognition sites relevant to the human population.

ACKNOWLEDGMENT

The author thanks Janice Hughes, Alan McLachlan, Ben Thornton and Ann Moriarty for their contributions and discussions. Research cited in this review was supported by NIH grants AI00585, AI18391, and AI20720, and a grant from the Johnson and Johnson Company.

REFERENCES

1.  L. F. Barker, F. V. Chisari, P. P. McGrath, D. W. Dalgard, R. L. Kirschstein, J. D. Almeida, T. S. Edgington, D. G. Sharp, and M. R. J. Peterson, Transmission of type B viral hepatitis to chimpanzees, J. Infect. Dis. 127:648 (1973).
2.  F. J. Dudley, R. A. Fox, and S. Sherlock, Cellular immunity and hepatitis-associated Australia antigen liver disease, Lancet I:723 (1972).
3.  W. Stibbe, and W. H. Gerlich, Structural relationships between minor and major proteins of hepatitis B surface antigen, J. Virol. 46:626 (1983).
4.  K. H. Heermann, U. Goldman, W. Schawartz, T. Seyffarth, H. Baumgarten, and W. H. Gerlich, Large surface proteins of hepatitis B virus containing the pre-S sequence, J. Virol. 52:396, 1984.
5.  A. Machida, S. Kishimoto, H. Ohnuma, K. Baba, Y. Ito, G. Miyamoto, G. Funatau, K. Oda, S. Usuda, S. Togami, T. Nakamura, M. Miyakawa, and M. Mayumi, A polypeptide containing 55 amino acid residues coded by the pre-S region of hepatitis B virus deoxyribonucleic acid bears the receptor for polymerized human as well as chimpanzee albumins, Gastroenterol. 86:910 (1984).
6.  D. R. Milich, and F. V. Chisari, Genetic regulation of the immune response to hepatitis B surface antigen (HBsAg). I. H-2 restriction of the murine humoral immune response to the a and d determinants of HBsAg, J. Immunol. 129:320 (1982).
7.  D. R. Milich, G. Leroux-Roels, and F. V. Chisari, Genetic regulation of the immune response to hepatitis B surface antigen (HBsAg). II. Qualitative characteristics of the humoral immune response to the a, d, and y determinants of HBsAg, J. Immunol. 130:1395 (1983).
8.  D. R. Milich, G. G. Leroux-Roels, R. E. Louie, and F. V. Chisari, Genetic regulation of the immune response to hepatitis B surface antigen (HBsAg). IV. Distinct H-2 linked Ir genes control antibody responses to different HBsAg determinants on the same molecule and map to the I-A and I-C subregions, J. Exp. Med. 159:41 (1984).
9.  D. R. Milich, G. B. Thornton, A. Neurath, S. Kent, M. Michel, P. Tiollais, and F. V. Chisari, Enhanced immunogenicity of the pre-S region of hepatitis B surface antigen, Science 228:1195 (1985).
10. D. R. Milich, A. McLachlan, F. V. Chisari, S. B. H. Kent, and G. B. Thornton, Immune response to the

pre-S(1) region of the hepatitis B surface
antigen (HBsAg): a pre-S(1)-specific T cell
response can bypass nonresponsiveness to the pre-
S(2) and S regions of HBsAg, <u>J</u>. <u>Immunol</u>. 137:315
(1986).

11.  D. R. Milich, M. K. McNamara, A. McLachlan, G. B.
     Thornton, and F. V. Chisari, Distinct H-2 linked
     regulation of T-cell responses to the pre-S and S
     regions of the same hepatitis B surface antigen
     polypeptide allows circumvention of nonrespon-
     siveness to the S region, <u>Proc</u>. <u>Natl</u>. <u>Acad</u>. <u>Sci</u>.
     82:8168 (1985).

12.  A. R. Neurath, S. B. H. Kent, and N. Strick, location
     and chemical synthesis of pre-S gene coded
     immunodominant epitope of hepatitis B virus,
     <u>Science</u> 224:392, 1984.

13.  G. N. Vyas, K. R. Rao, and A. B. Ibrahim, Hepatitis
     associated Australian antigen (HBsAg): a
     conformational antigen dependent on disulfide
     bonds, <u>Science</u> 178:1300 (1972).

14.  R. A. Lerner, N. Green, H. Alexander, F. -T. Liu, J. G.
     Sutcliffe, and T. M. Shinnick, Chemically
     synthesized peptides predicted from the nucleo-
     tide sequence of the hepatitis B virus genome
     elicit antibodies reactive with the native
     envelope protein of Dane particles, <u>Proc</u>. <u>Natl</u>.
     <u>Acad</u>. <u>Sci</u>. <u>USA</u> 78:3403 (1981).

15.  P. K. Bhatnagar, E. Papas, H. E. Blum, Immune response
     to synthetic peptide analogues of hepatitis B
     surface antigen specific for the a determinant,
     <u>Proc</u>. <u>Natl</u>. <u>Acad</u>. <u>Sci</u>. <u>USA</u> 79:4400 (1982).

16.  G. R. Dreesman, Y. Sanchez, I. Ionescu-Matiu, Antibody
     to hepatitis B surface antigen after a single
     inoculation of uncoupled synthetic HBsAg
     peptides, <u>Nature</u> 295:158 (1982).

17.  J. W. -K. Shih, R. J. Gerety, D. T. -Y. Liu, H. Yajima,
     N. Fujii, M. Nomizu, Y. Hayashi, and S. Katakura,
     Immunogenicity of the unconjugated synthetic
     polypeptides of Hepatitis-B surface antigen, <u>in</u>:
     "Modern Approaches to Vaccines", R. M. Chanock
     and R. A. Lerner, eds., Cold Spring Harbor
     Laboratory, Cold Spring Harbor, N.Y. (1985).

18.  J. L. Gerin, H. Alexander, J. W. -K. Shih, R. H.
     Purcell, G. Dapolito, R. Engle, N. Green, J. G.
     Sutcliffe, T. M. Shinnick, and R. A. Lerner,
     Chemically synthesized peptides of hepatitis B
     surface antigen duplicate the <u>d/y</u> specificities
     and induce subtype-specific antibodies in
     chimpanzees, <u>Proc</u>. <u>Natl</u>. <u>Acad</u>. <u>Sci</u>. <u>USA</u> 80:2365
     (1983).

19.  S. E. Brown, C. R. Howard, A. J. Zuckerman, and M. W.
     Steward, Determination of the affinity of
     antibodies to Hepatitis B surface antigen in
     human sera, <u>J</u>. <u>Immunol</u>. <u>Meth</u>. 72:41 (1984).

20.  F. R. Harmon, and J. L. Melnick, Synthetic vaccines for
     viral hepatitis, <u>in</u>: "Synthetic Vaccines, Vol.
     II", R. Arnon, ed., CRC Press Inc., Boca Raton,
     Fl. (1987).

21.  H. Okamoto, M. Imai, S. Usuda, E. Tanaka, K. Tachibana,
     S. Mishiro, A. Machida, Y. Miyakawa, and M.

Mayumi, Hemagglutination assay of polypeptide coded by the pre-S region of hepatitis B virus DNA with monoclonal antibody: correlation of pre-S polypeptide with the receptor for polymerized human serum albumin in serums containing hepatitis B antigens, J. Immunol. 134:1212 (1985).

22. D. R. Milich, A. McLachlan, F. V. Chisari, T. Nakamura, and G. B. Thornton, Two distinct but overlapping antibody binding sites in the pre-S(2) region of HBsAg localized within 11 continuous residues, J. Immunol. 137:2703 (1986).

23. A. R. Neurath, P. Adamowicz, S. B. H. Kent, M. M. Riottot, N. Strick, K. Parker, W. Offensperger, M. A. Petit, S. Wahl, A. Budkowska, M. Girard, and J. Pillot, Characterization of monoclonal antibodies specific for the preS2 region of the hepatitis B virus envelope protein, Molec. Immunol. 23:991 (1986).

24. A. R. Neurath, S. B. H. Kent, N. Strick, P. Taylor, and C. E. Stevens, Hepatitis B virus contains pre-S gene-encoded domains, Nature 315:154 (1985).

25. D. R. Milich, A. McLachlan, A. Moriarty, and G. B. Thornton, A single 10-residue pre-S(1) peptide can prime T cell for antibody production to multiple epitopes within the pre-S(1), pre-S(2), and S regions of HBsAg, J. Immunol. 138:4457 (1987).

26. K. -H. Heermann, F. Kruse, M. Seifer, and W. H. Gerlich, Immunogenicity of the gene S and pre-S domains in hepatitis B virions and HBsAg filaments, Intervirol. 28:14 (1987).

27. D. R. Milich, A. McLachlan, F. V. Chisari, and G. B. Thornton, Non-overlapping T and B cell determinants on an hepatitis B surface antigen pre-S(2) region synthetic peptide, J. Exp. Med. 164:532 (1986).

28. D. R. Milich, A. McLachlan, M. K. McNamara, F. V. Chisari, and G. B. Thornton, T-cell and B-cell recognition of native and synthetic pre-S region determinants on hepatitis-B surface antigen, in: "Vaccines 86", Cold Spring Harbor Laboratory, Cold Spring Harbor, N.Y. (1986).

29. J. A. Berzofsky, T-B reciprocity. An Ia-restricted epitope specific circuit regulating T cells-B cell interaction and antibody specificity, Surv. Immunol. Res. 2:223 (1983).

30. A. L. W. F. Eddleston, and R. Williams, Inadequate antibody response to HBAg or suppressor T-cell defect in development of active chronic hepatitis, Lancet II:1543 (1974).

31. M. Mondelli, G. M. Vergani, A. Alberti, D. Vergani, B. Portmann, A. L. W. F. Eddleston, and R. Williams, Specificity of lymphocyte cytotoxicity to autologous hepatocytes in chronic hepatitis B virus infection: evidence that T cells are directed against HBV core antigen expressed on hepatocytes, J. Immunol. 129:2773 (1982).

32. F. J. Gerety, E. Tabor, R. H. Purcell, and F. J. Tyeryar, Summary of an international workshop on

hepatitis B vaccines, J. Infect. Dis. 140:642 (1979).

33. K. Murray, S. A. Burce, A. Hinnen, P. Wingfield, P. M. van Erd, A. de Reus, and H. Schellekens, Hepatitis B virus antigen made in microbial cells immunize against viral infection, EMBO J. 3:645 (1984).

34. D. R. Milich, and A. McLachlan, The nucleocapsid of hepatitis B virus is both a T cell-independent and a T cell dependent antigen, Science 234:1398 (1986).

35. D. R. Milich, A. McLachlan, A. Moriarty, and G. B. Thornton, Immune response to hepatitis B virus core antigen (HBcAg): Localization of T cell recognition sites within HBsAg/HBeAg, J. Immunol. 139:1223 (1987).

36. D. R. Milich, R. E. Louie, and F. V. Chisari, Genetic regulation of the immune response to HBsAg. V. T cell proliferative response and cellular interactions, J. Immunol. 134:4194 (1985).

37. C. Ferrari, A. Penna, T. Sansoni, T. Giuberti, T. M. Neri, F. V. Chisari, and F. Fiaccadori, Selective sensitization of peripheral blood T lymphocytes to hepatitis B core antigen in patients with chronic active hepatitis type B, Clin. Exp. Immunol. 66:497 (1986).

38. D. R. Milich, J. L. Hughes, A. McLachlan, G. B. Thornton, and A. Moriarty, Hepatitis B synthetic immunogen comprised of nucleocapsid T-cell sites and an envelope B-cell epitope, Proc. Natl. Acad. Sci. USA 85:1610 (1988).

39. S. M. Russell, and F. Y. Liew, T cells primed by influenza virion internal components can cooperate in the antibody response to haemag-glutinin, Nature 147:280 (1979).

40. D. R. Milich, A. McLachlan, G. B. Thornton, and J. L. Hughes, A synthetic T cell site primes antibody production to both the nucleocapsid and envelope of the hepatitis B virus, Nature 329:547 (1987).

41. D. R. Milich, Immunological response to the pre-S antigens of the hepatitis B virus, Virol. Immunol. 1:83 (1987).

42. E. Takai, A. Machida, H. Ohnuma, H. Miyamoto, T. Tanaka, K. Baba, F. Tsuda, S. Usuda, T. Nakamura, Y. Miyakawa, and M. Mayumi, A solid phase enzyme immunoassay for the determination of IgM and IgG antibodies against translation products of the pre-S1 and pre-S3 regions of hepatitis B virus, Immunol. Meth. 95:23 (1986).

43. D. R. Milich, Genetic and molecular basis for T- and B-cell recognition of hepatitis B viral antigens, Immunol. Rev. 99:71 (1987).

44. Y. Itoh, E. Takai, H. Ohnuma, K. Kitajima, F. Tsuda, A. Machida, S. Mishiro, T. Nakamura, Y. Miyakawa, and M. Mayumi, A synthetic peptide vaccine involving the product of the pre-S(2) region of hepatitis B virus DNA: Protective efficacy in chimpanzees, Proc. Natl. Acad. Sci. 83:9174 (1986).

45. M. E. Walker, W. Szmuness, C. E. Stevens, and P. Rubinstein, Genetics of anti-HBs responsiveness:

I. HLA-DR7 and non-responsiveness to hepatitis vaccination (Abstract) <u>Transfusion</u> 21:601 (1981).

46. D. E. Craven, Z. L. Awdeh, L. M. Kunches, J. Yunis, J. L. Dienstag, B. G. Werner, B. Polk, D. R. Snydman, R. Platt, C. S. Crumpacker, G. F. Grady, and C. A. Alper, Nonresponsiveness to hepatitis B vaccine in health care workers, <u>Ann</u>. <u>Int</u>. <u>Med</u>. 105:356 (1986).

47. D. R. Milich, D. L. Peterson, G. G. Leoux-Roel, R. A. Lerner, and F. V. Chisari, Genetic regulation of the immune response to hepatitis B surface antigen (HBsAg). VI. T cell fine specificity, <u>J</u>. <u>Immunol</u>. 134:4203 (1985).

48. E. Celis, D. Ou, and L. Otvos, Jr., Recognition of hepatitis B surface antigen by human T lymphocytes. Proliferative and cytotoxic responses to a major antigenic determinant defined by synthetic peptides, <u>J.Immunol</u>. 140:1808 (1988).

# IMPORTANT DISEASES WITH A POSSIBLE VIRAL AETIOLOGY

C.A. Mims

Department of Microbiology, UMDS
Guy's Campus,
London Bridge, SE1 9RT U.K.

There can be few of us who have not at some time or another considered a virus to account for a disease of unknown origin.  Indeed as long as there are such diseases there will be suggestions that viruses are responsible. But we have to be careful because although the late Peter Medawar quite accurately described a virus as a "piece of bad news wrapped in protein" there is the alternative definition that virus is a Latin word used by doctors meaning "your guess is as good as mine".  However, viruses are unique in that they can do the following things.

1. They can infect tissues without inducing inflammation.

2. They invade any cell, at any site in the body.  This can include retinal neurones, B cells, T cells, ova, oligodendrocytes, adrenal cortical cells, or feather follicle epithelial cells.

3. They can replicate throughout life with no cell damage.

4. They can interfere with "luxury functions".  As an example of this, it has long been known that mice infected with LCM virus in utero or neonatally become virus carriers for life.  Cells in most organs of the body are infected without being damaged.  The carrier mice are runted, and it turns out that the cells in the anterior pituitary that produce growth hormone are infected and there are defects in the production of this hormone.  The blood contains half the normal amount of growth hormone, and if suckling mice are transplanted with growth hormone-producing cells the runting is prevented.  In other words, this is a virus-induced pituitary dwarfism (Rodriguez et al. 1983).

Today I am going to look at some of the chronic
diseases of obscure origin and discuss the evidence for
virus aetiology. Unfortunately 9 out of 10 reports linking
viruses with these diseases are inadequate and unacceptable,
so that the literature is burdened and the medical profession
confused by false alarms and unreproducible results. So we
have to be cautious, and practice a healthy scepticism,
always verging on disbelief. This exercises the critical
faculties and subjects any piece of research to the doubt
and criticism that is vital if anything is to be properly
established.

What sort of evidence for a viral aetiology might we
have? Epidemiological evidence can be very persuasive and
important (Gamble, 1980), but I shall be discussing laboratory
evidence. There are various possible mechanisms by which
viruses could cause chronic diseases and the more important
of these are listed in Table 1. The types of laboratory
evidence for a viral aetiology are as follows:-

1. Raised antibodies to a particular virus

This is the weakest type of evidence and by itself means
very little, although it may lead to more definitive
tests. For instance in multiple sclerosis and rheumatoid
arthritis there may be raised antibodies to certain
viruses, and local production of antibody in cerebro-
spinal fluid (multiple sclerosis) or joint fluid
(rheumatoid arthritis). In the multiple sclerosis brain
there are about $10^8$ plasma cells which produce about
100mg of IgG per day. The locally produced antibodies are
directed against up to 11 different viruses and also to
other common microbial agents such as mycoplasma and
tetanus (toxoid) (Salmi et al. 1983). In the rheumatoid
arthritis synovial fluid about a quarter of the immunoglobulin present is produced locally and is directed
against various common viruses. However in both these
diseases the locally produced antibodies are not thought
to give information about aetiology.

2. Electron microscopy

This is a difficult and often unsuitable test for the
presence of a virus. There are a small number of viruses
such as papovaviruses, adenoviruses where the ultrastructural appearances are unmistakable. Other viruses
are less characteristic in their appearance and many are
impossible to identify in thin sections. There have been
occasional incidences such as progressive multifocal
leucoencephalopathy, where electron microscopic studies
led the way to a proven virus aetiology, but in general
"virus like particles" are a snare and a delusion and can
be relied upon to raise the hackles of many a professional
virologist.

3. Local presence of viral antigens

Viral antigens can be detected with precision and
reliability and a positive result would mean that viral
antigen was present and possibly that local replication
was taking place. In the case of SSPE, presence of

measles virus antigen in brain cells gives reliable information about aetiology, and the presence of reverse transcriptase may give useful evidence about the presence of a retrovirus.

4.   Local presence of virus specific nucleic acid sequences

This can be tested for in homogenised tissues or at the histological level by in situ hybridisation. For instance it is now possible to examine routinely fixed and embedded paraffin sections 35 years old taken from a carcinoma of the cervix and left in the drawer of a pathologist, and show by in situ hybridisation that human papillomavirus type 18 sequences are present. There are pitfalls in this type of technology - herpes simplex virus and CMV DNA may hybridise with DNA from normal cells and retroviral nucleic acid probes may hybridise with endogenous (resident) retroviral nucleic acid. The human genome is heavily laden with retrovirus nucleic acid sequences.

5.   Local presence of infectious virus

This is clearly the best type of evidence for viral presence in a tissue although it is not often easily accomplished. Many viruses do not grow or grow very poorly in cells used for virus isolation (eg. hepatitis B, human parvoviruses, and papillomaviruses). Co-cultivation of cells to be tested together with known virus-susceptible cells may increase the likelihood of virus isolation but viruses often show a striking reluctance to reveal themselves. For instance in SSPE infectious measles virus is only recoverable from the brains of a small proportion of cases.

Even when any or indeed all of the above tests are positive, it may be difficult to conclude that the virus in question is the cause of the disease. The question of whether the viral presence is causal or casual (irrelevant) may still remain. Furthermore, many tissues contain resident persistent viruses that have nothing to do with the particular disease being studied.

Cancer

With all these considerations in mind let us look at some disease of unknown origin. Cancer is too big a topic to deal with on this occasion but I have included a list of possible viral culprits (Table 2). There is much current research activity on the role of human papillomaviruses in carcinoma of the cervix and virus types 16 and 18 in particular are implicated. Even Hela cells, originally derived from a human cervical carcinoma and passaged and used in laboratories throughout the world for the past 35 years, have recently been shown to contain nucleic acid sequences of human papillomavirus type 18. One notes that the evidence for EB virus in Burkitt's lymphoma and nasopharyngeal carcinoma is still not 100% although near it. There is still talk of possible co-factors but the final verdict will undoubtedly be clear when it is possible to vaccinate against EB virus and prevent the cancers. Similar remarks can also be made about hepatitis B virus, present as integrated DNA in liver cancer cells. Hundreds or thousands of humans are now being vaccinated against this virus and a study of their subsequent incidence of liver cancer will give an answer. It may be noted that there is no acceptable evidence for the viral aetiology of Hodgkin's lymphoma. EB virus in particular has been suggested for the aetiology of this disease but there has been a failure to detect EB DNA or the EB virus antigen EBNA in Hodgkin's tissues. The increased antibodies to EB virus seen in Hodgkin's disease appear to be without aetiological significance; indeed a few Hodgkin's patients have been shown to develop normal primary EB virus infection. On the other hand, it seems clear that EB virus is behind the lymphomas seen in immunosuppressed patients, such as renal transplant recipients and AIDS patients.

Table 1      Mechanisms by which viruses could cause chronic
             disease

| Mechanism | Example |
|---|---|
| 1. Very slow virus spread and low grade damage ⟶ disease later in life | SSPE |
| 2. Virus persistence in target cell with no overt damage but:<br>a. Interference with function | Pituitary dwarfism (LCM in mice) |
| b. Shortening of life span | ? Senile dementia |
| 3. Virus persistence ⟶ viral antigens ⟶ circulating immune complexes ⟶ eg. glomerulo-nephritis | LCM in mice |
| 4. Virus triggers damaging auto-immune response; strong host genetic determinants | Reovirus 1 in mice (onodera et al. 1981) |

## Neurological disease

There is no doubt that chronic diseases of the CNS can result from foetal infection with rubella or CMV and that SSPE and PML are chronic progressive virus infections. But it has proved exceedingly difficult to obtain acceptable evidence for other neurological conditions. In the case of multiple sclerosis serological studies, including the synthesis of anti-viral antibodies in the brains of MS patients (Salmi et al. 1983), have been inconclusive, as mentioned above. They have however provided a constant stimulus to fresh efforts, so that research grants have been given to fuel a vast amount of excellent work in neurobiology. Nevertheless there has been a complete failure to confirm reports of virus isolation from MS brain or the presence of virus-specific antigens or nucleic acid sequences. Work on the aetiology of multiple sclerosis has recently entered a more exciting phase with the increasing evidence that viruses can trigger damaging auto-immune-type responses. These responses may be mediated by antibodies or more probably by T cells, and it may be important to remember that lipids (which may be present on viral envelopes as well as on host target cells) may be involved as well as proteins. I have listed some of the autoimmune possibilities in Table 3 (see also Smith and Steinberg, 1983). The idea of a viral trigger or rather viral triggers for multiple sclerosis is still alive and encouraged by numerous excellent experimental models. But it can still be argued that the primary event in multiple sclerosis is exposure to an environmental toxin with secondary autoimmune events.

Table 2        Viruses and human cancer

| Cancer | Virus | Status |
|---|---|---|
| Certain T cell leukaemias | HTLV1, 2 | ++ |
| Burkitt's lymphoma | EB virus | + |
| Nasopharyngeal carcinoma | EB virus | + |
| Carcinoma of cervix | HPV16, 18 | + |
|  | Herpes simplex 2 | - |
| Carcinoma of penis | HPV16, 18 | ? |
| Hepatocellular carcinoma | Hepatitis B virus | + |

Parkinson's disease. The world wide outbreak of encephalitis lethargica between 1918 and 1930 gave rise to a series of patients with Parkinson's disease and the dramatic epidemiological features of this outbreak suggested an infectious aetiology. So far however there has been no acceptable evidence for a viral aetiology. In one report for instance neither influenza A nor herpes simplex type I nucleic acid sequences could be detected in 9 brains that were studied (Wetmur et al. 1979). Alternative aetiologies must be considered, especially environmental toxins, following the demonstration that chronic Parkinsonism with neuronal loss in the substantia nigra is caused by the substance MTPT (Langston et al. 1983). Parkinson's disease was recognised and described in 1817 shortly after the birth of industrial society.

Amyotrophic lateral sclerosis (motor neurone disease). Here too tests for poliovirus nucleic acid sequences have been negative and careful transmission experiments in primates have failed to provide evidence for a viral aetiology (Miller et al. 1980).

Schizophrenia. Reports of virus agents recovered from the cerebro-spinal fluid of patients with schizophrenia have not been confirmed.

Alzheimer's disease. A chronic virus infection could theoretically cause the dementia and degeneration of cortical neurones seen in this disease. So far however there is no acceptable evidence either for viruses present in Alzheimer brains (Rogo et al. 1987) or for transmission of the disease to primates.

Table 3        Damaging (a) autoimmune responses triggered by viruses

| Phenomenon | Status | Example |
|---|---|---|
| 1. Host antigens present on virus particle (helper determinant) | $\pm$ | Lindenmann & Klein 1967 |
| 2. Virus-infected cell displays its own antigens in immunogenic form | - | Autoimmune responses to heart or CNS antigens following heart or CNS disease |
| 3. Cross-reactive immune response recognises shared determinants on virus (b) and on host cell (molecular mimicry) | ++ | Srinivasappa et al. 1986 |
| 4. Virus infection interferes with suppressor mechanisms that normally restrain autoimmune responses | - | ? |
| 5. Virus infection causes polyclonal activation of B cells (or T cells?) | + | EB virus |
| 6. Antiviral antibody induces formation of antidiotype response which reacts with virus-specific receptor on normal cells | ? | Nepon et al. 1982 |

(a) Autoimmune responses, of course, are not necessarily harmful, and in the examples given there are no pathological consequences.

(b) Is also likely to occur with chlamydia, mycoplasma, bacteria etc.

Connective tissue diseases

    Systemic lupus erythematosus (SLE). This disease has so far resisted all efforts to incriminate viruses. Work was stimulated by the similar autoimmune disease of NZB strain mice which is closely associated with a retrovirus infection in a genetically predisposed host (Yoshiki et al. 1974). The increased expression of retroviruses however have now been dissociated from the development of the autoimmune disease (Datta et al. 1978). Evidence for SLE has remained at the level of raised antiviral antibodies which are probably a consequence of the immunoregulatory abnormalities seen in this disease. There have been no significant or reproducible reports of viral antigens in infectious virus or viral nucleic acid sequences.

Rheumatoid arthritis. The demonstration that mycoplasmas cause arthritis in experimental animals led to a great deal of work on mycoplasmas and viruses in rheumatoid arthritis but this has resulted in an impressive amount of negative evidence including the failure to find viral nucleic acid in rheumatoid cells (Norval and Smith, 1979) and unsuccessful transmission attempts with primates (Mackay et al. 1983). On the other hand there is no doubt that certain viruses cause arthritis in man, including rubella, the new human parvovirus and the Australian mosquito-transmitted Ross River Virus that causes epidemic polyarthritis. The latter disease interestingly is DR7-associated, and rheumatoid arthritis itself shows higher incidence in DR4 positive individuals. But in these viral arthritides we know very little about mechanisms and it is usually not clear even whether the virus invades the affected joint. A possible clue about rheumatoid arthritis comes from retroviruses. The goat retrovirus has been shown to cause an arthritis which is histologically indistinguishable from rheumatoid arthritis, after invading mononuclear cells in joints (Narayan et al. 1983). A similar type of arthritis is known to be caused by a cat retrovirus. As far as other viruses are concerned, reports about EB virus in rheumatoid arthritis are so far unacceptable and it seems that the phenomena observed reflect immunoregulatory disturbances in this disease. Reports of rubella and parvoviruses in chronic arthritis in man have not been confirmed and must be regarded with suspicion.

## Miscellaneous diseases

Juvenile diabetes. No one doubts that coxsackie B viruses cause diabetes in mice after growing in islet cells in the pancreas. On one dramatic occasion coxsackie virus B4 was not only isolated from the pancreas of a child with recent onset diabetes, but was also shown to cause a diabetic condition when inoculated into mice (Yoon et al. 1979). Otherwise however there is nothing but suggestive serological evidence for coxsackie B virus involvement in this disease (Gamble, 1980); only about 50% of patients, indeed, have antibodies to this virus. Rubella virus has also been clearly associated with diabetes in so far as about 20% of patients with congenital rubella who survive after the age of 20 have been shown in one study to develop diabetes (Menser et al. 1978). Also mumps virus is known to affect beta cells from islets of Langerhan in vitro (Prince et al. 1978) and has been associated with juvenile diabetes. CMV and varicella-zoster virus can invade islet cells in children suffering fatal infections with these viruses (Jenson et al. 1980). In summary, although there is no other evidence for viral aetiology in juvenile diabetes, viruses do look promising in this condition and more than one virus may be involved. There is a rich background of diabetes or diabetes-like conditions that can be produced experimentally in mice by viruses including coxsackie B (see above), mumps, EMC and reoviruses (Jenson et al. 1984). As if to confirm one's scepticism there is a recent report that viruses can prevent diabetes in mice! (Oldstone et al. 1986).

Crohn's disease.  Reports of the viral, or bacteria
aetiology of this disease have foundered "on the rocks of
reproducibility, uniformity, and specificity" (Janowitz 1981).

Ulcerative colitis, Paget's disease of bone, atherosclerosis.  There is no acceptable evidence for viral aetiology
in any of these conditions.  Reports that osteoblasts in
Paget's disease of bone contain nucleocapsids, measles
antigens, respiratory syncytial viral antigens, (Mills et al.
1981), await confirmation.  In the case of atherosclerosis
there is no doubt that certain viruses infect vascular
endothelium.  Atherosclerotic lesions appear in chickens
infected with Marek's disease virus and infected cells are
present in the lesions (Minick et al. 1970) and there is a
report that herpes simplex RNA has been detected in diseased
human aortas by in situ hybridisation (Benditt et al. 1983).
CMV also is known to infect vascular endothelium, but there
is no convincing evidence about a role in the aetiology of
atherosclerosis.

Post-viral fatigue syndrome (epidemic neuromyaesthenia;
myalgic encephalomyelitis).  This condition is included here
because it is a recent and interesting syndrome of possible
viral aetiology which illustrates many of the difficulties
that have been referred to above.  It occurs sporadically
but also as an apparently similar condition in dramatic
outbreaks which have been charted since 1934 and which
include Royal Free disease in 1955 and Lake Tahoe disease
in 1985.  Clinical descriptions include the presence of
general fatigue, depression, difficulty concentrating, fever,
headaches.  Neurologically there are no consistent abnormalities, and some of the alternative terms for the syndrome
(listed above) are misleading.  One cardinal feature appears
to be weakness, especially muscle weakness.  The condition
lasts more than a few months, may follow known viral infections, and females are more commonly affected.  The major
problem has been that a precise clinical definition has not
been made.  Patients with psychogenic symptoms are often
included, and indeed the condition may be commoner in such
individuals.  Inclusion of less well defined illnesses make
it very difficult to conduct a reliable survey for viruses
on sera and other clinical samples.  At least 4 different
viruses have been invoked as aetiological agents.  Some
cases (Krueger et al. 1987) are undoubtedly associated with
infection with human herpes virus type 6 - a recently isolated
virus which so far has no defined disease linked with it.
However, many patients with the post-viral fatigue syndrome
do not have antobidies.  EB virus certainly can cause this
clinical condition (Miller et al. 1987) and was responsible
for a proportion of the cases in the Lake Tahoe outbreak,
but evidence is confined to serological tests, as in the case
of human herpes virus type 6, and only a proportion of the
patients are positive.  The Inoue-Melnick virus (Melnick et
al. 1984) has also been suggested as an aetiological agent.
Unfortunately this virus has only been characterised by means
of a subtle cytopathic effect caused in cultivated cells
and although types 1, 2 and 3 have been distinguished in
antibody tests involving these cells, so far no virus particles
have been seen and no nucleic acid convincingly demonstrated.
Finally, there is recent evidence which sounds more convincing
that coxsackie B viruses may be involved in the post-viral

fatigue syndrome (Yousef et al. 1988). Viral antigens were
detected in blood in about half of patients and IgM antibodies
to these antigens in 90%. Enteroviruses were moreover recovered from 22% of faecal samples. Coxsackie B viruses are
known to persist in striped and cardiac muscle and would be
attractive aetiological agents. However this work awaits
confirmation. Endogenous interferon has been considered as a
cause of post-viral fatigue, but there is no evidence for
this (Lloyd et al. 1988).

It may turn out when the post-viral fatigue syndrome is
more clearly defined that there are multiple viral causes,
but exactly the same problems apply to this disease as to the
other diseases mentioned. We must maintain a healthy
scepticism as we await the reproducibility and specificity
that is needed.

Conclusions

With the exception of certain types of cancer there is
no acceptable evidence for viral aetiology in any of the
diseases mentioned in this article. Viral aetiology nevertheless remains a possibility. Absence of evidence is not
evidence of absence, as Professor Martin Ryle once observed
about extra-terrestrial life, but a healthy scepticism must
be maintained. It is quite probable that new human viruses
are waiting to be discovered and in addition old viruses may
behave in unexpected ways in the small proportion of affected
people who develop these diseases. The implications of a
given virus being essential for the production of one of
these diseases are immense. This is why we have been perhaps
too eager to reach out and grasp a virus as an aetiological
agent when the evidence is inadequate.

References

Benditt, E.P., Barrett, T. and McDougall, J.K. (1983)
Viruses in the aetiology of atherosclerosis. Proc. Natl.
Acad. Sci. 80:6386-6389.
Datta, S.K., McConahey, P.J., Manny, N. et al. (1978) Genetic
studies of autoimmunity and retrovirus expression in crosses
of New Zealand Black mice II The viral envelope glycoprotein
gp70. J. Exp. Med. 147:87-881.
Gamble, D.R. (1980) The epidemiology of insulin dependent
diabetes, with particular reference to the relationship of
virus infection to its aetiology. Epid. Rev. 2:49-70.
Janowitz, H.D. (1981) Crohn's disease - 50 years later. New
Engl. J. Med. 304:1600-1602.
Jenson, A.B., Rosenberg, H.S. and Notkins, A.L. (1980)
Pancreatic islet cell damage in children with fatal viral
infections. Lancet 2:354-358.
Jenson, A.B. et al. (1984) Multiple viruses in diabetes
mellitus. Prog. Med. Virol. 29:197-217.
Krueger, G.R.F., Koch, B. and Ablashi, D.V. (1987) Persistent
fatigue and depression in patients with antibody to human B
lymphotropic virus. Lancet 2:36.
Langston, J.W., Ballard, P., Tetrud, J.W. et al. (1983)
Chronic Parkinsonism in humans due to a product of meperidine-
analog synthesis. Science 219:879-880.

Lloyd, A., Hanna, D.A. and Wakefield, D. (1988) Interferon and myalgic encephalomyelitis. Lancet 1:471.

Mackay, J.M.K., Sim, A.K., McCormick, J.N. et al. (1983) Aetiology of rheumatoid arthritis: an attempt to transmit an infective agent from patients with rheumatoid arthritis to baboons. Ann. Rheum. Dis. 42:443-7.

Melnick, J.L., Wang, S., Seidel, E. et al. (1984) Characterisation of IM virus which is frequently isolated from cerebrospinal fluid of patients with multiple sclerosis and other chronic diseases of the central nervous system. J. Virol. 52:739-744.

Menser, M.A., Forrest, J.M. and Bransby, R.D. (1978) Rubella infection and diabetes mellitus. Lancet 1:57-60.

Miller, G., Grogan, E., Rowe, D. et al. (1987) Selective lack of antibody to a component of EB nuclear antigen in patients with chronic active Epstein-Barr virus infection. J. Inf. Dis.

Miller, J.R., Ramareddy, V.G. and Myers, J.C. (1980) Amyotrophic lateral sclerosis: search for poliovirus by nucleic acid hybridisation. Neurology 30:884-6.

Mills, B.G., Singer, F.R., Weiner, L.P. et al. (1981) Immunohistological demonstration of respiratory syncytial virus antigens in Paget's disease of bone. Proc. Natl. Acad. Sci. USA 78:1209-1213.

Mims, C.A. (1985) Viral aetiology of diseases of obscure origin. Brit. Med. Bull. 41:63-69.

Minick, C.R., Fabricant, C.G., Fabricant, J. et al. (1970) Atheroarterio-sclerosis induced by infection with a herpes virus. Am. J. Pathol. 96:673-706.

Narayan, O., Kennedy-Stoskpf, S., Sheffer, D. et al. (1983) Activation of caprine arthritis-encephalitis virus expression during maturation of monocytes to macrophages. Infect. Immunity 41:67-73.

Nepon, J.T., Weiner, J.L., Dichter, M.A. et al. (1982) Identification of a haemagglutinin-specific idiotype associated with reovirus recognition shared by lymphoid and neural cells. J. Exp. Med. 155:155-167.

Norval, M. and Smith, C. (1979) Search for viral nucleic acid sequences in rheumatoid cells. Ann. Rheum. Dis. 38:456-462.

Oldstone, M.B.A. (1988) Prevention of type 1 diabetes in non obese mice by virus infection. Science 239:500-501.

Onodera, T., Toniolo, A., Ray, U.R. et al. (1981) Virus induced diabetes mellitus. XX Polyendocrinopathy and autoimmunity. J. Exp. Med. 153:1457-1473.

Pogo, B.G.T., Casals, J. and Elizan, T.S. (1987) A study of viral genomes and antigens in brains of patients with Alzheimer's disease. Brain 110:907-916.

Prince, G.A., Jenson, A.B., Billups, L.C. et al. (1978) Infection of human pancreatic beta cells with mumps virus. Nature 271:158-161.

Rodriguez, M., von Wedel, R.J., Garrett, R.S. et al. (1983) Pituitary dwarfism in mice persistently infected with lymphocytic choriomeningitis virus. Lab. Investig. 49:48.

Salmi, A., Reunanen, M., Ilonen, J. et al. (1983) Intrathecal antibody synthesis to virus antigens in multiple sclerosis. Clin. Exp. Immunol. 52:214-249.

Smith, H.R. and Steinberg, A.D. (1983) Autoimmunity - a perspective. Ann. Rev. Immunol. 1:175-210.

Srinivasappa, J., Saegusa, J., Prabhakar, B.S. et al. (1986) Molecular mimicry: frequency of reactivity of monoclonal antiviral antibodies with normal tissues. J. Virol. 57:397-401.

Wetmur, J.G., Schwartz, J. and Elizan, T.S. (1979) Nucleic acid homology studies of viral nucleic acids in idiopathic Parkinson's disease. Arch. Neurol. 36:462-4.

Yoon, J.W., Austin, M., Onodera, T. et al. (1979) Virus-induced diabetes mellitus: isolation of a virus from the pancreas of a child with diabetic ketoacidosis. New Engl. J. Med. 300:1173-9.

Yoshiki, T., Mellors, R.C., Strand, M. et al. (1974) The viral envelope glycoprotein of murine leukaemia virus and the pathogenesis of immune complex glomerulonephritis of New Zealand mice. J. Exp. Med. 140:1011-1027.

Yousef, G.E., Ball, E.J., Mann, G.F. et al. (1988) Chronic enterovirus infection in patients with post-viral fatigue syndrome. Lancet i:146.

THE IMMUNOBIOLOGY OF RESPIRATORY SYNCYTIAL VIRUS:

PROSPECTS FOR A VACCINE

Erling Norrby[1], Britt Åkerlind[1] and Maurice A. Mufson[2]

[1]Department of Virology
Karolinska Institute
School of Medicine
Stockholm, Sweden

[2]Department of Medicine
Marshall University
School of Medicine and Veterans Administration
Medical Center
Huntington, West Virginia

Human Respiratory Syncytial Virus (RSV) is a member of the pneumovirus genus of the family Paramyxoviridae. This virus is the  most common
cause of severe lower respiratory infections among infants and children,
especially children between 2 and 6 months of age (Belshe et al., 1984).
All attempts that hitherto  have been made to develop a safe and effective
RSV vaccine have met with failure. During recent years there has been a
rapid accumulation of knowledge about the basic biochemistry and biology
of the virus. New possibilities for developing an effective vaccine have
become available.


THE STRUCTURE OF RSV

RSV is a non-segmented negative strand RNA virus wich codes for at
least 10 distinct gene products (Huang et al., 1985). Two of these gene
products are glycoproteins which are present on the surface of infected
cells and virions (Walsh et al., 1986; Walsh and Hruska, 1983; Gruber and
Levine, 1985). These two glycoproteins play a dominating role in humoral
immunoprotective reactions. Like in the case of other paramyxoviruses the
two glycoproteins serve functions of attachment and membrane fusion.
However the RSV attachment protein has some unique properties. This RSV
protein is denoted G for glycoprotein and it carries no detectable hemag-
glutinin or neuraminidase activity. It is a protein with a relatively
small molecular weight, 33 kDa, and it is heavily glycosylated not only
with N-linked but also unique to this paramyxovirus, O-linked sugars
(Gruber and Levine, 1985; Satake et al., 1985; Wertz et al., 1985). The
RSV attachment protein shows no homology with the corresponding protein of
other paramyxoviruses. In contrast, the RSV fusion (F) protein shows the
same basic size of its two cleavage products as other paramyxoviruses but
the homology to the corresponding protein of other paramyxoviruses is very
distant (Spriggs et al., 1986). The F protein is synthesized as an in-

active precursor, FO, which is activated by proteolytic cleavage into two disulphide linked polypeptide subunits, F1 and F2 (48 and 20 kDa, respectively; Fernie and Gerin, 1982; Gruber and Levine, 1983).

## PREVIOUS ATTEMPTS TO PRODUCE AN RSV VACCINE

Attempts have been made to produce both inactivated and live RSV vaccines. A formalin inactivated product was prepared from virus grown in African green monkey kidney tissue cultures. The virus was concentrated by centrifugation and absorbed to alum and the preparation was administered intramuscularly. This vaccine induced neutralizing antibodies against RSV in the immunized infants. However, the immunization did not confer protection against disease. Instead the vaccinees developed an aggravated form of disease when infected with the wild virus (Chanock et al., 1970; Fulginiti et al., 1969; Kapikian et al., 1969; Kim et al., 1969). The reason for this assumed accentuated immunopathological reactions to the infection is not known. However, it was shown in a more recent study (Prince et al., 1986) that prior intramuscular inoculation of formalin inactivated RSV in cotton rats enhances the immunopathology after exposure to infectious virus. It should be possible to use this experimental system for dissection of the relative role of selective immunity against native and formalin treated isolated G and F RSV components.

Attempts to develop a live RSV vaccine have aimed at generating a strain of RSV with a restricted capacity to replicate in the respiratory tract (Chanock and Murphy, 1980). A number of temperature sensitive (ts) RSV mutants showing reduced replication capacity in hamster lungs were isolated (Wright et al., 1970). One particular mutant, ts-1, was tested in children but was found to give a febrile rhinorrhea. In addition a reversion to virus producing plaques at the restricted temperature was seen. Further efforts were made to produce a genetically stable variant of ts-1, which could replicate in children without giving any symptoms. A second temperature sensitive mutant, ts-2, which in addition to being temperature sensitive also manifested altered plaque morphology was developed. This mutant was interpreted to have a lesion in the fusion glycoprotein. It was highly attenuated when administered to chimpanzees but when it was given to seronegative infants it was found to be overattenuated. Not even a high dose of virus gave infection in the majority of vaccinees (Wright et al., 1982). Thus to date no useful temperature sensitive mutants of RSV have been developed.

Live RSV has also been injected parenterally in an attempt to induce serum neutralizing antibodies (Buynak et al., 1978, 1979). Virus injected by this route into small experimental animals has provided protection against infection (Prince et al., 1979). These observations led to a clinical trial with a live RSV vaccine which was produced in WI-38 cells. When this virus material was used for parenteral immunization only low levels of antibodies were induced and no evidence for protection against disease was seen in the immunized children (Belshe et al., 1982).

## THE IMMUNOPROTECTIVE IMPORTANCE OF G AND F GLYCOPROTEINS OF RSV

In studies of paramyxoviruses other than RSV it has been found that both surface glycoproteins are required for efficient immunity but that quantitatively immunity to the attachment protein plays a dominating role. The situation in the case of RSV appears to be reversed in that antibodies to the fusion protein play a dominating role whereas antibodies to the G protein quantitatively have a somewhat lesser importance. Polyclonal as

148

well as monoclonal antibodies against both the F and the G protein can neutralize virus but in the case of monoclonal G-specific antibodies this generally is an indirect neutralization (Walsh et al., 1984a). Both monoclonal and polyclonal antibodies against the F and G proteins can protect small animals against RSV infections (Taylor et al., 1984; Walsh et al., 1984b). The relative immunoprotective importance of G and F proteins has also been evaluated by use of vaccinia vector transferred RSV genes specific for the two glycoproteins (Olmsted et al., 1986; Stott et al., 1986; Johnson et al., 1987). Again the dominating immunoprotective role of the F component was identified. In fact the G immunogen which gave an effective protection against deeper respiratory tract infection in cotton rats showed a reduced protective effect in the upper respiratory tract.

## TWO GROUPS OF RSV

RSV was originally considered to be an antigenically homogenous virus but already some early studies applying cross-neutralization tests with animal hyperimmune sera revealed slight antigenic differences among a few strains (Coates et al., 1963; Wulff et al., 1964; Doggett and Taylor-Robinson, 1965). However, not until the application of the monoclonal antibody technology did it become clear that two distinct subgroups of RSV, A and B, could be identified (Anderson et al., 1985; Mufson et al., 1985; Gimenez et al., 1986). Characterization of the F and G genes of representative subgroup A and B RSV strains by cDNA cloning techniques has revealed that the deduced amino acid homology is about 80% for the F protein but only about 30% for the G protein.

Studies in man identifying consecutive isolates with regard to subgroup characteristics showed that RS virus subgroup A gave a higher protection against infection with the homologous virus than against subgroup B virus infection (Mufson et al., 1987). In further studies (Hendry et al., 1988) the antibody response to homologous and heterologous glycoprotein components was determined in convalescent sera from subjects with primary RSV infections. The antibody responses to homologous and heterologous F glycoproteins were not significantly different. In contrast homologous versus heterologous antibodies responses to G glycoprotein were significantly different with 7.3% relatedness between the G glycoproteins of the two subgroups. Analysis of the neutralizing antibody response revealed a 31% relatedness. It still remains to explain why the neutralizing antibody response does not show a higher degree of relatedness considering the fact that the F protein has been concluded to be the dominating immunoprotective protein and further that no major subgroup differences between the immunogenic properties of this protein have been identified.

## POSSIBLE FUTURE AVENUES FOR DEVELOPMENT OF AN RSV VACCINE

The options for either developing an effective live vaccine which gives the appropriate restrictive replication of virus in the respiratory tract or designing an immunoprotective inactivated product which does not cause immunopathological complications still remains. One possibility would be to try the classical Jennerian approach that is to use an animal RSV for immunization of man. Attempts have already been made to use human RSV for immunization of cattle. A bovine and a human strain of RSV were adapted to in vitro growth in bovine cell culture and then inoculated into gnoto-biotic calves (Thomas et al., 1984). Virus strains replicated without causing symptoms and specific antibody responses were recorded. In further studies (Stott et al., 1984) an inactivated bovine virus vaccine was compared with two candidate live vaccines. These live vaccines were a modified bovine strain and a temperature-sensitive mutant of a human strain. Only the inactivated vaccine gave complete protection whereas the live immunogens gave an infection-permissive immunity. However, the mean peak titers and the mean duration of shedding was significantly reduced in the live vaccine candidate virus immunized animals compared to the non-immunized control animals. Thus there seems to be a cross-reactivity between human and bovine strains which might be exploited also in the human situation provided that it is possible to identify a bovine virus suitable for immunization of man. Potentially supplementary application of techniques of viral genetics might be considered in selecting the appropriate animal virus strain.

As concerns virus component immunization it is now clear that an RSV vaccine needs to include both the F and the G proteins and furthermore that in order to protect against members of the two subgroups, G proteins representing each of these need to be included in the product. Potentially this complex of antigens might be synthesized by use of a vaccinia vector or some other virus vector containing the appropriate selected viral genes. Since, however, the immunization needs to be performed at a very early stage in life and in the presence of maternal antibodies there is a necessity to guarantee absolute harmlessness of the vector used for the transfer of genes to be expressed. The alternative is to use purified virus components prepaired under conditions guaranteeing their native properties. This kind of combined inactivated product needs to be administered in the young infant in such a way that an efficient local immunity can be established. The alternate possibilities of immunizing pregnant women in order to raise the level of maternal antibodies to improve the efficacy of passive immune protection does not seem at this stage to be a realistic proposition.

In summary it appears that antibodies against both the F and the G glycoproteins of RSV are needed to give an efficient neutralization of virus infectivity. Further the low level of antigenic cross-reactivity between G glycoproteins of the two RSV subgroups indicate that the candidate future RSV vaccines need to include G proteins representing both subgroups in parallel with the F glycoprotein for amelioration of the morbidity and mortality associated with both primary and subsequent infections with RSV.

ACKNOWLEDGEMENTS

The authors' work cited in this review was supported by grants from the Swedish Medical Research Council (Project no B87-16X-00116) and the WHO programme for development of vaccines against respiratory infections.

REFERENCES

Anderson, L. J., Hierholzer, J. C., Tsou, C., Hendry, R. M., Fernie, B. F., Stone, Y., and McIntosh, K, 1985, Antigenic characterization of respiratory syncytial virus strains with monoclonal antibodies, J. Inf. Dis., 151:626.

Belshe, R. B., Van Voris, L. P., and Mufson, M. A., 1982, Parental administration of live respiratory syncytial virus vaccine: results of a field trial, J. Inf. Dis., 145:311.

Belshe, R. B., Bernstein, J. M., and Dansby, K. N., 1984, Respiratory syncytial virus, in: "Textbook of Human Virology", R. B. Belshe, ed., PSG Publishing Co., Littleton.

Buynak, E. B., Weibel, R. E., McClean, A. A., and Hilleman, M. R., 1978, Live respiratory syncytial virus vaccine administered parenterally, Proc. Soc. Exptl. Biol. Med., 157:636.

Buynak, E. B., Weibel, R. E., Carlson, A. J., McClean, A. A., and Hilleman, M. R., 1979, Further investigations of live respiratory syncytial virus vaccine administered parenterally, Proc. Soc. Exptl. Biol. Med., 160:272.

Chanock, R. M., Kapikian, A. Z., Mills, J., Kim, H. W., and Parrott, R. H., 1970, Influence of immunologic factors in respiratory syncytial virus disease, Arch. Environ. Health, 21:347.

Chanock, R. M., and Murphy, B. R., 1980, Use of temperature-sensitive and cold-adapted mutant viruses in immunoprophylaxis of acute respiratory tract disease, Rev. Inf. Dis., 2:421.

Coates, H. V, Kendrick, L., and Chanock, R.M., 1963, Antigenic differences between two strains of RS virus, Proc. Soc. Exptl. Biol. Med., 112:958.

Doggett, J. E., and Taylor-Robinson, D. T., 1965, Serological studies with respiratory syncytial virus, Arch. Ges. Virusforsch., 15:601.

Fernie, B. F., and Gerin, J. L., 1982, Immunochemical identification of viral and nonviral proteins of the respiratory syncytial virus virion, Inf. Immun., 37:243.

Fulginiti, V. A., Eller, J. J., Sieber, O. F., Joyner, J. W., Minamitani, M., and Meiklejohn, G., 1967, Respiratory virus immunization. I. A field trial of two inactivated respiratory virus vaccines: an aqueous trivalent parainfluenza virus vaccine and an alum-precipitated respiratory syncytial virus infection, Acta Med. Scand., 182:323

Gimenez, H. B., Hardman, N., Keir, H. M., and Cash, P., 1986, Antigenic variation between human respiratory syncytial virus isolates, J. Gen. Virol., 67:863.

Gruber, C., and Levine, S., 1983, Respiratory syncytial virus polypeptides. III. The envelope associated proteins, J. Gen. Virol., 64:825.

Gruber, C., and Levine, S., 1985, Respiratory syncytial virus polypeptides. IV. The oligosaccharides of the glycoproteins, J. Gen. Virol., 66:417.

Hendry, R. M., Burns, J. C., Walsh, E. E., Graham, B. S., Wright, P. F., Hemming, V. G., Rodriguez, W. J., Kim, H. W., Prince, G. A., McIntosh, K., Chanock, R. M., and Murphy, B. R., 1988, Strain specific serum antibody responses in infants undergoing primary infection with respiratory syncytial virus, J. Inf. Dis., 157:640.

Huang, Y. T., Collins, P. L., and Wertz, G. W., 1985, Characterization of the 10 proteins of human respiratory syncytial virus, identification of a fourth envelope associated protein, Virus Res., 2:157.

Johnson Jr., P. R., Olmsted, R. A., Prince, G. A., Murphy, B. R., Alling, D. W., Walsh, E. E., and Collins, P. L., 1987, Antigenic relatedness between glycoproteins of human respiratory syncytial virus subgroups A and B: Evaluation of the contributions of F and G glycoproteins to immunity, J. Virol., 61:3163.

Kapikian, A. Z., Mitchell, R. H., Chanock, R. M., Shvedoff, R. A., and Stewart, C. E., 1969, An epidemiologic study of altered clinical reactivity to respiratory syncytial (RS) virus infection in children previously vaccinated with an inactivated RS virus vaccine, Am. J. Epid., 89:405.

Kim, H. W., Canchola, J. G., Brandt, C. D., Pyles, G., Chanock, R. M., Jensen, K., and Parrott, R. H., 1969, Respiratory syncytial virus disease in infants despite prior administration of antigenic inactivated vaccine, Am. J. Epid., 89:422.

Mufson, M. A., Örvell, C., Rafnar, B., and Norrby, E., 1985, Two distinct subtypes of human respiratory syncytial virus, J. Gen. Virol., 66:2111

Mufson, M. A., Belshe, R. B., Örvell, C., and Norrby, E., 1987, Subgroup characteristics of respiratory syncytial virus strains recovered from children with two consecutive infections, J. Clin. Microbiol., 25:1535.

Olmsted, R. A., Elango, N., Prince, G. A., Murphy, B. R., Johnson, P. H., Moss B., Chanock, R. M., and Collins, P. L., Expression of the F glycoprotein of respiratory syncytial virus by a recombinant vaccinia virus: Comparison of the individual contributions of the F and G glycoproteins to host immunity, Proc. Natl. Acad. Sci. USA, 83:7462.

Prince, G. A., Potash, L., Horswood, R. L., Camargo, E., Suffin, S. C., Johnson, R.A., and Chanock, R. M., 1979, Intramuscular inoculation of live respiratory syncytial virus induces immunity in cotton rats, Inf. Immun., 23:723.

Prince, G. A, Jenson, A. B., Hemming, V. G., Murphy, B. R., Walsh, E. E., Horswood R. L., and Chanock, R. M., 1986, Enhancement of respiratory syncytial virus pulmonary pathology in cotton rats by prior intramuscular inoculation of formalin inactivated virus, J. Virol., 57:721.

Satake, M., Coligan, J. E., Elango, N., Norrby, E., and Venkatesan, S., 1985, Respiratory syncytial virus envelope glycoprotein (G) has a novel structure, Nucl. Acid. Res., 13:7795.

Spriggs, M. K., Olmsted, R. A., Venkatesan, S., Coligan, J. E., and Collins, P. L., 1986, Fusion glycoprotein of human parainfluenza virus type 3, nucleotide sequence of the gene. Direct identification of the cleavage activation site and comparison with other paramyxoviruses, Virology, 152:241.

Stott, E. J., Thomas, L. H., Taylor, G., Collins, A. P., Jebbett, J., and Crouch, S., 1984, A comparison of three vaccines against respiratory syncytial virus in calves, J. Hyg. Camb., 93:251.

Stott, E. J., Bell, L. A., Young, K. K., Furze, J., and Wertz, G. W., 1986, Human respiratory syncytial virus glycoprotein G expressed from a recombinant vaccinia virus vector protects mice against live virus challenge, J. Virol., 60:607.

Taylor, G., Stott, E. J., Bew, M., Fernie, B. F., Cote, P. J., Collins, A. P., Hughes, M., and Jebbett, J., 1984, Monoclonal antibodies protect against respiratory syncytial virus infection in mice, Immunology, 52:137.

Thomas, L. H., Stott, E. J., Collins, A. P., Crouch, S., and Jebbett, J., 1984, Infection of gnotobiotic calves with a bovine and human isolate of respiratory syncytial virus. Modification of the response by dexamethasone, Arch. Virol., 79:67.

Walsh, E. E., and Hruska, J., 1983, Monoclonal antibodies to respiratory syncytial virus proteins: Identification of the fusion protein, J. Virol., 47:171.

Walsh, E. E., Schlesinger, J. J., and Brandriss, M. W., 1984a, Purification and characterization of GP90 one of the envelope glycoproteins of respiratory syncytial virus, J. Gen. Virol., 65:761.

Walsh, E. E., Schlesinger, J. J., and Brandriss, M. W., 1984b, Protection from respiratory syncytial virus infection in cotton rats by passive transfer of monoclonal antibodies, Inf. Immun., 43:756.

Walsh, E. E., Cote, P. J., Fernie, B. F., Schlesinger, J. J., and Brandriss, M. W., 1986, Analysis of the respiratory syncytial virus fusion protein using monoclonal and polyclonal antibodies, J. Gen. Virol., 67:505.

Wertz, G. W., Collins, P. L., Huang, Y., Gruber, C., Levine, S., and Ball, L. A., 1985, Nucleotide sequence of the G protein gene of human respiratory syncytial virus reveals an unusual type of viral membrane protein, Proc. Natl. Acad. Sci. USA, 82:4075.

Wright, P. F., Woodend, W. G., and Chanock, R. M., 1970, Temperature-sensitive mutants of respiratory syncytial virus: in vivo studies in hamsters. J. Inf. Dis., 122:501

Wright, P. F., Belshe, R. B., Kim, H. W., Van-Voris, L. P., and Chanock, R. M., 1982, Administration of highly attenuated live respiratory syncytial virus vaccine to adults and children, Inf. Immun., 37:397.

Wulff, H., Kidd, P., and Wenner, H. A., 1964, Respiratory syncytial virus: Observation on antigenic heterogeneity, Proc. Soc. Exptl. Biol. Med., 115:240.

PATHOGENESIS OF DENGUE HAEMORRHAGIC FEVER:

CURRENT PERSPECTIVES

Tikki Pang

Department of Medical Microbiology
University of Malaya
59100 Kuala Lumpur
Malaysia

INTRODUCTION

Dengue viruses belong to the family Flaviviridae and the four recognized serotypes (dengue-1, dengue-2, dengue-3, dengue-4) cause a spectrum of disease ranging from an undifferentiated febrile illness to the severe, life-threatening dengue haemorrhagic fever/dengue shock syndrome (DHF/DSS). DHF/DSS remains as a major public health problems in various parts of Southeast Asia, the Pacific region and, more recently, the Caribbean. The disease syndromes pose special problems for health authorities because of their epidemic potential, the often high case-fatality rate and difficulties associated with treatment and prevention. As part of the overall effort at disease control and prevention, a comprehensive understanding of the pathogenesis of DHF/DSS is vital in the quest for more rational approaches to treatment and the design of appropriate and effective vaccines.

Much has been said previously about the immunopathologic basis of DHF/DSS (Halstead, 1980, 1981, 1982; Bhamarapravati, 1981; Pang, 1983). However, it seems timely and appropriate that a more current perspective of the problem be put forward, taking into account recent advances in molecular virology, plus a resurgence of interest in the overall biology of flaviviruses and the diseases they cause (Monath, 1985). What follows is thus an attempt to present a current view of the problem, taking into consideration the immunopathological aspects, other host related factors and the importance of natural virus variation in disease pathogenesis.

ASSOCIATION OF DHF/DSS WITH SECOND INFECTIONS

The definition of risk factors through long-term epidemiological studies of DHF/DSS are invaluable for a better understanding of pathogenesis and ultimate application of vaccines.

Extensive seroepidemiological studies, mainly in Thailand, have strongly established the fact that a large majority of DHF/DSS cases occurs in persons undergoing a secondary infection with a heterologous dengue serotype (Halstead, 1980, 1981). In one study, the relative risk of developing the shock syndrome was 160 times greater for children with  secondary infection than for those with primary infection (Halstead, 1980). It is important to point out, however, that a small minority of primary infections can result in DHF/DSS (Scott et al, 1976). It has also been established that DSS is seen in two groups of children : with primary infection in children younger than one year and with secondary (sequential) infection in children one year of age and older (Halstead, 1984).

The sequence of dengue infections also appears to be an important risk factor. In one study (Sangkawibha et al, 1984), the highest risk factor were infections in the sequence dengue-1 followed by dengue-2. Although a second infection with dengue-2 appears to be crucial in the causation of DHF/DSS in Thailand, dengue-3 virus has been implicated in Indonesia and Malaysia (Gubler  et al, 1979; Norazizah et al, 1983).

The proposal that DHF/DSS is strongly associated with a second infection was also demonstrated during an outbreak of DHF/DSS in Cuba in 1981, the first time the disease has occurred outside of Southeast Asia. No dengue transmission had occurred in Cuba for over 30 years between World War II and 1977-1978. During 1977-78 an outbreak caused by dengue-1 virus occurred with a predominantly mild disease picture. This was followed by a dengue-2 outbreak in 1981 where 10,000 cases of DHF/DSS with 158 deaths were reported. A large majority of cases showed a secondary-type antibody response (Halstead, 1988).

PATHOPHYSIOLOGY OF DHF/DSS

The usual observation in fatal cases of DHF/DSS is the absence of sufficiently severe pathological manifestations to explain the death of patients (Bhamarapravati et al, 1967) thus suggesting a physiologic mode of death most probably involving short-acting chemical mediators. The major pathophysiological changes that occur result in essentially two effector pathways. One leads to haemorrhage which is due to thrombocytopenia, coagulation defects, some degree of vasculopathy and, in a number of cases, disseminated intravascular coagulation. The other effector pathway results in increased vascular permeability with resultant leakage of water, low molecular weight proteins, and electrolytes from the vascular compartment ultimately resulting in hypovolemic shock. Haemorrhage does not appear to contribute significantly to shock.

SITES OF DENGUE VIRUS REPLICATION

Dengue viruses appear to have a predilection for lymphoid tissues and the central role of mononuclear phagocytes as target cells and principal sites of replication of dengue viruses seems well established (Halstead, 1980). Studies on human cases of DHF/DSS have demonstrated the presence of viral antigens in peripheral blood monocytes and virus has also been recovered from these cells (Scott et al, 1980). Viral antigens have also been detected in mononuclear phagocytes in skin, spleen, thymus, lymph nodes, and in Kupffer cells of the liver and alveolar macrophages in the lung (Bhamarapravati, 1981). Dengue-like viral particles have also been obseved in renal glomerular macrophages.

Additionally, Boonpucknavig et al (1976) also demonstrated the presence of viral antigen on the surface of circulating B lymphocytes and dengue-2 virus is capable of in vitro growth in human endothelial cells (Andrews et al, 1978). There is, however, no evidence of viral antigens in cells of the vascular endothelium in vivo (Sahaphong et al, 1980).


PATHOGENESIS OF DHF/DSS

Current knowledge about DHF/DSS pathogenesis is compatible with the concept that DHF/DSS is the result a pathological immune elimination response directed against dengue-infected monocytes/macrophages during an exacerbated secondary infection with a heterologous dengue virus.

At the mechanistic level, the underlying pathogenic mechanisms of the DHF/DSS syndrome appears to involve all the important components of the host immune response : antigen-antibody complexes, monocytes/macrophages, cell-mediated immune mechanisms, complement, and chemical mediators. These will now be discussed in more detail.

Antigen-antibody complexes and monocytes/macrophages

A key role has been proposed for the phenomenon of antibody dependent enhancement (ADE) in the pathogenesis of DHF/DSS (Halstead 1980, 1981). The key observation was made that dengue virus showed enhanced replication in human and simian peripheral blood monocytes in the presence of subneutralizing concentrations of antibody. These "enhancing" antibodies are of the IgG class and appear to promote viral entry via Fc receptors on the membranes of monocytes/macrophages (Peiris & Porterfield, 1979). As a corollary to this, possession of neutralizing, non-enhancing IgM class antibodies may serve to reduce the risk of DHF/DSS (Pang, 1987). The in vivo correlate of ADE was obtained when monkeys, previously treated with dengue-immune sera and then inoculated with dengue-2 virus, developed virus titres which were three- to 51-fold higher compared to animals which received normal sera prior to virus inoculation (Halstead, 1980).

On the basis of the above observations, the immunological enhancement hypothesis has been put forward as the pathogenetic basis of DHF/DSS (Halstead 1980, 1981). The hypothesis proposes that, in the presence of enhancing antibodies, virus is engulfed and then replicates within monocytes. Due to the mobility of the monocytes, the infection may then spread to other parts of the body. The infected monocytes then become the targets of an immune elimination response which activates the monocytes to release a variety of chemical mediators which then produce shock and haemorrhage.

Alternatively, dengue-infected monocytes may act as antigen-presenting cells to induce the release of lymphokines and other factors from activated T cells (Pang, 1983).

One word of caution seems warranted. Although there is good in vitro and some in vivo evidence for the ADE phenomenon, the studies to date have not conclusively established the relative importance of this phenomenon in the causation of DHF/DSS. For example, with regard to the in vivo evidence for ADE, the dosage, timing, and relative affinities of the antibodies tranferred may be quite out of proportion to the natural situation. Also, the transferred antibodies may have a second, more vital effect or function in vivo.

## Cell-mediated immunity

Given the mode of dengue virus replication (within monocytes/macrophages), it would seem very likely that cell-mediated immune mechanisms play an important role in the immune elimination of infected cells and viral clearance. Such an immune elimination response could then result in the release of various chemical mediators which in turn causes shock and haemorrhage. Little is known, however, about which mechanisms may be important and operating in the in vivo situation.

Although a T cell response, as manifested by a delayed-type hypersensitivity reaction (Pang et al, 1982) and leukocyte migration inhibition (Nagarkatti et al, 1978), has been demonstrated in dengue-infected mice, its relevance in DHF/DSS in humans is not known.  A likely candidate in the immune elimination of dengue-infected monocytes/macrophages are cytotoxic T lymphocytes (CTL). The existence of these cells in dengue-infected mice has been demonstrated recently in our laboratory. The CTL were generated in an in vitro system by spleen cells from immune animals with dengue-infected "stimulator" cells. Once generated they were tested against histocompatible dengue-infected target cells utilizing a chromium 51 release assay. A significant amount of specific lysis (between 17-26 %) was consistently observed in several experiments and the cytotoxic cells appear to be T cells possessing both Lyt 1 and Lyt 2 antigens (Pang et al, 1988). Some cross-reactivity in cytotoxic activity was noted against JEV-infected target cells but none against Sindbis-infected targets. Cytotoxicity was restricted by the H-2 gene complex (Pang et al, 1988).

It has also been reported that NK (natural killer) cells and K cells (which mediate antibody-dependent cell-mediated cytotoxicity, ADCC) from human peripheral blood can cause significant lysis of dengue-infected Raji cells in vitro (Kurane et al, 1984). These cells appear to be heterogeneous with regards to surface markers and peripheral blood lymphocytes active in ADCC seem to belong to the same subset as those active in the NK cell assay (Kurane et al, 1986). It has also been shown that NK cell cytotoxicity was increased on the first day of shock, although the total numbers of NK cells (and T cells) were depressed during the shock phase (Suvatte et al, 1988).

## Complement

It has been shown that in the acute stage of DHF/DSS, the serum levels of C1q, C3, C4, C5-8, and C3 proactivator are depressed and more so in severe grades of DHF (W.H.O., 1985a, 1985b). This would seem to indicate the existence of a complement consumption process via the classical pathway and perhaps via the alternative pathway as well. The activation of complement could conceivably result in the formation of C3a and C5a, potent anaphylotoxins which may increase vascular permeability and cause shock. In a recent study, it was shown that both C3a and C5a were significantly increased at the time of shock in DHF/DSS patients (Suvatte et al, 1988). Other studies have demonstrated the presence of C3 split products during the shock phasein severely ill patients with DHF/DSS, which rapidly disappeared during the convalescent phase (Churdboonchart et al, 1983).

Despite these observations, it has been shown that wide variability exists in complement levels among DHF/DSS cases and that normal levels of complement have been found in some patients with DHF/DSS.

## Chemical Mediators

As mentioned above, short-acting chemical mediators are likely to be the effectors of the major pathogenetic features of DHF/DSS. Despite their obvious importance, the identities of the various chemical mediators involved in the pathogenesis of DHF/DSS remain unknown. The important findings can be summarized in the following manner :

a. One study suggested that the kinin system is not involved (Edelman et al, 1975). This was supported by the observation that bradykinin-kallikrein levels are normal in DHF/DSS patients (Rohde, 1978).

b. The importance of histamines is still unresolved. Russell and Brandt (1973) proposed that histamine is released from basophils/mast cells in response to complement activation and generation of C3a anaphylatoxin in DHF/DSS. In support of this concept, Bhamarapravati et al (1967) have shown the presence of cells in perivascular infiltrates of skin from fatal cases which resemble degranulated mast cells, and Tuchinda et al (1977) have reported increased urinary excretion of histamine by DHF patients. Pavri et al (1979) have also found increased levels of IgE in sera from DHF/DSS cases and proposed that IgE-mediated histamine release plays a role in the pathogenesis of shock. On the other hand, it has been found that shock in DHF/DSS does not respond to antihistamines and steroids (Rohde, 1978; Halstead, 1982) and that DHF/DSS is not accompanied by cutaneous signs such as urticaria (Halstead, 1982).

c. The potential role played by arachidonic acid metabolites, leukotrienes, prostacyclins and prostaglandins, is not known but should be considered (Pang, 1983). These highly potent mediators, leoktriene C is 1000-times more potent than histamine, are released largely by macrophages following various stimuli. They have been shown to induce systemic hypotension, cause haemodynamic and permeability changes, and affect the cutaneous microcirculation. An attempt to detect the presence of these metabolites in sera from DHF/DSS patients was unsuccessful (D.S. Burke, personal communication). This failure could perhaps be the result of lability of the metabolites, interference from other serum proteins and difficulties with the assay methods used, rather than a reflection of the absence of the metabolites in the sera.

d. The potential role of other molecules produced by monocytes/macrophages should also be mentioned. These include various antiproteases (e.g. alpha-1-antitrypsin, alpha-2-antiplasmin), procoagulants, tumour necrosis factor and plasminogen activator.

OTHER IMMUNOLOGICAL ASPECTS

Other aspects of the host immune response should also be considered, including:

State of activation of monocytes/macrophages

Experimental studies have shown that activated macrophages with increased phagocytic capacity are more permissive to flavivirus replication in the presence of specific antiviral antibody than are resting macrophages. This could suggest that non-antibody activation of these cells may also result in increased replication of dengue viruses. For example, it has been shown that mouse peritoneal macrophages treated with various macrophage-activating substances of bacterial origin (e.g. LPS, bacterial toxins, cell wall peptidoglycans) and parasite extracts produced significantly more dengue virus compared to untreated control cells (Hotta & Hotta, 1982; Wiharta et al, 1985). The phenomenon has also been observed in vivo in monkeys pretreated with Corynebacterium parvum and pertussis vaccine (Halstead, 1983). The importance and significance of these observations are unclear at this stage but the findings of Wiharta et al (1985) may have some relevance to a previous hypothesis that an underlying parasitic infection is a risk factor in DHF/DSS pathogenesis (Pavri & Prasad, 1980).

Interferons

Despite its obvious potential importance, very little is known about the role of interferons in dengue virus infections. Some investigators reported suppression of dengue-2 virus replication in human peripheral blood monocytes in vitro when cells were treated with alpha- and beta-interferon prior to virus inoculation (Hotta et al, 1984). Other workers reported marginal and variable effects of alpha- and beta-interferon on dengue-2 replication in vitro (Vithanomsat et al, 1984). Kurane et al (1984) reported the detection of alpha-interferon in the culture supernatant of human peripheral blood mononuclear cells and dengue-infected Raji cells, perhaps indicating that interferons are produced during immune elimination of dengue-infected cells. These workers also found that pretreatment with exogenous alpha-interferon enhanced the cytotoxic activity of NK cells against dengue-infected target cells (Kurane et al, 1984).

## Immunosuppression

A considerable amount of data has been accumulated in experimental infection of mice suggesting that infection with dengue virus results in immunosuppression to homologous and heterologous antigens (Chaturvedi et al, 1978; Tandon et al, 1979; Wong et al, 1984). Suppression may involve the production of cytotoxic factors (Chaturvedi et al, 1980) or the generation of T suppressor cells (Wong et al, 1984; The significance and relevance of these findings to DHF/DSS is unknown but it is virtually certain that the regulation of the human immune response to dengue virus infection will involve T suppressor cells. Consequently, any transient episode of immunosuppression could result in immunoregulatory disturbances which may form the basis for a later pathological reaction e.g. favoring formation of enhancing antibodies (Pang, 1983).

## GENETIC FACTORS

In addition to the immunological factors discussed above, it would seem appropriate to point out that genetic factors, appear to be involved in the pathogenesis of DHF/DSS. In a study involving 87 unrelated patients with DHF/DSS, and 138 non-dengue control subjects, a positive DSS association was found with HLA-A2 and HLA-blank and a negative relationship with HLA-B13 (W.H.O., 1985a). That genetic factors may be involved was further suggested by the significantly higher incidence of DHF in whites than in blacks during the DHF/DSS epidemic in Cuba in 1981; of 124 Havana children with DHF/DSS, only 14% were blacks and mulattos (which make up 34% of the population) (Guzman et al, 1985; Halstead, 1988).

## NATURAL VIRUS VARIATION

One of the basic tenets in the study of pathogenesis of viral infections is the fact that the characteristics of the virus and of the host contribute to the outcome of infection and either can exercise a determining influence (Mims & White, 1984). In contrast to the proposal that DHF/DSS has an immunopathological basis, Rosen (1977) has proposed that DHF/DSS is caused by dengue virus variants of increased pathogenic potential. The potential importance of natural variation among the dengue viruses has been recently restated by several authors (Igarashi, 1984; Monath, 1986) and the suggestion made that such variation, which may influence the virulence of dengue strains for human hosts and its infectivity for mosquito vectors, is probably more important than generally appreciated. The high mutation rates among RNA viruses is a well known fact (Holland et al, 1982; Holland, 1984) and flaviviruses are unlikely to be exempt.

For example, strain variation correlated with human pathogenicity have been documented recently with Japanese encephalitis virus (JEV) infections in Thailand (Burke et al, 1985). In this study, virus strains isolated from an area where transmission occurs, but where human encephalitis is virtually absent, differed in their oligonucleotide fingerprints from strains obtained from an area where JEV causes annual epidemics (Burke et al, 1985). With tick-borne encephalitis (TBE) virus, monoclonal antibodies were used to distinguish between strains prevalent in Far-Eastern USSR and in Europe and a correlation found with human pathogenicity (Heinz et al, 1982; Stephenson et al, 1984). It was found that the Far Eastern strain caused infection of the brainstem and upper cervical chord with a high case-fatality rate (up to 35 %) whereas the European strain caused mainly meningoencephalitis with a lower lethality (1-5 %).

With regard to the dengue viruses, a lack of suitable and convenient laboratory models has hampered studies to correlate strain characteristics with biological properties. However, several studies have suggested that strain variation does occur. Earlier studies (reviewed by Schlesinger, 1980) suggested that new isolates of dengue contain mixtures of virion populations having different virulence properties. A mixture of large and small plaque variants were also detected in a dengue-2 strain isolated from a human case (Eckels et al, 1976). Furthermore, the small plaque variant appeared to correlate with mouse virulence , temperature sensitivity and attenuation for nonhuman primates (Eckels et al, 1976; Harrison et al, 1977).

The existence of dengue variants can also be inferred from epidemiological observations. Studies in Thailand have shown that a second infection with dengue-2 virus following primary infection with dengue-1 represents an important risk factor for DHF/DSS (Sangkawibha et al, 1984). A similar sequence may also be operating in Cuba during the 1981 DHF/DSS epidemic (Guzman et al, 1984). In contrast, secondary infections with dengue-3 virus were associate with DHF/DSS in Indonesia (Gubler et al, 1979) and Malaysia (Norazizah et al, 1983). These observations are suggestive of a difference in virulence potential between dengue-2 and dengue-3 strains in the three countries. Furthermore, the possibility that some strains of dengue may have enhanced pathogenic potential is suggested by the various reports of occasional cases of DHF/DSS occurring during primary infection with dengue-2 virus in Niue (Barnes & Rosen, 1974), dengue-2 and dengue-3 viruses in Thailand (Scott et al, 1976) and with dengue-1 virus in Puerto Rico and the Netherlands Antilles (CDC, 1985).

Despite the lack of definitive data correlating strain variation with human pathogenicity, there is hope that recent advances in monoclonal antibody technology, molecular virology and antibody enhancement assays, will result in more conclusive studies in the near future. The recent report of Halstead et al (1984), for example, used dengue-specific and flavivirus group-reactive monoclonal antibodies to analyze enhancing epitopes in seven dengue-2 strains. In accordance with the findings of natural variation mentioned above, it was found that the virus strains were heterogeneous with respect to distribution of such enhancing epitopes (Halstead et al, 1984).

Oligonucleotide fingerprinting studies of dengue-1 and dengue-2 viruses have also demonstrated major differences between strains of different geographic origin and also within strains from the same region isolated at different times (Repik et al, 1983; Trent et al, 1983). However, no correlation has been found so far with disease severity or immune enhancement.

Most importantly, genomic sequence data have recently been published for dengue-1 (Mason et al, 1987), dengue-2 (Hahn et al, 1988) and dengue-4 (Zhao et al, 1987) viruses. Such analysis is also being carried out on a wide variety of dengue virus isolates from human cases and also candidate vaccine strains. It is expected that these studies will provide important clues to the molecular basis of dengue virus pathogenicity.

Finally, what are the selective forces operating in nature which are responsible for the appearance of variants ?. It has been suggested that the high mutation rates of RNA viruses combined with alternative growth in two phylogenetically remote hosts (ie. vertebrates and arthropods) may exert strong selective pressures for selection of virus variants (Igarashi, 1984; Monath, 1986). Such a selection may also be closely related to changes in the susceptibility of the vertebrate hosts (e.g. acquisition of immunity) or within arthropod populations (e.g. genetically determined changes in vector competence) (Monath, 1986).


CONCLUSIONS

From the previous discussion it would seem inescapable that the pathogenesis of DHF/DSS may well turn out to be multifactorial in nature and more complex than previously imagined. A wide range of factors may be important, genetic and non-genetic, immunological and non-immunological. Although the weight of evidence points to the strong immunopathologic correlates, it would seem prudent to bear in mind other factors which may influence the overall picture, especially the pathobiological significance of natural virus variation and heterogeneity for

".....indeed, the virulence of a virus and the susceptibility/ resistance of the host cannot be considered in isolation-it is their interaction that is relevant..." (Mims & White, 1984).

It is the author's hope that this more integrated view of the pathogenesis of DHF/DSS will prevail in our future endeavours at elucidating and ultimately controlling this unique and important human disease.


ACKNOWLEDGEMENT

I would like to acknowledge various esteemed colleagues for advice and discussions. In particular I am indebted to Bob Blanden, Alastair Cunningham, Cedric Mims, Tom Monath, Bob Shope, Scott Halstead, Peter MacDonald, Joel Dalrymple, Don Burke, Phil Russell and Skip Fournier.

REFERENCES

Andrews, B.S., Theofilopoulos, A.N., Peters, C.J., Loskutoff, D.J., Brandt, W.E., Dixon, F.J. (1978). Replication of dengue and Junin viruses in cultured rabbit and human endothelial cells. Infect.Immun. 20 : 776-781.

Barnes, W.J.S., Rosen, L. (1974). Fatal haemorrhagic disease and shock associated with primary dengue infection on a Pacific island. Am.J.Trop.Med.Hyg. 23 : 495-506.

Bhamarapravati, N. (1981). Pathology and pathogenesis of DHF. in "Dengue Haemorrhagic Fever 1981" (S. Hotta, ed.), ICMR, Kobe, pp. 207-214.

Bhamarapravati, N., Tuchinda, P., Boonyapaknavik, V. (1967). Pathology of Thailand haemorrhagic fever : A study of 100 autopsy cases. Ann.Trop.Med.Parasit. 61 : 500-510.

Boonpucknavig, S., Bhamarapravati, N., Nimmannitya, S., Phalavadhtana, A., Siripont, J. (1976). Immunofluorescent staining of the surfaces of lymphocytes in suspension from patients with dengue haemorrhagic fever. Am.J.Pathol. 85 : 37-47.

Burke, D.S., Schmaljohn, C.S., Dalrymple, J.M. (1985). Strains of Japanese encephalitis virus isolated from human brains have a highly conserved genotype compared to strains isolated from other natural hosts. Abstr.Ann.Meet.Am.Soc. Virol., Univ. New Mex., Albuquerque.

CDC (Centers for Disease Control), Dengue Surveillance Summary, no. 26, September 1985, pp. 1-2

Chaturvedi, U.C., Tandon, P., Mathur, A., Kumar, A. (1978). Host defense mechanisms against dengue virus infection in mice. J.Gen.Virol. 39 : 293:302.

Chaturvedi, U.C., Bhargava, A., Mathur, A. (1980). Production of cytotoxic factor in the spleen of dengue virus-infected mice. Immunology 40 : 653-658.

Churdboonchart, V., Bhamarapravati, N., Futrakul, P. (1983). Crossed immunoelectrophoresis for the detection of split products of the third complement in dengue haemorrhagic fever. I. Observations in patient's plasma. Am.J.Trop.Med. Hyg. 32 : 569-576.

Eckels, K.H., Brandt, W.E., Harrison, V.R., McConn, J.M., Russell, P.K. (1976). Isolation of a temperature-sensitive dengue-2 virus under conditions suitable for vaccine development. Infect.Immun. 14 : 1221-1227.

Edelman, R., Nimmannitya, S., Colman, R.W., Talamo, R.C., Top, F.H. (1975). Evaluation of the plasma kinin system in dengue haemorrhagic fever. J.Lab.Clin.Med. 86 : 410-421.

Gubler, D.J., Suharyono, W., Sumarmo, Wulur, H., Jahja, E., Sulianti Saroso, J. (1979). Virological surveillance for dengue haemorrhagic fever in Indoneis using the mosquito inoculation technique. Bull. WHO 57 : 931-936.

Guzman, M.G., Kouri, G., Morier, L., Soler, M., Fernandez, A. (1984). A study of fatal haemorrhagic dengue cases in Cuba. Bull.PAHO 18: 213-220.

Guzman, M.G., Kouri, G., Bravo, J., Soler, M., Vasquez, S. (1985). Dengue haemorrhagic fever in Cuba, 1981. II. Study of patients clinically diagnosed with dengue haemorrhagic fever and dengue shock syndrome. Trans.Roy.Soc.Trop.Med. Hyg. 78 : 239-247.

Hahn, Y.S., Galler, R., Hunkapiller, T., Dalrymple, J.M., Strauss, J.H., Strauss, E.G. (1988). Nucleotide sequence of dengue 2 RNA and comparison of the encoded proteins with those of other flaviviruses. Virology 162 : 167-180.

Halstead, S.B. (1980). Immunological parameters of togavirus disease syndromes. In "The Togaviruses" (R.W. Schlesinger, ed.), Academic Press, New York, pp. 107-173.

Halstead, S.B. (1981). The pathogenesis of dengue : Molecular epidemiology in infectious disease. Am.J.Epidemiol. 114 : 632-648.

Halstead, S.B. (1982). Dengue : Haematologic aspects. Sem. Haematol. 19 : 116-131.

Halstead, S.B. (1983). Pathogenesis of dengue : New knowledge depends upon epidemiological and clinical studies. In "Proceedings of the International Conference on Dengue/ Dengue Haemorrhagic Fever (T. Pang, R. Pathmanathan, eds.), University of Malaya, pp. 34-44.

Halstead, S.B. (1984). Selective primary health care : Strategies for control of disease in the developing world. XI. Dengue. Rev.Infect.Dis. 6 : 251-264.

Halstead, S.B. (1988). Pathogenesis of dengue : Challenges to molecular biology. Science 239 : 476-481.

Halstead, S.B., Venkatesan, C.N., Gentry, M.K., Larsen, L.K. (1984). Heterogeneity of infection enhancement of dengue 2 strains by monoclonal antibodies. J.Immunol. 132 : 1529-1532.

Harrison, V.R., Eckels, K.H., Sagartz, J.W., Russell, P.K. (1977). Virulence and immunogenicity of a temperature-sensitive dengue 2 virus in lower primates. Infect.Immun. 18 : 151-156.

Heinz, F.X., Berger, R., Najdic, O., Knapp, W., Kunz, C. (1982). Monoclonal antibodies to the structural glycoprotein of tick-borne encephalitis virus. Infect.Immun. 37 : 869-874.

Holland, J.J. (1984). Continuum of change in RNA virus genomes. In "Concepts in Viral Pathogenesis" (A.L. Notkins, M.B.A. Oldstone, eds.), Springer-Verlag, New York, pp. 137-143.

Holland, J.J., Spindler, V., Horodyski, F., Grabau, E., Nichol, S., vandePol, S. (1982). Rapid evolution of RNA genomes. Science 215 : 1577-1585.

Hotta, H., Hotta, S. (1982). Dengue virus multiplication in cultures of mouse peritoneal macrophages : Effects of macrophage activators. Microbiol.Immunol. 26 : 665-676.

Hotta, H., Hotta, S., Homma, M. (1984). Effect of interferons on dengue virus multiplication in cultured monocytes/macrophages. Biken J. 27 : 189-193.

Igarashi, A. (1984). A hypothesis on the geographical distribution of arboviruses. Trop.Med. 26 : 173-180.

Kurane, I., Hebblewaite, D., Brandt, W.E., Ennis, F.A. (1984). Lysis of dengue-infected cells by natural cell-mediated cytotoxicity and antibody-dependent cell-mediated cytotoxicity. J.Virol. 52 : 223-230.

Kurane, I., Hebblewaite, D., Ennis, F.A. (1986). Characterization with monoclonal antibodies of human lymphocytes active in natural killing and antibody-dependent cell-mediated cytotoxicity of dengue virus-infected cells. Immunology 58 : 429-436.

Mason, P.W., McAda, P.C., Mason, T.L., Fournier, M.J. (1987). Sequence of the dengue-1 virus genome in the region encoding the three structural proteins and the major nonstructural protein NS1. Virology 161 : 262-267.

Mims, C.A., White, D.O. (1984). Viral Pathogenesis and Immunology. Blackwell Scientific Publications, Oxford.

Monath, T.P. (1986). Pathobiology of flaviviruses. In "The Togaviruses and Flaviviruses" (S. and M. Schlesinger, eds), Plenum Press, New York, pp. 375-440.

Nagarkatti, P.S., D'Souza, M.B., Rao, K.M. (1978). Use of sensitized spleen cells in capillary tube migration inhibition test to demonstrate cellular sensitization to dengue virus in mouse. J.Immunol.Meth. 23 : 241-348.

Norazizah Md. Taib, Chan, S.P., Rethwan, T., Pang, T., Lam, S.K. (1983). The 1982 dengue outbreak in Malaysia : The University Hospital experience. In "Proceedings of the International Conference on Dengue/Dengue Haemorrhagic Fever" (T.Pang, R. Pathmanathan, eds.), University of Malaya, pp. 88-94.

Pang, T. (1983). Delayed-type hypersensitivity : Probable role in the pathogenesis of dengue haemorrhagic fever/dengue shock syndrome. Rev.Infect.Dis. 5 : 346-352.

Pang, T. (1987). Dengue-specific IgM and dengue haemorrhagic fever/shock. Lancet 1 : 988.

Pang, T., Wong, P.Y., Pathmanathan, R. (1982). Induction and characterization of delayed-type hypersensitivity to dengue virus in mice. J.Infect.Dis. 146 : 235-242.

Pang, T., Devi, S., Blanden, R.V., Lam, S.K. (1988). T cell-mediated cytotoxicity against dengue-infected target cells. Microbiol. Immunol. 32 (5), in press.

Pavri, K.M., Prasad, S.R. (1980). T suppressor cells : Role in dengue haemorrhagic fever and dengue shock syndrome. Rev.Infect.Dis. 2 : 142-146.

Pavri, K.M., Swe Than, Ramamoorthy, C.L., Chodankar, V.P. (1979). Immunoglobulin E in dengue haemorrhagic fever (DHF) cases. Trans.Roy.Soc.Trop.Med.Hyg. 73 : 451-452.

Peiris, J.S.M., Porterfield, J.S. (1979). Antibody mediated enhancement of flavivirus replication in macrophage-like cell lines. Nature 282 : 509-511.

Repik, P.M., Dalrymple, J.M., Brandt, W.E., McCown, J.M., Russell, P.K. (1983). RNA fingerprinting as a method for distinguishing dengue 1 virus strains. Am.J.Trop.Med.Hyg. 3 : 577-589.

Rohde, J. (1978). Clinical management of severe dengue. Trop.Doct. 8 : 54-61.

Rosen, L. (1977). The emperor's new clothes revisited, or reflections on the pathogenesis of dengue haemorrhagic fever. Am.J.Trop.Med.Hyg. 26 : 337-343.

Russell, P.K., Brandt, W.E. (1973). Immunopathologic processes and viral antigens associated with sequential dengue virus infection. Persp.Virol. 7 : 263-277.

Sahaphong, S., Riengrojpitak, S., Bhamarapravati, N., Chirachariyavej, T. (1980). Electron microscopic study of the vascular endothelial cell in dengue haemorrhagic fever. S.E.Asian J.Trop.Med.Publ.Hlth. 11 : 194-204.

Sangkawibha, N., Rojanasuphot, S., Ahandrik, S. (1984). Risk factors in dengue shock syndrome : A prospective epidemiologic study in Rayong, Thailand. I. The 1980 outbreak. Am.J.Epidemiol. 120 : 653-669.

Schlesinger, R.W. (1980). Virus-host interactions in natural and experimental infections with alphaviruses and flaviviruses. In "The Togaviruses" (R.W. Schlesinger, ed.), Academic Press, New York, pp. 83-106.

Scott, R.M., Nimmannitya, S., Bancroft, W.H., Mansuwan, P. (1976). Shock syndrome in primary dengue infection. Am.J. trop.Med.Hyg. 25 : 866-874.

Scott, R.M., Nisalak, A., Cheamudon, U., Seridhoranakul, S., Nimmannitya, S. (1980). Isolation of dengue viruses from peripheral blood leukocytes of patients with haemorrhagic fever. J.Infect.Dis. 141 : 1-6.

Stephenson, J.R., Lee, J.M., Wilton-Smith, F.D. (1984). Antigenic variation among members of the tick-borne encephalitis complex. J.Gen.Virol. 65 : 81-89.

Suvatte, V., Malasit, P., Sarasombath, S., Wasi, C. (1988). Immunopathogenesis of dengue haemorrhagic fever/dengue shock syndrome. In Proceedings of the First International Congress of Tropical Paediatrics, in press.

Tandon, P., Chaturvedi, U.C., Mathur, A. (1979). Dengue virus induced thymus-derived suppressor cells in the spleens of mice. Immunology 38 : 653-658.

Trent, D.W., Grant, J.A., Rosen, L., Monath, T.P. (1983). Genetic varaiation among dengue 2 viruses of different geographic origin. Virology 128 : 271-286.

Tuchinda, M., Dhorranintra, B., Tuchinda, P. (1977). Histamine content in 24-hour urine in patients with dengue haemorrhagic fever. S.E.Asian J.Trop.Med.Pub.Hlth. 8 : 80-83.

Vithanomsat, S., Wasi, C., Harinasuta, C., Thongcharoen, P. (1984). The effect of interferon on flaviviruses in vitro : A preliminary study. S.E.Asian J.Trop.Med.Publ.Hlth. 15 : 27-31.

Wiharta, A.S., Hotta, H., Hotta, S., Matsumura, T., Sujudi, Tsuji, M. (1985). Increased multiplication of dengue virus in mouse peritoneal macrophage cultures by treatment with extracts of Ascaris-Parascaris parasites. Microbiol. Immunol. 29 : 337-348.

Wong, P.Y., Devi, S., McKenzie, I.F.C., Yap, K.L., Pang, T. (1984). Induction and Ly phenotype of suppressor T cells in mice during primary infection with dengue virus. Immunology 51 : 51-56.

World Health Organization (WHO) (1985a). Arthropod-borne and rodent-borne viral diseases. Tech.Rep.Ser. no. 719, WHO, Geneva.

World Health Organization (WHO) (1985b). Viral haeomorrhagic fevers. Tech.Rep.Ser. no. 721, WHO, Geneva.

Zhao, B., Mackow, E., Buckler-White, A., Markoff, L., Chanock, R.M., Lai, C.J., Makino Y. (1986). Cloning full-length dengue type 4 viral DNA sequences : Analysis of genes coding for structural proteins. Virology 155 : 77-88.

# PERSISTENCE OF HUMAN PAPILLOMAVIRUSES

Judit Acs[1], William C. Reeves[2] and William E. Rawls[1]

[1]Molecular Virology and Immunology Program
Department of Pathology
McMaster University
Hamilton, Ontario, Canada

[2]Division of Epidemiology
Gorgas Memorial Laboratories
Republic of Panama

## Introduction

The role of viruses in the induction of papillomas which may persist and give rise to malignant lesions was demonstrated in animals a number of years ago (Rous and Beard, 1935). More recent observations suggest that a similar relationship could exist in humans infected with certain types of papillomaviruses (HPV) (Gissmann, 1984). The nature of the host-virus interactions leading to persistence of papillomavirus infections are poorly understood as are the factors leading to the development of virus-associated malignant disease. In addition to intracellular events controlling virus replication, lesion formation appears to be controlled, in part, by host immune responses. It is our intent in this paper to portray the nature of persistence of human papillomaviruses and to review the role of immune factors in controlling the diseases induced by these viruses.

## Nature of Papillomaviruses

Papillomaviruses are widely distributed and species specific viruses have been obtained from rabbits, hamsters, sheep, goats, deer, cattle, dogs, monkeys as well as humans (Lancaster and Olson, 1982). The viruses are relatively simple. The genome is double-stranded, circularly closed DNA of about 7-8 kb which is complexed with histones, condensed into nucleosomes and encapsidated into icosahedral virions (Broker and Botchan, 1986). The genomes replicate as plasmids and are normally found as supercoiled circles when extracted from virions or infected cells. The genes of the viral proteins are encoded on only one of the DNA strands and have been divided into early and late domains on the basis of the time in the replicative cycle when they are thought to be expressed (see figure 1).

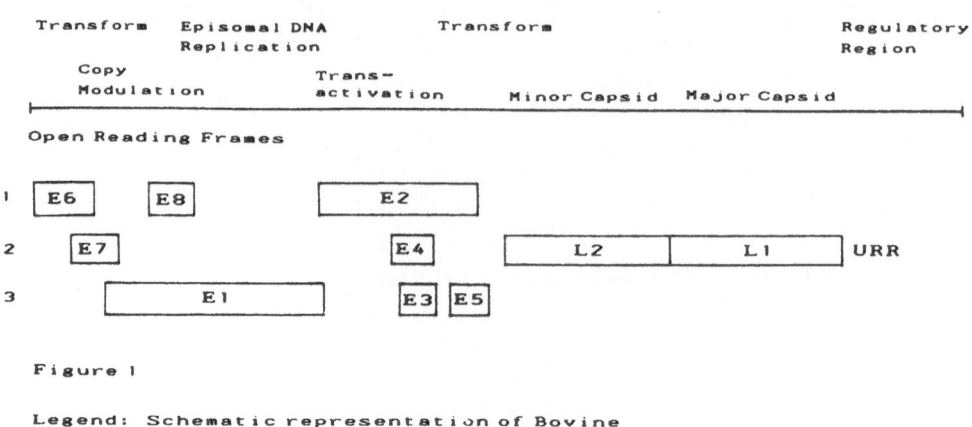

Figure 1

Legend: Schematic representation of Bovine
        Papillomavirus 1 (BPV-1) genome

Figure 1

Although relatively simple in structure, the genome is
fairly complex with respect to expression.  Most of the
knowledge of the molecular biology of these viruses has been
obtained from studies of BPV.  A non-coding region located
upstream from the first early gene (E6) appears to control
transcription and this segment has been labeled the upstream
regulatory region (URR).  The URR of BPV has been found to
contain regions with characteristics of transcriptional
promoters.  In addition the  URR contains the origin of
replication, plasmid maintenance sequences and transcriptional
enhancer elements responsive to the E2 gene product (Baker and
Howley, 1987; Lusky and Botchan, 1984; Spalholz et al., 1987;
Waldeck et al., 1984).  Similar promoter and enhancer elements
have been found in the URR of the genomes of other
papillomaviruses including those isolated from humans where
tissue specific enhancer sequences have also been identified
(Cripe et al., 1987; Hirochika et al., 1987).

Several open reading frames (ORF) of the genome appear to
code for regulatory proteins.  The E1 ORF is associated with
replicative functions of the plasmid DNA.  The 3' end of the
gene is required for transient DNA replication while the 5' end
has a negative regulatory function required for the establishment of the viral genome as a stable plasmid (Berg et al.,
1986; Lusky and Botchan, 1986).  Domains of the E2 ORF product
interact with URR sequences to enhance or repress genome
transcription.  The E5 and E6 genes of BPV are capable of
transforming cells in vitro (Schiller et al., 1986; Yang et
al., 1985), possibly through a  mechanism involving the
stimulation of host cell DNA synthesis.  Rather elegant studies
using chemically synthesized oligopeptides have shown that only
13 amino acids at the C-terminus of BPV E5 are necessary to
induce DNA synthesis in growth arrested cells (Green and
Loewenstein, 1987).  Human papillomavirus type 16 (HPV 16) has
been implicated in the genesis of cervical cancer (Gissmann,
1984) and in vitro transformation has been accomplished with
the E6/E7 region of this virus.  These genes are also expressed
in several cell lines derived from human cervical carcinomas
(Androphy et al., 1987; Seedorf et al., 1987).  The E6 and E7

proteins contain a regularly spaced pattern of cystine residues (cys-X-X-cys) similar to those found in cellular DNA binding proteins (Giri and Danos, 1986). Thus, these proteins may alter cellular DNA transcription by binding to promoter or enhancer elements of cellular genes.

The late genes, L1 and L2, are thought to code for viral structural proteins. L1 ORF is highly conserved and codes for the major 55 kilodalton capsid protein (Meinke and Meinke, 1981; Tomita et al., 1987) while the L2 gene apparently codes for a minor structural protein. The E4 gene is also expressed late in the replicative cycle but is not incorporated into virions. In HPV 1 induced lesions, E4 proteins are among the most abundant viral proteins found in the cytoplasm of stratified keratinocytes and may play a role in virus maturation (Doorbar et al., 1986).

Papillomavirus Replication

For most papillomaviruses, expression of the various genes, viral DNA synthesis and virion production occur in squamous epithelial cells and are related to the stage of differentiation of the keratinocyte. Since keratinocyte differentiation has not been successfully duplicated in vitro, concepts regarding papillomavirus replication have been extrapolated from observations of lesion formation in vivo. In lesions, DNA synthesis is readily detected in cells of the stratum spinosum and granular layers but not in the cells of the basal layer or underlying fibroblast. Synthesis of viral capsid proteins and virion assembly take place in cells of the upper stratum spinosum and in the granular layers. Cells of the basal layer are thought to be infected and contain the genome which is minimally expressed.

The early events of BPV infection have been examined in vitro. Following infection the genome was rapidly amplified. The copy number then stabilized at about 50 to 100 per cell and each copy was there after replicated once per cell division in synchrony with the replication of cellular DNA replication (Broker and Botchan, 1986). Analysis of mutants have identified 4 viral genes involved in the regulation of early viral DNA synthesis and control of plasmid copy number. These include the E6-E7 genes, the positive and negative regulatory domains of E1 and the transactivation domain of E2 (Berg et al., 1986; Lusky and Botchan, 1986).

It can be envisioned that basal layer reserve cells acquire papillomavirus genetic information during the initial infections in vivo and that the host cell/virus interaction is similar to that of the in vitro BPV model. The virus genome is maintained in a stable, regulated state and is transmitted to daughter cells during cell division. The expression of the viral DNA remains regulated in the daughter cells retaining the properties of reserve cells. As daughter cells are committed to differentiation, regulation of viral DNA replication is lost and expression of virus genomes culminates in virion production in terminally differentiated keratinocytes. This model predicts that yet to be identified factors specific for the differentiation of keratinocytes are involved in complete expression of the virus genome (Cripe et al., 1987).

## Papillomavirus Pathogenesis

The events following initial infection with papillomaviruses are poorly understood. The recently observed association of human genital papillomaviruses with human cancers has stimulated several studies characterizing genital HPV infections. The results of these studies provide clues to the host-virus relationship of these agents in humans. For example, observation of individuals exposed to condylomata acuminata revealed an incubation period of about 6 weeks to 3 months before the development of genital warts (Oriel, 1971). Genital warts are normally caused by HPV 6 or 11 (Gissmann et al., 1982) and the other HPV types infecting the genital tract probable have similar incubation periods. HPV infections of the cervix seldom give rise to exophytic condylomas but rather are manifested as flat or inverted condylomas (Meisels et al., 1981). Certain epithelial abnormalities found in the flat lesions are similar to those belonging to the spectrum called intraepithelial neoplasia (CIN) (Koss, 1987). A number of studies have shown that 50-70% of condylomatous lesions of the cervix are associated with CIN. In women followed prospectively, approximately 10% of the lesions progressed to more severe stages of CIN while the remainder regressed or persisted (Reid et al., 1982; Syrjanen et al., 1985).

The papillomaviruses display remarkable tissue tropism. HPV types 6 and 11 are found in about 95% of condylomata acuminata (Gissmann et al., 1982). These types are also associated with CIN but their occurrence is inversely related to the severity of such lesions and these virus types are seldom found in invasive cervical cancers. In a recent report, for example, HPV 6 or 11 DNA was found in 33% of CIN I lesions, 15% of CIN II lesions and 4.5% of CIN III lesions (Reid et al., 1987). In contrast, the other types of HPVs infecting the genital tract seldom produce exophytic growths on the external genitalia but are associated with epithelial lesions of mucosal surfaces. The prevalence of HPV 16 DNA varies directly with the severity of CIN and is commonly found in invasive lesions (Gissmann, 1984; Reid et al., 1987). Thus, the human genital papillomaviruses appear capable of inducing lesions which may persist and undergo malignant conversion as was demonstrated a number of years ago in rabbits infected with Shope papillomavirus (Rous and Beard, 1935).

Studies of the association between HPV and cervical cancer have revealed HPV DNA not only in lesions but also in the genital tract of women without apparent disease. For example, HPV 16 was detected in biopsies from 31 of 47 (66%) patients with invasive cancer and in cervical epithelial samples from 9 of 26 (35%) control women (Meanwell et al., 1987). In a second case-control study HPV 16 was found among 31 of 46 (67%) cases and 22 of 51 (43%) controls (Reeves et al., 1987). The rates of detection HPV DNA among control women of these studies are somewhat higher than noted by most investigators studying women without proliferative diseases of the cervix. Analysis of exfoliated cervical cells by filter in situ hybridization detected HPV DNA among 2.2% to 11% (DeVilliers et al., 1987; Schneider et al., 1985) of women with normal cytological smears. Southern blot analysis of exfoliated cells yielded positivity rates of 11% among cytologically normal women in one study (Lorincz et al., 1986) and 12.5% and 28%, respectively,

among non-pregnant and pregnant women in another study
(Schneider et al., 1987).  Southern blot analysis of DNA
extracted from biopsies of normal cervices also yielded
positivity rates of about 10% (Wickenden et al., 1985).  These
observations suggest that the genital HPVs persists in
apparently normal epithelium, a concept supported by analysis
of tissue adjacent to neoplastic or condylomatous lesions of
the cervix (Ferenczy et al., 1985; MacNab et al., 1986).

The nature of the persistence of genital HPVs can be
inferred from the results of the studies of women at high risk
of being infected with these agents. We followed a cohort of
181 Panama City prostitutes at monthly intervals for eight
months and collected cervical samples for cytology and HPV DNA
analysis by filter in situ hybridization (Reeves et al., in
preparation).  Samples were obtained at 741 visits and HPV-
16/18 and HPV-6/11 were detected in 25% and 12% of the samples,
respectively.  Between visit rates varied from 17 to 41% for
HPV-16/18 and from 10 to 24% for HPV-6/11.  The proportion of
women positive for HPV DNA increased as the women were sampled
more frequently.  Almost 80% of the women were negative after a
single visit but only 57% of those sampled twice were negative
on both occasions.  Twenty nine percent of women sampled 5
times were negative on all 5 occasions and about 20% remained
negative after 8 samplings.  Examples of the patterns of
reactivity are illustrated in table 1.  Some women had HPV DNA
detected on a single visit while others were positive on
several visits with or without varying numbers of intervening
negative test.  The observed periodic positivity could
represent scattered foci of persistent infection which were
inconsistently included in the sample or periodic reactivation
and shedding of latent virus.  Interestingly, we found a poor
correlation between the presence of detectable HPV DNA on a
visit and cytological abnormalies associated with
papillomavirus infections which favors the existence of
inconsistently sampled persistent focal disease.

Table 1.   Patterns of Detection of HPV 16/18 DNA
           in Cervical Sample Obtained from a
           High Risk Population

| Subject No. | Visit Number | | | | | | | |
|---|---|---|---|---|---|---|---|---|
| | 1 | 2 | 3 | 4 | 5 | 6 | 7 | 8 |
| 009 | - | - | - | - | - | + | + | + |
| 029 | - | - | - | - | - | - | - | - |
| 037 | + | - | - | + | - | + | - | - |
| 042 | + | - | - | + | - | + | + | + |
| 058 | - | - | - | - | - | + | - | - |
| 069 | - | - | - | + | - | - | - | + |
| 102 | - | - | + | + | + | - | - | - |

The samples were analyzed by the filter in situ
hybridization assay using a combined HPV 16 and
18 DNA probe and assessed as positive (+) or
negative (-) according criteria described elsewhere (Reeves et al., 1987).

## Immune Responses to Papillomaviruses

Available evidence indicates that proliferative aspects of papillomavirus infections are influenced by host immune responses. Early experimental studies in rabbits provided data which supported immune mechanisms in the control of lesions induced by Shope papillomavirus. The warts tended to undergo spontaneous regression and the frequency of regression could be increased by inoculating the rabbit with homologous or autologous tissue derived vaccines (Evans et al., 1962). In reinfection experiments, no warts developed in rabbits in which the primary lesions had regressed while papillomas developed at the site of injection in 90% of normal rabbits and in 75% of rabbits in which the primary lesion had persisted (Evans and Ito, 1966). Immunity to lesion formation could not be passively transferred with sera from regressor rabbits and antibody titers to the virus were unrelated to the fate of the lesions suggesting that regression of the papillomas was not related to a humoral immune response (Kreider and Bartlett, 1987). However, high doses of methylprednisolone were found to strongly suppress the frequency of lesion regression supporting a role for cell-mediated immunity in controlling lesion progression and regression (McMichael, 1967).

Recently, attempts have been made to define the antigenic determinants of the papillomavirus proteins. Initially, virus particles purified from extracts of warts were used as antigens while current studies use proteins produced in engineered vectors or synthetic peptides. Antisera raised against intact virions were found to be type specific while antisera raised by injecting disrupted particles recognize epitopes shared by other types of papillomaviruses, i.e. genus specific antigens (Jenson et al., 1980). These findings suggest that the major capsid protein contains type specific epitopes on the virion surface and type common epitopes hidden within the virion. The conclusion that the major capsid protein contain both type common and type specific epitopes has been confirmed by expressing in E. coli L1 ORFs cloned into plasmids (Pilacinski et al., 1984; Tomita et al., 1987). The L1 ORFs of BPV 1, HPV 1, HPV 6 and HPV 16 have all been expressed and the protein products have been used to raise antisera as has a 14 amino acid oligopeptide representing the N-terminus of HPV 6 L1 (Doorbar and Gallimore, 1987). The HPV 6 and HPV 16 L1 ORF products reacted with antisera raised against BPV 1 disrupted virions and antiserum to the HPV 6 L1 protein reacted with the HPV 16 L1 protein. The anti-HPV 6 L1 serum recognized antigens in condyloma acuminata which contained HPV 6 DNA but did not react in HPV 11 lesions (Li et al., 1987). Analysis of monoclonal antibodies derived from mice immunized with disrupted BPV 1 virions revealed some with type specific reactivity while others cross reacted with types of papillomaviruses other than BPV (Nakai et al., 1986). Thus, the major capsid protein appears to contain both genus specific and type specific epitopes capable of inducing antibody responses.

Less is known about the antigenicity of other HPV proteins. The L2 ORF of HPV 1 has been partially cloned and expressed as a fusion protein in bacteria and this protein is not recognized by a genus specific antiserum (Komly et al., 1986). Antisera raised against the L2 proteins made in

bacteria have generally reacted in a type specific manner
(Doorbar and Gallimore, 1987; Komly et al., 1986) and appear to
recognize a minor 76 kd protein detectable in virions.
Bacterial products of cloned ORFs of the early region genes
have also been used to induce polyclonal and monoclonal
antibodies (Matlashewski et al., 1986). The greatest interest
has focused on the E6 and E7 genes which are thought to be
involved in cell transformation. A comparison of HPV 16 and 18
E6 ORFs suggest that their products are antigenically unique
(Banks et al.,, 1987). Using appropriate antisera, small
amounts of these proteins can be detected in cell lines derived
from cervical cancers as can the proteins encoded by E1 and E4
(Banks et al., 1987; Seedorf et al., 1987). However, the role
of these proteins in the induction of antibodies during a
natural infection is not known nor has the nature of the
antigenic determinants involved in cell-mediated immune
reactions been determined.

Characterization of antibody responses to papillomavirus
infections has been hampered by the lack of a ready source of
well defined viral antigens. Early studies of the humoral
response in humans were conducted using virions purified from
pooled wart preparations which probably contained different
types of HPVs (Pass and Maizel, 1973). These studies clearly
demonstrated the production of antibodies to HPVs but the
varying sensitivity and specificity of the techniques used to
measure antibody makes the results of the different studies
difficult to compare. Information is available from
experiments in which animals were infected with
papillomaviruses but the experimental results did not always
agree with observations made in naturally infected animals.
For example, cattle infected with various types of BPV became
heavily tumored and did not produce detectable antibodies to
the viruses although naturally occurring cases had high titers
of circulating antibody (Jarrett, 1985). Recently, an E. coli
derived L1 protein of BPV was used to immunize cattle and fewer
vaccinated cattle developed warts on challenge than animals
receiving placebo (Pilacinski et al., 1986) confirming the
ability of the animals to respond to the viral antigens under
experimental conditions.

Antibodies to papillomavirus antigens have been found in
human sera and the occurrence of antibodies usually correlates
poorly with a history of warts (Morison, 1975). For example,
IgG and IgM antibodies against disrupted HPV were detected in
70.4% and 40.7%, respectively, of women with warts while 54.6%
and 24.3%, respectively, of control women possessed detectable
IgG and IgM antibodies to the virus. With the BPV preparation,
IgG antibodies were detected in 70.4% of the patient group and
in 45.6% of the control group (Portolani et al., 1987). The
investigators noted that a recent onset of lesions and the
presence of exophytic warts correlated with the absence of
antibodies. In another study, antibodies to HPV 1 were
measured in sera from 162 patients possessing common warts,
flat warts, genital warts or deep plantar warts (myrmecia).
The highest occurrence of antibodies was found among the
patients with myrmecia and no significant differences were
found between patients afflicted with the other types of warts
and a control group of subjects (Kienzler et al., 1983).
Interestingly, a booster effect was observed in patients with
myrmecia after injection with purified HPV 1 capsid protein.

Attempts are presently being made to overcome the lack of
specific HPV antigens by using fusion proteins of HPV ORFs
produced in bacteria.  In a preliminary report, a protein of
HPV 6 L1 reacted with 18 of 20 sera from women being
investigated for cervical neoplastic disease and 2 of 20
children by Western blot technique (Li et al., 1987).  Although
infected individuals produce antibodies to papillomavirus
antigens, the specificity of these responses against specific
virus types and the role of humoral immune responses in the
pathogenesis of infection remain unknown.


## Cell-mediated Immunity in Papillomavirus Infections

The importance of cell-mediated immune functions in
controlling the lesions induced by papillomaviruses has been
argued from three lines of evidence.  These include the nature
of cells found in warts at different stages of evolution;
observations made in patients with epidermodysplasia
verruciformis (EV), who appear incapable of eliminating
papillomas; and the fate of papillomas in immunosuppressed
patients.  Lymphoid tissues are associated with the skin (SALT)
and mucosal surfaces (MALT) and these elements are thought to
provide an immune surveillance mechanism.  Langerhans' cells
(LC) and intraepithelial lymphocytes (IEL) are part of this
system and may interact with adjacent keratinocytes (Volz-
Platzer et al., 1984).  Several investigators have attempted to
quantitate various cell types involved different types of human
papillomas.  Most studies have found a decreased number of LC
in warts, increased numbers in condyloma acuminata and in
cervical intraepithelial neoplasia (CIN) (Bhawan et al., 1986;
Caorsi et al., 1986).  The increase in LC in CIN lesions has
generally been reported to be greater in more severe lesions.
Attempts to correlate LC numbers with the presence of
detectable viral genus specific antigens in warts induced by
HPV 1 or HPV 2 revealed the reduced numbers to be associated
with antigens but not with HPV type or location of the warts
(Chardonnet et al., 1986).  Somewhat different results were
recently reported (Tay et al., 1987). In this study, reduced
numbers of LC were found in both cervical atypical condyloma
and CIN lesions.  This study also revealed changes in the
subpopulations of LC with a 60% reduction in lesions of T6,
ATPase and MHC II positive cells and an almost complete absence
of S-100 positive cells.  One reason for the discrepancies in
this report could be the use of different markers of LC
although an increase in S-100 containing LC were found in CIN
by a second study using monoclonal antibodies to enumerate the
cells containing this protein (McArdle and Muller, 1986).

Several investigators have reported minimal differences
between patients with HPV related lesions and controls with
respect to the numbers of T-lymphocytes or the T4/T8 ratios in
cervical mucosa (Vayrnen et al., 1985).  However, Tay and co-
workers (Tay et al., 1987) noted a depletion in the number of
intraepithelial T-lymphocytes, especially of the T4 subset, in
cervical condylomas and CIN.  The number of T-lymphocytes and
the T4/T8 ratio were lower in cases of condyloma than in CIN
while in the subepithelial stroma the numbers and phenotypes of
these cells were not different between diseased and normal
specimens.  Activated T and B-lymphocytes were rarely found in
either normal or abnormal epithelium.

Another indication of the role of cell-mediated immunity can be implied from the nature of cells infiltrating a lesion. Regressing warts characteristically display an intense infiltration of mononuclear cells compatible with a T-lymphocyte mediated reaction (Tagami et al., 1977). _In vitro_ experiments comparing the behavior of mononuclear cells from regressing warts and normal skin showed that mononuclear cells migrating from wart explants attached to the wart-derived cells rather than floating free in the medium as observed in explants of normal skin (Tagami et al., 1985). As a continuation of these experiments, the presence of intranuclear HPV particles and the subpopulation of mononuclear cells were examined in warts (Aiba et al., 1986). The genus-specific antigens were demonstrated in 12 of 19(63%) ordinary flat warts but not in 31 regressing warts containing inflammatory infiltrates. The connective tissue underlying the warts was infiltrated with lymphocytes and the T-helper type were in excess to T-suppressor cells. In contrast, the numbers of T-helper and T-suppressor lymphocytes were similar in the infiltrates of the tumor tissue. The majority of the infiltrating lymphocytes expressed HLA-DR antigen suggesting that they were activated. These findings are compatible with a lymphocyte mediated destruction of the tumor cells in the regressing warts, a concept supported by the demonstration of tumor specific transplantation antigens and immune responses to these antigens in Shope papillomavirus infected cells in rabbits (Evans et al., 1962; Seto et al., 1977).

Epidermodysplasia verruciformis (EV) is a rare disorder characterized by the development of extensive cutaneous verrucae which appear to be induced by papillomaviruses (Orth et al., 1979). The lesions usually develop at a young age and persist for life without undergoing spontaneous regression. DNA of numerous types of HPV have been cloned from the lesions and several types have only been found in EV lesions (Pfister et al., 1985). These patients also experience a high rate of epidermoid carcinomas in which HPV 5 or 8 DNA have been detected. EV is thought to represent a genetic disease with autosomal or X-linked recessive inheritance (Androphy et al., 1985; Lutzner, 1978). It has been suggested that EV patients possess a defect in cell-mediated immunity which prevents immune-induced spontaneous regression of the viral induced lesions (Glinski et al., 1976). This concept is based on decreased responses of peripheral blood lymphocytes from EV patients in non-specific assays of cell-mediated immunity (Glinski et al., 1976; Glinski et al., 1981). In addition, these patients reportedly have decreased numbers of total circulating T-lymphocytes and of T-helper cells as well as increased NK cell activity (Majewski et al., 1986). However, these abnormalities have not been found by all investigators (Androphy et al., 1985; Claudy et al., 1982) and EV patients do not experience high frequencies of infections with other agents commonly seen in people with impaired cellular immunity. The so called EV-specific HPV types are rarely found in other immunosuppressed patients suggesting that immune defects in the EV patients may be virus type specific and that the non-specific abnormalities of peripheral T-cells could be secondary to the skin disease rather than causal.

Stronger evidence implicating cellular immunity in the control of the HPV proliferative lesions comes from

observations of patients on immunosuppressive therapy.  High
rates of both squamous cell carcinomas and warts have been
observed in such patients and the rates increase with the
duration of immunosuppression.  For example, in one study 15%
of renal transplant patients had warts by the end of the first
year after surgery while 87% of those who survived 5 or more
years developed warts (Rudlinger et al., 1986).  In a another
study, 43% of Australian renal transplant patients surviving
more than 14 years developed cutaneous neoplasms (Sheil et al.,
1985) and neoplastic or preneoplastic lesions were found in 25%
to 38% of patients surviving more than 80 months after
transplantation in New Wales (Shuttleworth et al., 1987).
Elevated rates of CIN have also been noted in transplant
recipients.  Interestingly, DNA of HPV types (Aiba et al.,
1986; Androphy et al., 1985; Androphy et al., 1985; Androphy et
al., 1987; DiMaio et al., 1986) usually found in the commonly
encountered wart lesions have generally been detected in
immunosuppressed these patients although DNA related to HPV 5
(an EV associated type) is occasionally encountered (Rudlinger
et al., 1986).  These findings suggest that mechanisms involved
in transplant rejection may be operative in controlling the
proliferative phase of virus induced papillomas and of squamous
cell neoplasia.

Worthy of mention is the observed increased prevalence of
warts among patients with systemic lupus erythematosus (SLE).
In one study, warts were found in 45% of SLE patients and in
12% of a control group (Johansson et al., 1977).  The
prevalence of warts was unrelated to corticosteriod therapy but
did increase with age in the patient group and correlated
inversely with rheumatoid factor activity.  The significance of
the association of warts with this disease is unclear but could
reflect primary or secondary immunological abnormalities of SLE
that prevent normal regression of warts.

Comments

Although our understanding of the pathogenesis of
papillomavirus infections is far from complete, it is clear
that these viruses can persist both in the presence of
papillomas and in the absence of clinically or histologically
apparent lesions.  As exemplified by laryngeal papillomas, the
persisting lesions, although benign, may produce significant
morbidity.  In addition, available evidence suggest that clones
of malignant cells may arise with increased frequency in
persisting papillomas (Rous and Beard, 1935).  Thus,
elucidating the mechanisms of persistence of papillomas has
practical implications with respect to controlling disease.

There appears to be two phases in the pathogenesis of
papillomavirus infections.  During the incubation period,
infection of cells of the basal layer of the epithelium seems
to occur and this is followed by epithelial cell proliferation
which occurs focally within infected areas (Reid et al., 1987).
An initial rapid proliferative phase is followed by a slowing
of lesion growth and subsequent persistence or regression
(Syverton et al., 1950).  Spontaneous regression or eradication
of the focal lesion is not necessarily accompanied by
elimination of virus (Murdoch et al., 1988).  Thus,
establishment and expression of latent virus in the epithelium
appears to represent one phase while the induction of

epithelial cell proliferation represents another.

Classically, persistent and latent infections have been defined by the ability to detect infectious virus; persistent infections being those in which infectious viruses are continually present while infectious virus can only periodically be demonstrated in latent infections.  Since papillomavirus infectivity is not readily demonstrable, classification of the prolonged relationship between the host and these viruses is difficult.  The detection of viral DNA in histologically normal epithelium argues strongly for the persistence of papillomavirus genetic information in the absence of proliferative lesions.  This would also occur if the epithelium were sampled early in the incubation period of the infection, however, the results of our study of high risk women followed for a number of month as well as the reports of others (Ferenczy et al., 1985; MacNab et al., 1986) make this explanation unlikely.  Thus, papillomaviruses appear to remain associated with the host in a latent or persisting form either in the presents or absence of an apparent hyperplastic lesion.

The factors involved in the persistence of papillomaviruses and the associated lesions appear to differ. The expression and replication of the virus genomes are to be regulated in part by cellular factors involved in the differentiation of keratinocyte and a model of persistence of the virus in epithelial reserve cells not yet committed to differentiation can be constructed from knowledge regarding the structure and functions of the viral genes.  This model is illustrated in figure 2.

The model predicts that the papillomavirus DNA replicates as a plasmid with a controlled copy number in the reserve cells.  Alternatively, the viral DNA could persist in an integrated form in the host cell genome as suggested by recent studies of cervical neoplastic disease (Durst et al., 1987; Murdoch et al., 1988).  However, this is an unlikely functional form of persistence since the integration sites in the viral DNA are not constant.  In the model, viral plasmid DNA is transmitted to daughter cells during cell division.  The daughter cell maintaining reserve cell potential would receive a complement of viral plasmids which would remain regulated. However, in the daughter cells committed to differentiation the cellular factors associated with keratinocyte maturation would alter the regulatory state of viral DNA.  This would result in an increase of viral genome copy numbers and an increase in the number of genes that are functionally transcribed.  As the keratinocyte terminally differentiate, capsid proteins would be produced and progeny virions would be ultimately assembled. Only viruses produced from the keratinocytes derived from a single reserve cell might be shed at any one time as an expression of the viral persistence.

Papillomas appear to represent the product of hyperplasia of epithelial cells as well as the supporting stromal elements. The stimuli for increased cell division could represent growth factors transcribed from the viral genome or cell-derived growth factors induced by viral gene products.  These possibilities are not mutually exclusive and the functions assigned to several early ORFs are compatible with their participation in stimulating cell DNA synthesis (Green and

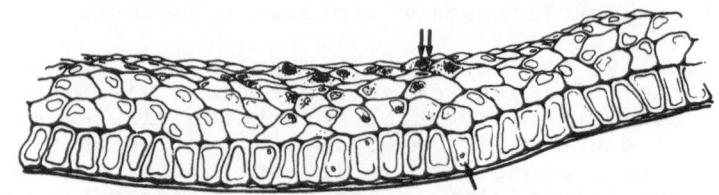

Fig. 2. Model of papillomavirus persistence.
The viral DNA exist as plasmids in a
regulated state in the reserve cells
of the epithelium (single arrow).
During cell division each daughter
cell receives a complement of plasmids
which remain regulated in terms of copy
number and gene expression in the
daughter cell retaining reserve cell
properties.  In the daughter cell
committed to differentiation, viral
genomes copy numbers increase and gene
transcription increases as the cell
matures into a keratinocyte where
intact virions ultimately develop
(double arrow).

Loewenstein, 1987) or altering cell transcription (Durst et
al., 1987).  As reviewed above, spontaneous regression of
papillomas appears to be immune mediated by non-humoral
mechanisms.  Immune functions may also prevent persisting
viruses from inducing clinically evident papillomas although
the genome of the virus would be fully expressed in certain
clones of differentiating keratinocytes.  Thus, it appears that
the clinical manifestations of papillomavirus infections are
controlled by immune responses while the virus genome persist
in a regulated state unaffected by the host immune status.

Clearly, studies defining the mechanisms of papilloma
formation and the role of the immune system in the pathogenesis
of papillomavirus infections should enhance our understanding
of skin and mucosal immunity.  The results of such studies
might also indicate new approaches for controlling malignant
diseases at these sites.

References

Aiba, S., Rokugu, M., and Tagami, H., 1986, Immunohistologic
        analysis of the phenomenon of spontaneous regression of
        numerous flat warts, Cancer, 58:1246.
Androphy, E.J., Dvoretzky, T., and  Lowy, D.R., 1985, X-linked
        inheritance of epidermodysplasia verruciformis, Arch.
        Dermatol., 121:864.
Androphy, E.J., Schiller, J.T., and Lowy, D.R., 1985,
        Identification of the protein encoded by E6 transforming
        gene of bovine papillomavirus, Science, 230:442, 1985
Androphy, E.J., Hubbert, N.L., Schiller, J.T., and Lowy, D.R.,
        1987, Identification of the HPV-16 E6 protein from
        transformed mouse cells and human cervical cell lines,
        EMBO J, 6:989.

Baker, C.C., and Howley, P.M., 1987, Differential promoter utilization by the bovine papillomavirus in transformed cells and productively infected wart tissue, EMBO J, 6:1027.

Banks, L., Spence, P., Androphy, E., Hubbert, N., Matlashewski, G., Murray, A., and Crawford, L., 1987, Identification of human papillomavirus type 18 E6 polypeptide in cells derived from human cervical carcinomas, J. gen. Virol. 68:1351.

Berg, L.J., Singh, K., and Botchan, M., 1986, Complementation of a bovine papillomavirus low-copy-number mutant: evidence for a temporal requirement of the complementing gene, Mol. Cell. Biol., 6:859.

Bhawan,J., Dayal, Y., and Bhan, T.K., 1986, Langerhans' cells in molluscum contagiosum, verruca vulgaris, plantar wart and condyloma acuminatum, J. Am. Acad. Derm., 15:645.

Broker, T.R., and Botchan, M., 1986, Papillomaviruses: retrospectives and prospectives, in: "Cancer Cell 4/ DNA Tumor Viruses", pp. 17-36, Cold Spring Harbor Laboratory, Cold Spring Harbor, New York.

Caorsi, T., and Figueroa, C.D., 1986, Langerhans' cell density in normal exocervical epithelium and in the cervical intraepithelial neoplasia, Br. J. Obst. Gynecol., 93:993.

Chardonnet, Y., Viac, J., and Thivolet, J., 1986, Langerhans' cells in human warts, Br. J. Derm., 115:669.

Claudy, A.L., Touraine, J.L., and Mitanne, D., 1982, Epidermodysplasia verruciformis induced by a new human papillomavirus (HPV-8) (Report of a case without immune dysfunction. Effect of a treatment with an aromatic retinoid. Arch. Dermatol. Res., 274:213.

Cripe, T.P., Haugen, T.H., Turk, J.P., Tabatabai, F., Schmidt, P.G., Durst, M., Gissmann, L., Roman, A., and Turek, L.P., 1987, Transcriptional regulation of the human papillomavirus-16 E6-E7 promoter by a keratinocyte-dependent enhancer, and by viral E2 trans-activator and repressor gene products: implications for cervical carcinogenesis, EMBO J, 6:3745.

De Villiers, E.-M., Schneider, A., Milklaw, H., Papendick, U., Wagner, D., Wesch, H., Wahrendorf, J., and zur Hausen, H., 1987, Human papillomavirus infections in women with and without abnormal cervical cytology, Lancet, ii:703.

DiMaio, D., Guralski, D., and Schiller, J.T., 1986, Translation of open reading frame E5 of bovine papillomavirus is required for its transforming activity, Proc. Natl. Acad. Sci., 83:1797.

Doorbar, J., Campbell, D., Grand, R.J.A., and Gallimore, P.H., 1986, Identification of the human papillomavirus-1a E4 gene product, EMBO J, 5:355.

Doorbar, J., and Gallimore, P.H., 1987, Identification of proteins encoded by the L1 and L2 open reading frames of human papillomavirus 1a, J. Virol., 61:2793.

Durst, M., Croce, C.M., Gissmann, L., Schwarz, E., and Huebner, K., 1987, Papillomavirus sequences integrate near cellular oncogenes in some cervical carcinoma, Proc. Natl. Acad. Sci., 84:1070.

Evans, C.A., Gorman, L.R., Ito, Y., and Weiser, R.S., 1962, Antitumor immunity in the Shope papilloma-carcinoma complex of rabbits. I Papilloma regression induced by homologous and autologous tissue vaccines. JNCI, 29:277.

Evans, C.A., and Ito, Y., 1966, Antitumor immunity in the Shope papilloma-carcinoma complex of rabbits. III Response to reinfection with viral nucleic acid, JNCI, 36:1161.

Ferenczy, A., Mitao, M., Nagai, N., Silverstein, S.J., and Crum, C,P., 1985, Latent papillomavirus and recurring genital warts. N. Engl. J. Med., 313:784.

Giri, T., and Danos, O., 1986, Papillomavirus genomes: from sequence data to biological properties, Trends in Genetics, 2:227.

Gissmann, L., de Villiers, E.-M., and zur Hausen, H., 1982, Analysis of human genital warts (condylomata acuminata) and other genital tumors for human papillomavirus type 6 DNA, Int. J. Cancer, 29:143.

Gissmann, L., 1984, Papillomaviruses and their association with cancer in animals and in man, Cancer Surv., 3:162.

Glinski, W., Jablonska, S., Langer, A., Obalek, S., Haftek, M., and Proniewska, M., 1976, Cell-mediated immunity in epidermodysplasia verruciformis, Dermatologica, 153:218.

Glinski, W., Obalek, S., Jablonska, S., and Orth, G., 1981, T cell defect in patients with epidermodysplasia verruciformis due to human papillomavirus type 3 and 5, Dermatologica, 162:141.

Green, M., and Loewenstein, P.M., 1987, Demonstration that chemically synthesized BPV-1 oncoprotein and its C-terminal domain function to induce cellular DNA synthesis, Cell, 51:705.

Hirochika, H., Broker, T.R., and Chow, L.T., 1987, Enhancers and trans-acting E2 transcriptional factors of papillomaviruses, J. Virol., 61:2599.

Jarrett, W.F.H., 1985, The natural history of bovine papillomavirus infections, Adv. Viral Oncol., 5:83.

Jenson, A.B., Rosenthal, J.D., Olson, C., Pass, F., Lancaster, W.D., and Shah, K., 1980, Immunologic relatedness of papillomaviruses from different species, JNCI, 64:495.

Johansson, E., Pyrhonen, S., and Rostila, T., 1977, Warts and wart virus antibodies in patients with systemic lupus erythematosus, Br. Med. J., 1:74.

Kienzler, J.L., Th.Lemoine, M., Orth, G., Jibard, N., Blanc, D., Laurent, R., and Agache, P., 1983, Humoral and cell mediated immunity to human papillomavirus type 1 (HPV-1) in human warts, Br. J. Derm., 108:665.

Komly, C.A., Beitburd, F., Croissant, O., and Streeck, R.E., 1986, The L2 open reading frame of human papillomavirus type 1a encodes a minor structural protein carrying type-specific antigens, J. Virol., 60:813.

Koss, L.G., 1987, Cytologic and histologic manifestations of human papillomavirus infections of the female genital tract and their clinical significance, Cancer, 60:1942.

Kreider, J.W., and Bartlett, G.L., 1987, Shope papilloma-carcinoma complex of rabbits, Adv. Cancer Res., 35:81.

Lancaster, W.D., and Olson, C., 1982, Animal Papillomaviruses, Microbiol. Rev., 46(2):191.

Li, C.H., Shah, K.V., Seth, A., and Gilden, R., 1987, Identification of the human papillomavirus type 6b L1 open reading frame protein in condylomas and corresponding antibodies in human sera, J. Virol., 61:2684.

Lorincz, A.T., Lancaster, W.D., Kurman, R.J., Jenson, A.B., and Temple, G.F., 1986, Characterization of the human papillomaviruses in cervical neoplasias and their detection in routine clinical screening, in: "Viral

Etiology of Cervical Cancer", R. Peto and H. zur Hausen
ed., Branbury Report 21, pp. 225-237, Cold Spring Harbor
Laboratory, Cold Spring Harbor, NY.

Lusky, M., and Botchan, M.R., 1984, Characterization of the
papillomavirus plasmid maintenance sequences, Cell,
36:391.

Lusky, M., and Botchan, M.R., 1986, A bovine papillomavirus
type 1-encoded modulation function is dispensable for
transient viral replication but is required for
establishment of the stable plasmid state, J. Virol.,
60:729.

Lutzner, M.T., 1978, Epidermodysplasia verruciformis: an
autosomal recessive disease characterized by viral warts
and skin cancer, Bull. Cancer, 65:169.

MacNab, J.C.M., Walkinshaw, S.A., Cordiner, J.W., and Clements,
J.B., 1986, Human papillomavirus in clinically and
histologically normal tissue of patients with genital
cancer, N. Engl. J. Med., 315:1052.

Majewski, S., Sopinska-Rozewska, E., Jablonska, S., Wasik, M.,
Misiewic, J., and Orth, G., 1986, Partial defects of
cell-mediated immunity in patients with
epidermodysplasia verruciformis, J. Am. Acad. Dermatol.,
15:966.

Matlashewski, G., Banks, L., Wu-Liao, J., Spence, P., Pim, D.,
and Crawford, L., 1986, The expression of human
papillomavirus type 18 E6 protein in bacteria and
production of anti-E6 antibodies, J. gen. Virol.,
67:1909.

McArdle, J.P., and Muller, H.K., 1986, Quantitative assessment
of Langerhans' cells in human cervical intraepithelial
neoplasia and wart virus infection, Am. J. Obst.
Gynecol., 154:509.

McMichael, H., 1967, Inhibition by methylprednisolone of
regression of the Shope rabbit papilloma, JNCI, 39:55.

Meanwell, C., Cox, M.F., Blackledge, G., and Maitland, N.J.,
1987, HPV 16 DNA in normal and malignant cervical
epithelium: implications for the etiology and behavior
of cervical neoplasia, Lancet, i:703.

Meinke, W., and Meinke, C.G., 1981, Isolation and
characterization of the major capsid protein of bovine
papillomavirus type 1, J. gen. Virol., 52:15.

Meisels, A., Roy, M., Fortier, M., Morin, C., Casas-Corero, M.,
Shah, K.V., and Turgeon, H., 1981, Human papillomavirus
infection of the cervix. The atypical condyloma, Acta
Cytol., 25:7.

Morison, W.L., 1975, In vitro assay of immunity to human wart
antigens, Br. J. Dermatol., 93:545.

Murdoch, J.B., Cassidy, L.J., Fletcher, K., Cordiner, J.W., and
Macnab, J.C.M., 1988, Histological and cytological
evidence of viral infection and human papillomavirus
type 16 DNA sequences in cervical intraepithelial
neoplasia and normal tissue in the west of Scotland:
evaluation of treatment policy, Br. Med. J., 296:381.

Nakai, Y., Lancaster, W.D., Lim, L.Y., and Jenson, A.B., 1986,
Monoclonal antibodies to genus- and type-specific
papillomavirus antigens, Intervirol., 25:30.

Oriel, J.D., 1971, Natural history of genital warts, Br. J.
Vener. Dis., 46:1.

Orth, G., Jablonska, S., Jarzabek-Chorzelska, M., Obalek, S.,
Rzesa, G., Favre, M., and Croissant, O., 1979,
Characteristics of the lesions and risk of malignant

conversion associated with the type of human papillomavirus involved in epidermodysplasia verruciformis, <u>Cancer Res</u>., 39:1074.

Pass, F., and Maizel, J.B., 1973, Wart-associated antigens. II Human immunity to viral structural proteins, <u>J. Invest. Dermatol</u>., 60:307.

Pfister, H., Iftner, T., and Fuchs, P.G., 1985, Papillomaviruses from epidermodysplasia verruciformis patients and several allograft recipients, <u>in</u>: "Papillomaviruses: molecular and clinical aspects", UCLA Symposia on molecular and cellular biology, new series, P.M. Howley, T.R. Broker, eds., vol. 32:85-100, Alan R. Liss, NY.

Pfister, J.C., Krubke, J., Dietrich, W., Iftner, T., and Fuchs, P.G., 1986, Classification of the papillomaviruses-mapping of the genome, <u>in</u>: "Papillomaviruses", Ciba Foundation Symposium 120, pp.3-22, Wiley, Chichester.

Pilacinski, W.P., Glassman, D.L., Krzyzek, R.A., Sadowski, P.L., and Robbins, A.K., 1984, Cloning and expression in Escherichia coli of the bovine papillomavirus L1 and L2 open reading frames, <u>Biotech</u>., 1:356.

Pilacinski, W.P., Glassman, D.L., Glassman, K.F., Reed, D.E., Lum, M.A., Marshall, R.F., Muscoplat, C.C., and Faras, A.J., 1986, Immunization against bovine papillomavirus infection, <u>in</u>: "Papillomaviruses", pp.136-156, Ciba Foundation Symposium 120, Wiley, Chihester.

Portolani, M., Mantovani, G., Pietrosemoli, P., Cermelli, C., and Boselli, F., 1987, Antibodies to papillomavirus genus-antigens in women with genital warts, <u>Microbiol</u>., 10:271.

Reeves, W.C., Caussy, D., Brinton, L.A., Brenes, M.M., Montalvan, P., Gomez, B., de Britton, R.C., Morice E., Gaitan, E., Loo de Lao, S., and Rawls, W.E., 1987, Case-control study of human papillomaviruses and cervical cancer in Latin America, <u>Int. J. Cancer</u>, 40:450.

Reeves, W.C., Arosemena, J.R., Garcia, M., Loo de Lao, S., de Torraza, I., Cuevas, M., Quiroz, E., Willett, J., Guerrero, D.M., Brenes, M.M., Caussy, D., Kurman, R., and Rawls, W.E., Cervical cytopathology and genital papillomavirus (HPV) infection in Panamanian prostitutes (in preparation).

Reid, R., Stanhope, C.R., Herschman, B.R., Booth, E., Phibbs, G., and Smith, J.P., 1982, Genital warts and cervical cancer. I. Evidence of an association between subclinical papillomavirus infection and cervical malignancy, <u>Cancer</u>, 50:377.

Reid, R., Greenberg, M., Jensen, B., Husain, M., Willett, J., Daoud, Y., Temple, G., Stanhope, C.R., Sherman, A.I., Phibbs, G.D. and Lorincz, A.T., 1987, Sexually transmitted papillomaviral infections. I The anatomic distribution and pathologic grade of neoplastic lesions associated with different viral types, <u>Am. J. Obstet. Gynecol</u>., 156:212.

Rous, P., and Beard, J.W., 1986, The progression to carcinoma of virus-induced rabbit papilloma (Shope), <u>J. Exp. Med</u>., 62: 523.

Rudlinger, R., Smith, J.W., Bunney, M.H., and Hunter, J.A.A., 1986, Human papillomavirus infections in a group of renal transplant recipients, <u>Br. J. Dermatol</u>., 115:681.

Schiller, J.T., Vass, W.C., Vousden, K.H., and Lowy, D.R., 1986, E5 open reading frame of bovine papillomavirus type 1 encodes a transforming gene, <u>J. Virol.</u>, 57:1.

Schneider, A., Kraus, H., Schuhmann, R., and Gissmann, L., 1985, Papillomavirus infection of the lower genital tract: detection of viral DNA in gynecological swabs, <u>Int. J. Cancer</u>, 35:443.

Schneider, A., Hotz, M., and Gissmann, L., 1987, Increased prevalence of human papillomaviruses in the lower genital tract of pregnant women, <u>Int. J. Cancer</u>, 40:198.

Seedorf, K., Oltersdorf, T., Krammer, G., and Rowekamp, W., 1987, Identification of early proteins of the human papillomaviruses type 16 (HPV 16) and type 18 (HPV 18) in cervical cancer cells, <u>EMBO J</u>, 6:139.

Seto, A., Notake, K., Kawanishi, M., and Ito, Y., 1977, Development and regression of Shope papillomas induced in newborn domestic rabbits, <u>Proc. Soc. Exp. Biol. Med</u>,. 156:64.

Sheil, A.G., Flavel, S., Disney, A.P.S., and Mathew, T.H., 1985, Cancer development in patients progressing to dialysis and renal transplantation, <u>Transplant. Proc.</u>, 17:1685.

Shuttleworth, D., Marks, R., Griffin, P.J.A., and Salaman, J.R., 1987, Dysplastic epidermal change in immunosuppressed patients with renal transplants, <u>Quart. J. Med.</u>, 64:609.

Spalholz, B.T., Lambert, P.F., Yee, C.L., and Howley, P.M., 1987, Bovine papillomavirus transcriptional regulation: localization of the E2-responsive elements of the long control region, <u>J. Virol.</u>, 61:2128.

Syrjanen, K., Varynen, M., Saakikosoi, S., Mantyjarvi, R., Parkkinen, S., Hippelainen, M., and Castren, O., 1985, Natural history of cervical human papillomavirus (HPV) infection based on prospective follow-up, <u>Br. J. Obstet. Gynecol.</u>, 92:1086.

Syverton, J.T., Dascomb, H.E., Koomen, J. Jr, Wells, E.B., and Berry, G.P., 1950, The virus-induced papilloma-to carcinoma sequence. I. The growth pattern in natural and experimental infections, <u>Cancer Res.</u>, 10:379.

Tagami, H., Tarigawa, M., Ogino, A., Tinamura, S., and Ofuji, S., 1977, Spontaneous regression of plane warts after inflammation: Clinical and histologic studies in 25 cases, <u>Arch. Dermatol.</u>, 113:1209.

Tagami, H., Oku, T., and Jwatsuki, K., 1985, Primary tissue culture of spontaneously regressing flat warts, <u>Cancer</u>, 55:2437.

Tay, S.K., Jenkins, D., Maddox, P., Campion, M., and Singer, A., Subpopulation of Langerhans' cells in cervical neoplasia, <u>Br. J. Obst. Gynecol.</u>, 94:10.

Tay, S.K., Jenkins, D., Maddox, P., and Singer, A., 1987, Lymphocyte phenotypes in cervical intraepithelial neoplasia and human papillomavirus infection, <u>Br. J. Obst. Gynecol.</u>, 94:16.

Tomita, Y., Shirasawa, H., and Simizu, B., 1987, Expression of human papillomavirus types 6b and 16 L1 open reading frames in Escherichia coli: detection of a 56,000-dalton polypeptide containing genus-specific (common) antigens, <u>J. Virol.</u>, 61:2389.

Vayrnen, M., Syrjanen, K., Mantyjarvi, R., Castren, O., and Saarikoski, S., 1985, Immunophenotypes of lymphocyte in prospectively followed up human papillomavirus lesions of the cervix, <u>Genitourin. Med.</u>, 61:190.

Volz-Platzer, B., Majdic, O., Knapp, W., Wolff, K.,

Hinterberger, W., Lechner, K., and Stingi, G., 1984,
Evidence of HLA-DR antigen biosynthesis by human
keratinocytes in disease, <u>J. Exp. Med</u>., 159:1784.

Waldeck, S., Rosl, F., and Zentgraf, J.C., 1984, Origin of
replication in episomal bovine papillomavirus type 1 DNA
isolated from transformed cells, <u>EMBO J</u>, 3:2173.

Wickenden, C., Steele, A., Malcolm, A.D., and Coleman, D.V.,
1985, Screening for warts virus infection in normal and
abnormal cervices by DNA hybridization of cervical
scrapes, <u>Lancet</u>, i:65.

Yang, Y., Okayama, J.C., and Howley, P.U., 1985, Bovine
papillomavirus contains multiple transforming genes,
<u>Proc. Natl. Acad. Sci</u>., 82:1030.

GENETIC ENGINEERING AND PROPERTIES OF NOVEL HERPES SIMPLEX
VIRUSES FOR USE AS POTENTIAL VACCINES AND AS VECTORS OF
FOREIGN GENES

Bernard Meignier[$] and Bernard Roizman[*]

Institut Mérieux, 69280 Marcy L'Etoile, France[$] and
The Marjorie B. Kovler Viral Oncology Laboratories
The University of Chicago, Chicago, IL 60637[*]

INTRODUCTION

The pathology associated with viral infections as well
as the attempts to prevent them are closely dependent on the
biology of virus-host interactions. A unique feature of
herpes simplex viruses 1 and 2 (HSV-1 and HSV-2) is that, in
the course of the primary infection, the viruses multiply at
the portal of entry, infect sensory or autonomic nerve
endings, ascend through the axon to the neuronal nucleus and
establish a latent infection that is shielded from the immune
defenses of the host. In a significant fraction of infected
individuals, the latent viruses are reactivated by a variety
of stimuli and cause clinically discernible lesions. Indeed,
most of the morbidity associated with HSV is the result of
such recrudescences. Prevention of morbidity caused by HSV,
unlike that caused by other viruses, requires blocking the
establishment of latency. Inasmuch as establishment of
latency results from multiplication at the portal of entry,
such prevention requires a state of immunity induced by
immunization that is potent enough to preclude this initial
multiplication, a very tall order. On the grounds that
inactivated or subunit vaccines are not likely to generate
both the required duration and level of immunity, we chose to
construct a live attenuated HSV vaccine by the techniques of
genetic engineering described in detail elsewhere (Roizman
and Jenkins, 1985).

Our objective was to design a virus with specific
characteristics, i.e. (i) the virus should retain the ability
to replicate in the recipient but not that of invading the
central nervous system (CNS) or of causing recurrences, (ii)
it should be genetically stable, (iii) it should protect
against both HSV-1 and HSV-2, (iv) it should be easily
differentiable from wild type virus, (v) it should be capable
of acting as vector for non HSV genes whose products are able
to induce protection against other viral diseases, and (vi)
the immune response to vaccination should be differentiable
from that resulting from wild type infection.

# THE CONSTRUCTION OF RECOMBINANT R7020

The details of the design and construction of the recombinant were published elsewhere (Meignier et al, 1988). HSV-1 strain F [HSV-1(F)] was selected as the backbone of the construction to avoid the presence of the sequences of HSV-2 genomes reported to be associated with transformation in cell cultures (reviewed in Tevethia, 1985).

As a first step, two deletions were introduced in the genome of the parent virus. The first, 700 bp long, was in the domain of the thymidine kinase (TK) gene. The second, approximatively 14.5 Kbp long, removed nearly all of the internal inverted repeat sequences of the L and S components. These deletions were made to remove some of the genetic loci implicated in neurovirulence (Centifanto-Fitzgerald et al, 1982), and to create convenient sites and space within the genome for insertion of non HSV genes (Shih et al, 1984). Into the genome of this virus (R3410) were inserted two sets of genes. The first set inserted in place of the inverted repeat sequences, consisted of an HSV-2 DNA fragment encoding the glycoproteins D, G and I and a truncated portion of glycoprotein E. The function of glycoproteins G, I and E in conferring immunity is not known, but glycoprotein D has been recognized as a major determinant of the immune response to infection. The other set was the TK gene, inserted next to the fragment expressing the HSV-2 glycoproteins, to enable the virus to multiply efficiently at the site of inoculation and to render the virus susceptible to antiviral chemotherapy. To preclude the TK gene from recombining at its natural position, it was placed under the control of an α rather than under its natural, β promoter.

As a consequence of the alterations made in the course of its construction, the genome of R7020 is frozen in the prototype orientation and does not invert ; it has sufficient space for insertion of additional 10 Kbp of DNA. Because of the insertions, the restriction endonuclease patterns are unique and different from those of the wild type viruses (Buchman et al, 1978).

# THE PROBLEM OF EVALUATING BIOLOGICAL PROPERTIES OF PROTOTYPE HSV VACCINES

Virulence, latency or the induction of protection are the result of so many complex interactions between the virus and the organism that they can only be evaluated in vivo by the use of animal models. Even though a number of animal species can be experimentally infected with HSV, none of them mimics the spectrum of clinical presentations of human infections with this virus. The evaluation of a vaccine candidate must therefore rely on a combination of tests carried out with the most adequate substitutes of each of the major kinds of pathology encountered by patients. The models can be used to compare the behavior of the recombinant and wild type viruses to assess the degree of attenuation or ability to establish latency. They may also be used for the estimation of the protection conferred by immunization with the candidate vaccine. We selected the mouse inoculated by the intracerebral route to asses the neurovirulence. Even though this model is highly artificial, it has the advantage of allowing relatively precise measurement coupled with statistical analysis of the mortality pattern by the use of serial dilutions of both recombinant and wild type viruses.

The rabbit and the guinea pig were chosen for their longstanding reputation as suitable models to study ocular and genital infections respectively. The development and the evolution of the local symptoms and lesions following infection with wild type virus have been well characterized and can be scored according to a reference scale of severity for comparison of results. It is important to stress here that notwithstanding the similarity with analogous infections in humans, local infections in the rabbit and the guinea pig are frequently complicated by CNS involvement and that death is not uncommon. Also noteworthy is the observation that the reactivations of latent virus, should they occur, do not have all the attributes of recurrent lesions in humans.

Two species of monkeys were also included in our panel of animal models because of their propensity to develop systemic infections upon inoculation, even with very small quantities of virus administered by peripheral routes such as the eye or the vagina. The susceptibility of and the pattern of disease seen in <u>Aotus trivirgatus</u>, (the owl monkey) or in <u>Callithrix jacchus</u> (the common marmoset) are reminiscent of the generalized infections that occur in the newborn.

ATTENUATION OF R7020

R7020 was highly attenuated in rodents (Meignier et al, 1988). In mice injected by the intracerebral route, it required approximatively 10,000 fold more virus than the wild type HSV-1 parent virus to obtain 50 % lethality. In guinea pigs or rabbits the highest inocula ($10^8$ pfu) were not lethal by peripheral routes whereas $10^5$ pfu of wild type virus caused 10 to 80 % mortality. The lesions caused by the recombinant virus was localized, superficial and healed much more rapidly than those caused by wild type virus. We have not seen either the ulcerative lesions of rabbit cornea or of the genital tract of guinea pigs.

Two common marmosets and fourteen owl monkeys were inoculated with a maximum of $10^8$ pfu of R7020, i.e. quantities 1000 to 100,000 times higher than the lethal doses of wild type HSV-1 or HSV-2. All the monkeys survived the inoculations. They showed none or minimal to mild local inflammation and swelling after inoculation in the vagina (owl monkeys and marmosets) or on the cornea (owl monkeys). There were no side effects after subcutaneous or intramuscular injections, including booster immunizations. It is particularly significant that no secondary lesions were observed at sites distant from the site of inoculation indicating that the recombinant is not prone to disseminate in the body of even the most susceptible species.

# ABILITY OF R7020 TO ESTABLISH LATENCY AND TO REACTIVATE

R7020 established latency in all the models tested. However, to establish latency it was necessary to use very high amounts of virus (1,000 fold higher than those required for wild type strains) and the most efficient routes of inoculation, such as the eye in the mouse and rabbit or the vagina in the guinea pig and in monkeys. The same dosages, given by subcutaneous or intra-muscular routes failed to produce any detectable latent infection. It should be noted that among rodents the highest incidence by any route yielded only a small proportion of latently infected animals and that the amount of recombinant virus recovered from individual ganglia was usually smaller than that obtained from the tissues of wild type infected animals.

In Callithrix and Aotus monkeys latent virus could only be detected in one or very few sensory ganglia, all of which were afferent to the anatomical site of inoculation. This is an indirect confirmation that the recombinant virus did not spread from the site of inoculation.

Aotus monkeys were monitored by routine cultures of the tear film and vaginal secretions for several months after inoculation. Asymptomatic reactivation was detected occasionally in genital secretions of some monkeys. Immunosuppression by repeated total gamma irradiation slightly increased the frequency of recovery but did not induce any pathology related to herpes infection.

## R7020 AS AN IMMUNOGEN

In mice, immunization with R7020 effectively reduced mortality and the incidence of latency caused by wild type challenge infections. The level of protection conferred by R7020 was similar to that obtained with the parent virus HSV-1(F) but his required larger doses of virus. Protection could be partially overcome by use of greater challenge doses.

In guinea pigs and in rabbits, a single immunization by subcutaneous or intramuscular routes did not block the infection with the challenge virus but it did prevent the diseases and the mortality associated with the progression of the virulent virus to the CNS. Local lesions developed at the site of challenge, the eye and the perineum respectively, though milder in "vaccinees" than those observed in naive animals. It was remarkable that, after immunization with the recombinant and virulent challenge, the incidence of latent infections was markedly reduced, not only in terms of number of infected animals, but also in terms of number of infected dorsal root ganglia in the positive individuals.

Eight monkeys, previously inoculated with R7020 to determine the level of its attenuation, were challenged with the wild type HSV-2(G) to determine if the exposure to the recombinant viruses had induced protective immunity. Of two females challenged by the vaginal route, one was completely protected and the other experienced a severe infection but recovered completely. Of six males challenged by the intradermal route, one died, one was severely ill and the other four showed lesions at the site of challenge and a few days of hyperthermia but otherwise maintained a normal pattern of behavior. The two severe infections followed the typical course of disease seen in nonimmunized owl monkeys, but at a slower pace and the one death occurred on day 25 compared to days 6 to 12 seen in non immunized animals. After recovery from the acute challenge infection, all but one monkey experienced spontaneous clinical recurrences at various sites of the body, most of which when assayed yielded isolates of the challenge HSV-2 strain.

GENETIC STABILITY OF R7020

Genetic stability was tested by serially passing R7020 in mouse brain and analyzing the virus recovered after the passages for virulence and for restriction enzymes cleavage sites. Earlier studies have shown that this procedure made it possible to select revertants in 4 to 6 mouse brain passages (Roizman et al, 1988 and unpublished results). The R7020 virus isolated after nine serial passages in the mouse brain could not be differentiated from the parent virus with respect to the pfu/LD$_{50}$ ratio or the sequence arrangements of its DNA as determined from tests with restriction endonucleases. It is noteworthy that serial passage of HSV-1 (F) in the same fashion resulted in the selection of mutants 40 fold more virulent that the parent virus.

CONCLUSIONS

R7020 showed a grossly reduced virulence and ability to establish latency in sensory ganglia of the rodents and monkeys tested. After challenge with wild type viruses, animals immunized with R7020 were not protected against infection but were protected from severe disease, from mortality and, for a significant proportion of them, against the establishment of latency by the wild type HSV-2 challenge virus.

The two most significant findings were that in no species did the recombinant virus spread from the site of inoculation to cause systemic or CNS infections and that the efforts to select R7020 revertants or viruses exhibiting compensatory mutations with increased virulence were not successful. R7020 has not yet been used for the expression of non HSV genes but a similar construct has been successfully used as expression vector (Arsenakis et al, 1987).

REFERENCES

Arsenakis M, Poffenberger KL and Roizman B. Novel herpes simplex virus genomes : Construction and application. In : Gallo RC, Haseltine W, Klein G and Zur Hausen H, eds. : "Viruses and human cancer". Proceedings of the UCLA Symposia on Molecular and Cellular Biology, New series. 1987. Vol 43, pp 427-441. Alan R. Liss, Inc., New york.

Buchman TG, Roizman B, Adams G and Stover H. Restriction endonuclease fingerprinting of herpes simplex virus DNA: A novel epidemiological tool applied to a nosocomial outbreak. J. Infect. Dis. 1978;138:488-498.

Centifano-Fitzgerald YM, Yamaguchi T, Kaufman M, Tognon M, Roizman B. Ocular disease pattern induced by herpes simplex virus is genetically determined by a specific region of viral DNA. J. Exp. Med. 1982;155:475
Meignier B, Longnecker R and Roizman B. In vivo behavior of genetically engineered herpes simplex viruses R7017 and R7020. I. Construction and evaluation in rodent animal models. J. infect. Dis. 1988. accepted for publication.

Roizman B, Meignier B, Norrild B, Wagner JL. Bioengineering of herpes simplex virus variants for potential use as live vaccines. 1984. In : "Modern approaches to vaccines. Molecular and chemical basis of virus virulence and immunogenicity". Edited by RM Chanock and RA Lerner. Cold Spring Harbor Laboratory. pp 274-281,1984

Roizman B and Jenkins FJ. Genetic engineering of novel genomes of large DNA viruses. Science 1985;229:1208-1214.

Shih MF, Arsenakis M, Tiollais P and Roizman B. Expression of Hepatitis B virus S gene by Herpes simplex virus 1 vectors carrying α and β regulated gene chimeras. Proc. Nat. Acad. Sci. (USA) 1984;81:5867-5870.

Tevethia MJ. Transforming potential of herpes simplex viruses and human cytomegalovirus. In : B Roizman and C Lopez, eds, The herpesviruses, 1985, Vol 3, 257-313. Plenum Press, New york, N.Y.

ADENOVIRUSES AS VECTORS FOR THE TRANSFER OF GENETIC INFORMATION AND FOR

THE CONSTRUCTION OF NEW TYPE VACCINES

T.I.Tikchonenko

Institute of Agricultural Biotechnology
Academy of Agricultural Sciences, Moscow
127253, USSR

## BIOLOGICAL AND MEDICAL ASPECTS

Members of adenoviral family represent middle-sized viruses composed
of a complex protein capsid and a linear double-stranded DNA of molecular
weight 20-22 mD or 35-36 kb coding for about 30 virion and virus-specific
proteins. The ends of adenoviral genome have inverted terminal repeats 100-
200 nucleotide long, each of which contains identical replication origins.
These sites represent the cis-acting element indispensable in any virus
vector construction. The second cis-element of adenoviral genome absolu-
tely necessary in the capsid type of viral vectors is the package signal
located in the left noncodig part of the DNA molecule 200-300 nucleotides
long between the terminal repeat and the E1 cistron (Doerfler, 1983/1984;
Ginsberg, 1984).

Adenoviral genome contains 5 early and 5 late regions (Fig.1), of
which only two important for vector developing are of interest for us
now. First of all it is the E1 cistron occupaying the region with coordinates 1.3-11.2 map units (m.u.) and consisting of two independently
transcribed units: E1A and E1B. The E1 plays a crucial role in the start
and development of adenoviral infection. Its products, especially E1A,
serve as transactivators launching expression of other viral and some
cellular genes. Besides that both these regions are typical viral oncogenes, thereby the first one is responsible for cell immortalization and
the second one stabilizes the transformed cells and confers tumorigenicity to them (Doerfler, 1983/1984; Ginsberg,1984; Struk et al. 1987).

Of course,it goes without saying that a real oncogenic virus can not
be used as vector for human beings. Not going far into discussion of the
complex problem of viral tumorigenicity, it is reasonable to limit
ourself to two points. First,being a lytic agent adenovirus kills the infected cells, even under semi-permissive conditions when infectious viral
progeny is not formed. The cells previously transformed by adenoviral oncogenes, as a rule, sustain the subsequent adenovirus replication better
in comparison with original cell types. As a result medical oncology does
not consider the adenoviral family as real tumorigenic viruses which
could be responsible for tumors in humans. In accordance to numerous investigations the adenoviral transforming region is totally absent in human
tumors of different origins (Green & Mackey,1977; Green et al. 1979; Mackey
et al. 1976, 1979). Therefore, the doubts which were expressed by some

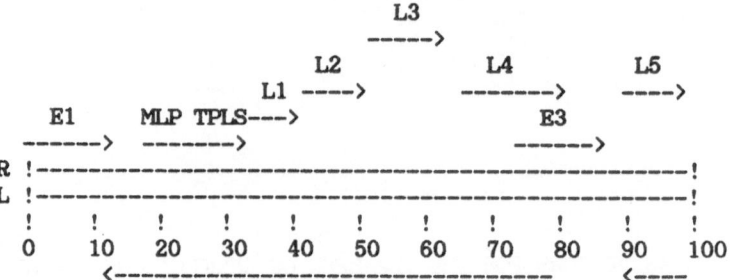

Fig.1. <u>Functional and genetic map of human adenovirus
genome</u>. Early and late cistrons are abbreviated
as E1A,E1B,E2,E3,E4, and L1,L2,L3,L4,L5 respectively.
Directions of transcription are indicated by arrows; R and L are right and left polynuleotide
chains respectively; MLP-major late promoter;
TPLS-tripartite leader sequence; numbers 0 to
100 are map units (m.u.)

authors on several occasions concerning possible oncogenic danger of
adenoviral infections in humans are of a purely speculative nature
(Ibelgaufts, 1982; Berenci & Nasz, 1983).

Second, one may object that the absence of oncogenicity in natural
adenoviral infections can not automatically guarantee the safety of adenoviral vectors or recombinants whose genomes were subjected to serious
changes and which can be applied in much higher concentrations in comparison with natural ones. In both cases officially recognised tests for evaluation of tumor-inducing ability can be recommended to answer the problem:
injection of virus or virus oncogenes into newborn hamsters and immune-
deficient rodents. These two systems are the most sensitive to the neoplastic developments, especially the last one. Of course, positive results
with adenoviral or some other oncogenes in these systems do not necesarilly
mean any real oncogenic danger in other less sensitive or immunocompetent
organisms.

From this point of view the so-called C group of nononcogenic human
adenoviruses is quite safe as vectors. The members of this group or their
isolated oncogenes do not induce tumors in newborn hamsters and do it
only in immunodeficient nude rats or mice (Graham, 1984; van der Eb &
Bernards, 1984; Struk et al. 1987). The last purely theoretical suspicions
concerning the safety of adenoviral vectors can be cleared up using adenoviral mutants of the cyt type. These mutants are defective in the 19K protein coded by the E1B region and responsible for tumorigenicity. In spite
of this defect cyt mutants retain the ability to autonomous replication
(Chinandurai, 1983; Shenk & Williams, 1984; Young et al. 1984). From the
technological point of view adenoviruses with cyt phenotype are characterized usually by lower yield in comparison with wild type, but this defect
can be compensated by replicating the cyt vector in some special cell lines
(Young et al. 1984). Thus, the E1 region is indispensable for construction
of non-defective type of adenoviral vector, but it can be omitted from defective helper-dependent adenoviral vector (see below).

Next point of interest is the E3 region with coordinates 76.8-86.0
m.u. So far only this part of adenoviral genome has been identified as
being dispensable for autonomous replication of virus in permissive cell
lines. Therefore the E3 cistron can be removed during vector construction.

For many years the natural recombinants of human adenovirusesxSV40 have been know capable of replication in simian cells where wild type human adenovirus do not replicate. The SV40 genome or its part occupies in these recombinants almost the entire E3 region without hampering viral infectivity (Jones & Shenk, 1978; Klessig, 1984).

In general, members of adenoviral family belong to lytic viruses, therefore the nondefective adenovectors also launch in infected cells productive infection ending in cell lysis. Contrary to oncornaviral vectors adenoviruses do not permit to establish permanently infected cell lines producing products of interest for long period of time. Nevertheless, recombinant adenoviruses do induce in infected cells high-level synthesis of foreign proteins of commercial importance. At the same time use of some types of defective adenovectors (devoid, as a rule, of late functions harmfull to the resident cell) allows to get permanently transformed cell lines producing high levels of recombinant proteins (see below).

But at the present time the main perspective in adenovectors is their use for the construction of live or killed gene-engineered vaccines designed for oral or respiratory administration. All data concerning adenoviral vectors will be considered mainly from this point of view. There is some positive experience in the use of adenoviral live oral vaccines of classic type (Dudding et al., 1972; Takafuji et al., 1979; Meikleojohn, 1983). As a rule, in all these cases under conditions of oral administration the epidemic strains of adenoviruses were used straight without the development of conventional attenuated live vaccines. Nevertheless, such vaccination of military recruites against acute respiratory disease has been successful, quite safe and effective.

The only known patented gene-engineered adenovaccine against hepatitis B and rotaviruses (Davis & Huang, 1987) just uses this simple principle. It is obvious that civil human patients could strongly object to this rather drastic treatment, and for the wide use of adenoviral vectors it will be necessary to develop nonpathogenic attenuated conventional adenovaccines.

## THE CAPACITY OF CORPUSCULAR ADENOVIRAL VECTORS

Because our present interest in adenoviruses is connected mainly to their use in constructing gene-engineered live and inactivated vaccines, our consideration will be limited to the corpuscular expressing types of vectors. One of the most important characteristics of any virion vector is how much foreign DNA can be put inside the capside or in other words what is its packaging limit or genetic capacity? The answer to this question, first of all, depends on the type of the constructed vector: is it a nondefective helper-free vector capable of autonomous replication and productive infection or is it a defective or conditional-defective vector which needs some sort of helper for normal multiplication. It is obvious that in the case of defective vectors the genetic capacity could be much higher in comparison with nondefective autonomously replicating ones.

In all wild type adenoviruses studied so far there exists a small spare volume available for packing some extra DNA amounting to 5% of the total genome (Jones & Shenk, 1978; Tikchonenko, 1985). Thus, any wild type adenovirus can be considered as a low capacity virion vector which can incapsidate additionally about 1.0-1.1 mD or 1.5 kb of foreign DNA (Table 1). The vector capacity may be increased at the expense of the E3 region mentioned earlier. In the case of Ad5 and some other members of the C group of human adenoviruses, the most part of this cistron can be removed using restriction endonuclease XbaI (Berkner & Sharp, 1982; Gluzman et al., 1982). The deleted fragment with coordinates 78.5-84.5 m.u. comprizes about 65% of the E3 region, instead of which any foreign DNA can be inserted in this or some other part of the viral genome. Keeping in mind the above-discussed 5% limit for extra DNA packing and the size of the deleted E3 fragment, this simplest type of nondefective adenovector can incapsidate additionally 2.2-2.4 mD DNA.

If necessary, the capacity of the vector can be increased additionally at the expense of small deletions in different parts of adenoviral genome: in the inverted terminal repeats, in the left noncoding part before the E1 region, in the rest of the E3 cistron and so on (Hay & McDougall, 1986). Approximate calculations show that this genetic economizing can save about 1 additional mD of packing limit. As a result, from the theoretical point of view it is possible to construct a nondefective infectious adenoviral vector with a genetic capacity of about 3 mD or 4.5 kb which is enough to code for proteins of molecular weight about 150 kD. This estimate seems to be the upper limit capacity for nondefective adenoviral vectors. There seems to be no lower limit on the size of DNA molecule to be packed into the capside since natural defective (incomplete) adenoviral particles are known carrying only a third to a half of the genome (Doerfler, 1983/1984).

Further increase in size limit for packing up foreign DNA into an adenoviral capside can be achieved only in the case of defective vectors at the expense of essential part of genome whose removal can be complemented by a helper virus or by transformed cell lines expressing the deleted viral genes. Gluzman et al. (1982) was the first to construct the Ad5 vector with the most part of the E1 region deleted (see Table 1). The removal of the essential part of the adenoviral genome deprives it off the ability to replicate autonomously in the usual types of cell culture. At the same time this defect can be complemented by multiplication of AdE1⁻ mutants in transformed cell culture expressing the deleted E1 genes. One of these cultures able to support replication of AdE1 mutants is the well known cell line 293 which represents human embryonic kidney cell transformed by the Ad5E1 region (Graham et al., 1977). Usually this type of vector is called conditional-defective to stress the difference from other types of defective helper-dependent vectors.

Of course, the complementation of the AdE1⁻ vector can be achieved just as well with a wild type adenovirus helper, but use of helpers has its own drawbacks. First of all, in mixed populations of normal (helper) and defective (vector) virions the proportion of the second virus rarely exceeds 5-10% and often drops to 1-2% (Gluzman et al. 1982).

Important point in manipulating the E1 region is the safety of the gene coding for polypeptide IX. The last one is located on the right shoulder of the E1B region close to the site of cutting by restricting endonucleases and can be accidentaly destroid. Polypeptide IX controls the termostability of assembled virions and completeness of packing full-sized DNA into the capsid. When this polypeptide is deficient or lacking in infected cells only a part of the adenoviral DNA molecule is packed into the capsid (Ghosh-Choudhry & Graham,1987).

The gain in capacity of the adenoviral vector due to removal of the most part of the E1 cistron is equal to 1.7-1.8 mD. This value may be practically doubled by combining both types of deletions considered above (in E1 and E3)(Ghosh-Choudhry & Graham, 1987). As a matter of fact AdE1⁻ virions multiplied in cell line 293 may be considered as safe killed adenoviral vaccine. Contrary to conditional-defective E1⁻type of vector,in helper-dependent defective vectors any essential genes can be deleted. Usually in these constructions late genes with coordinates 26-29 to 41-59

Table 1.Capacity of nondefective and defective adenoviral vectors

| Type of vector | Function[a] Inf. Tumor. | | Deletion m.u.[b] | Capacity mD | References |
|---|---|---|---|---|---|
| Wild type | + | + | none | 1.0-1.1 | Jones & Shenk,1978 |
| Nondefective | + | + | E3 78.5-84.5 | 1.2-1.4 | Gluzman et al.,1982; Berkner & Sharp,1982 |
| -"- cyt | + | - | E1B 19K | -"- | Shenk & Williams,1984; Fukui et al.,1984 |
| Conditional-defective | - | - | E1 1.4-9.1 | 1.7-1.8 | Gluzman et al.,1982 |
| -"- | - | - | E1,E3 1.4-9.1 78.5-84.5 | 4.4-4.6 | Ghosh-Choudhury & Graham, 1987 |
| Replication-defective | - | + | L1-L3 29.0-59.0 | 6.0-6.5 | Yamada et al.,1985 |
| -"- | - | - | 26.0-41.0 | -"- | Mansour et al.,1985 |
| Defective helper-dependent | - | - | E-L 1.0-99.0 | 20 | Mansour et al.,1985 Tikchonenko,1985 |

[a]Inf. - infectivity, Tumor. - tumoroginecity, [b]m.u. - map units.
Ad - adenovirus

m.u. controlling capsid proteins are deleted providing the vector capacity of 7.0-7.5 mD (Yamada et al.,1985;Mansour et al.,1985).

From the theoretical point of view it possible to construct a repli-cation-defective helper-dependent adenovector with a maximal genetic capa-city equal to 20 mD (Tikchonenko,1985). This supervector must contain prac-tically only cis-acting regulatory elements of adenoviral genome required for autonomous replication and for packing up viral DNA. Both indispen-sable cis-acting elements (origin of replication and packaging signal) are placed in short terminal fragments of DNA of molecular weight within 1 mD taken together (Ginsberg, 1984;Doerfler,1983/1984). As a matter of fact, these short fragments are similar to lambda phage shoulder DNA preparations used in different types of phage cloning vectors.

All types of defective vectors can be used without helper for highly efficient transfer of large pieces of foreign DNA into animal cells. Compa-risons between efficacy of transformation of animal cells by genes intro-duced by conventional Ca phosphate precipitation and by genes packed in adenoviral capsid usually ended with a score at 1:10-1:100 in favour of the vector (Karlsson et al.,1985). In the absence of a helper the AdE1⁻ recom-binant virions, for example, do not multiply in transfected cells,provi-ding only a safe transport of genes of interest into the cells.In some cases for easier selection of transformed cells an antibiotic-resistance marker can be introduced into the vehicle (see below). In trasformed cells vector DNAs were integrated in 1-3 copies per cell genome (van Doren et al.,1984).

SELECTION PROBLEMS

A necessary step in any gene engineering manipulation including con-struction of recombinant viruses is a selection which allows to get the product of interest free of wild type virus. Two approaches to this goal are available. The first one, of general nature, is based on introducing a gene controlling resistance to antibiotic G418 into the vector, the second one can be applied only to human adenoviruses unable to multiply in monkey cells.

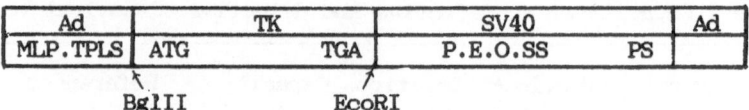

| Ad | TK | | SV40 | | Ad |
|---|---|---|---|---|---|
| MLP.TPLS | ATG | TGA | P.E.O.SS | PS | |

BglII             EcoRI

Fig.2. <u>Gene engineering cassete for "late" expression
and selection of human adenovirus recombinants
in monkey cells</u> (Yamada et al.,1985). MLP-major
late promoter; TPLS-tripartite leader sequence;
P-promoter; E-enhancer; O-origin; SS-splicing
site; PS-polyadenylation signal. Other explana-
tins in the text.

In several studies the gene controlling the neo-resistance marker
has been built into a conditional-defective adenoviral vector of the
E1⁻ type. This recombinant adenovirus transferred resistance to G418
into many types of animal cells where this vector does not multiply (van
Doren et al.,1984). Along with the marker gene any neutral gene of interest may be built in and selected later among antibiotic-resistant colonies growing in the presence of G418.

The second approach is based on the SV40 early A gene introduced
into a human adenovector together with any other gene. The fragment of
cassette used by some authors to provide a selection procedure for the
recombinant virus can be seen in Fig.2 (Yamada et al.,1985). The fragment
of SV40 DNA carring the origin of replication, enhancer, promoter, A coding region with intron and polyadenylation signal has been joined covalently with a gene of interest (TK gene of herpes virus in this case) and
the whole construction has been built into the Ad5 genome behind the
"late" promoter. Joining the above-described recombinant DNA fragment and
a wild type adenovirus is achieved usually by overlapping recombination
when the plasmid and the virus are transfected together into cells. The
desired site of recombination is predetermined by a common fragment of DNA
in the cassette and in the virus. The resulting stock is a mixture of wild
type and recombinant Ad5 which can be selected by passaging through monkey
cell line CV1 where only the recombinant adenovirus can multiply.

EXPRESSING FOREIGN GENES IN ADENOVIRAL VECTORS

In early works with adenoviral vectors there were some problems with
expressing foreign genes built into the recombinant virus (Table 2). For
example, some authors reported that the gene of hepatitis B virus surface
antigene (HBsAg) being built in behind its own or some other promoters in
the Ad5 vector is easily transcribed in transfected cells but HBsAg mRNA
is often badly translated (Tiollais et al., 1981; Saito et al., 1985). At
the same time there are several publications reporting positive results
with expressing the HBsAg gene, in particular, the genes built in the E3
region of the adenovector behind SV40 (Naroditsky et al., 1983, 1986) or
E3 promoters (Morin et al., 1987).

As proved later, the discrepancy was brought about by some peculiar
features of protein synthesis at the late stages of adenoviral infection
(Doerfler,1983/1984; Ginsberg,1984). Contrary to the early phase of adenoviral infection, all late transcripts initiate at single place from the
strong major late promoter (MLP) positioned at 17.6 m.u. (see Fig.1).
Besides that, almost all late mRNAs have at their 5'-end a common noncoding
structure, so called tripartite leader sequence (TPLS) 200 nucleotide long.
TPLS is homologous to three noncontiguous regions of the viral genome.

Table 2. Some recombinant adenoviruses expressing different foreign genes.

| Type of vector | Promoter | Genes | Location m.u. | References |
|---|---|---|---|---|
| Ad1 defective | MLP TPLS | Herpes virus TK | 29-59 | Yamada et al.,1985; Haj-Ahmad & Graham,1985 |
| Ad5  -"- | -"- | $\alpha$-Chorionic gonadotropin | -"- | Yamada et al.,1985 |
| Ad5  -"- | -"- | Polyomavirus early antigen | 26-41 | Mansour et al.,1985 |
| Ad5 conditional -defective | -"- | SV40 T antigen | 1-9 | Logan & Pilder,1985 |
| -"- | -"- | $\gamma$-Interfron | -"- | Kaufman et al.,1985 |
| -"- | -"- | Polyomavirus TM antigen | -"- | Berkner & Sharp,1985; Berkner et al.,1985; Davidson & Hassel,1987 |
| Ad2 conditional -defective | -"- | HBsAg | -"- | Davis et al.,1985,1986 |
| Ad2/Ad1 non-defective | SV40 | -"- | 78-84 | Naroditsky et al.,1983, 1986 |
| Ad5  -"- | HBV | -"- | 99 | Saito et al.,1985 |
| Ad5 conditional -defective | SV40 | Dihydrofolat reductase (DHFR) | 1-9 | Ghosh-Choudhury & Graham,1987 |
| Ad5 nondefective | E3 | HBsAg | 78-84 | Morin et al.,1987 |
| Ad5 conditional -defective | SV40 | neo | 1-9 | Karlsson et al.,1985 |
| -"- | MLP TPLS | DHFR | -"- | Berkner & Sharp,1985 |
| Ad5 non-defective | SV40 | Herpes virus gB protein | 78-84 | Jonson et al.,1988 |

Together with short adenovirus-specific VA RNAs which are usually accumulated in infected cells at late stages, all these elements efficiently
prevent translation of other type of normal cell mRNAs or early viral mRNAs
devoid of TPLS. At the same time there are several exceptions from this
block in translation of mRNAs lacking the adenospecific leader sequence.
For example, translation of influenza virus mRNAs is known to proceed at
a normal level in adenoinfected cells (Katze et al., 1981). Other cases of
escaping the late transcription block (Jonson et al., 1988) will be considered below.

The adenospecific TPLS element joined by recombinant DNA technology
to any foreign mRNAs confers to them the capability of highly efficient
translation at the late stages of adenovial infection (Table 2). As a
result, such foreign protein can accumulate in infected cells in large
quantities frequently amounting to 10% of newly synthesized polypeptides,
which is equal to 1% of total cell protein (Yamada et al., 1985; Logan &
Pilder, 1985; Kaufman et al., 1985). Therefore most of adenoviral vectors
and recombinant adenoviruses created in the last two or three years were
based on the use of MLP and TPLS regulatory elements.

Fig. 3 shows the fragment of a convenient gene engineering cassette
(Logan & Pilder, 1985) which provides an easy way to put any gene of interest behind MLP and to confer to any mRNA the adenovirus-specific TPLS.
The construction contains adenoviral packaging signal, E1 enhanser, MLP,
TPLS, signals for transcription termination and for polyadenylation. The
presence of E1 enhancer stimulates the transcription from MLP, bacause
this late promoter has no enhancer. This or a similar construction can
be put in different parts of adenoviral genome: instead of the E1

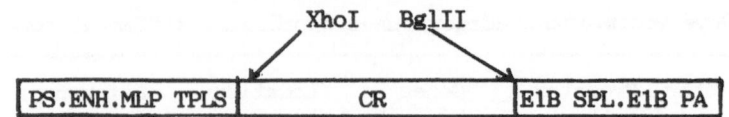

Fig.3.Gene engineering cassete for "late" expression of
foreign genes in adenoviral vectors (Logan & Pilder,
1985). PS-packing signal; ENH-E1A enhanser; MLP-major
late promoter; TPLS-tripartite leader sequence; CR-
coding region flanked by XbaI and BglII sites;
E1B SPL-E1B gene with splicing sequence; E1B PA- E1B
polyadenylation signal.

region (Logan & Pilder, 1985; Kaufman et al., 1985; Davis et al., 1985), in
the middle part of the genome instead of L1-L3 late genes (Yamada et al.,
1985; Mansour et al., 1985). In all these instances the system consisting
of MLP and TPLS elements provided high levels of transcription and translation of built-in foreign genes. Reported values of different proteins
synthesized under these conditions in transfected cells usually amount to
1-5 $\mu$g per 10 cells.

From this point of view the above-mentioned negative results on expression of the HBsAg gene in some adenoviral recombinants can be logicaly explained by the block in translation of HBsAg-specific mRNAs at the late
stages of adenovirus infection owing to the absence of TPLS regulatory elements (Saito et al., 1985). In the last case the HBsAg gene has been placed
at the extreme right noncoding part of adenoviral genome (99.3 m.u.) where
no interference with transcription from other promoters can take place. But
nevertheless, Naroditsky et al. (1983, 1986) and Morin et al. (1987) reported the high-level expression of HBsAg gene placed in the early E3 region.
Both groups reported the synthesis of 1.0-3.0 $\mu$g of HBsAg per $10^6$ cels in
infected cell culture. Jonson et al. (1988) recently reported successful
synthesis of herpes virus gB protein, whose gene was also built in the E3
region. Thus, one is led to believe that the insertion of foreign genes in
the E3 cistron makes their mRNAs somehow insensitive to the above-discussed
block of translation at the late stages of adenoviral infection.

This mystery was solved by Morin et al. (1987) who presented a very
simple and plausible explanation based on position effect and dual behaviour of the E3 cistron. The juice of the matter is that the E3 region
belonging to early genes is located just between late genes, on the same
R chan and is transcribed in the same direction left to right (see Fig.1).
At the early stage of infection when late genes are silent the E3 transcripts are typical early messengers initiated independently at their own
promoters, devoid of any late regulatory elements and insensitive to cyto-
sine-arabinoside which prevents transcription of late adenoviral mRNAs by
inhibiting adenovirus DNA replication. At the late stages of adenoviral
infection the E3 mRNA or, better to say, the transcripts started from the
E3 promoter may function having acquired the features characteristic of late
mRNAs. This hypothesis is supported by the observation that primary late
transcripts originated from MLP contain the E3 sequence which is subsequently removed during mRNA splicing. When the downstream splice site was deleted from the E3 region, mature late mRNAs proved to contain the intact E3
sequence (Bhat et al., 1986).

As a whole this hypothesis is sustained by the results obtained
earlier by us (Naroditsky et al., 1983, 1986) and recently by Jonson et al.
(1988) on successful expression of foreign genes put into the E3 region
behind the SV40 promoter. At the same time some experimental detailes regi-
stered by these authors do not agree wiht the proposed mechanism. In parti-

cular, the expression of the herpes virus gB gene inserted in the E3 region proceeded independently of its orientation left to right or right to left. Anyway, the plausible hypothesis put forward by Morin et al. (1987) needs direct evidence, that is analysis of "late" HBsAg-specific mRNA transcribed from the E3 region.

One can find in Table 2 some other results which need an additional explanation. For example, it is not clear why E1A products do not inhibit under these conditions the expression of foreign genes placed behind the SV40 promoter (Ghosh-Choudhury & Graham,1987; Jones et al.,1988; Naroditsky et al.,1986), although this inhibition was well documented by many authors (Borelli et al., 1984; Velchich & Ziff, 1985).

From our point of view the most interesting results are the expression of capsid proteins of HBV, rotaviruses and herpes virus. These recombinant adenoviruses can be used as live or killed gene-engineered vaccines.

Indeed, nondefective human vectors with the HBsAg gene were able to replicate in lower respiratory tracts and lungs of hamsters. This replication caused efficient seroconversion and appearance of high levels of neutralizing HBV antibodies (Morin et al., 1987). These results create an impression that there appeared, at last, an HBsAg-adenovirus recombinant which can be considered as a candidate for the role of safe live antihepatitis gene-engineered vaccine. The impression is sustained by the patent "Oral vacine" granted recently to Davis and Huang (1987) on the above-described constructions with capside viral genes under the E3 promoters in nondefective Ad5 vector. The patent describes vaccines against HBV and rotaviruses. These live recombinant vaccines might have in future some advantage in comparison with the well-known live hepatitis recombinant vaccine virus in one important point - oral administration in the form of enteric coated tablets. Besides, the recent test performed in the USSR for HBsAg-vaccine virus vector in humans gave negative results: vaccination did not induce HBV-neutralizing antibodies, although exerted some priming effect (V.I.Chernos, personal communication).

Concluding the review I would like to stress that virology needs pluralistic scientific policy towards the types of viral vectors designed for future gene engineering of vaccines. Irrespective of the fact how good the vaccine vector is it cannot be the only answer to all vaccination problems. The above-mentioned negative results in vaccination of humans against hepatitis B with HBsAg-recombinant vaccine virus confirm the necessity to have at our disposal many different viral vectors. The more - the better. Anyway, in accordance to the English proverb it is reasonable not to put all eggs in one basket, the more so that the eggs are of gene engineering origin.

ACKNOWLEDGMENT

The author wish to thank Mr. V.A.Karpov for helpful technical assistance in preparation of this manuscript.

SUMMARY

At present many types of corpuscular nondefective, conditional-defective and helper-dependent expressing adenoviral vectors are available which can be used in constructing gene-engineered live or inactivated viral vaccines. In particular, promising results have been obtained with live recombinant human adenoviruses expressing the S antigen of hepatitis B virus, capsid protein of rotaviruses and gB protein of herpes virus. These recombinants are proper candidates for testing as corresponding vaccine strains, a good alternative to well-known recombinant vaccine virus.

Berencsi G. & Nasz I.,1983, Adenoviruses and human tumors: regulation of eukaryotic chromatin structure, Arch.Geschwulstforsch. 53:239.

Berkner K. L. & Sharp P.A., 1982, Preperation of adenovirus recombinants using plasmid of viral DNA, in "Eukaryotic viral vectors", Y.Gluzman, ed., Cold Spring Harbor Laboratory, Cold Spring Harbor.

Berkner K. L. & Sharp P. A., 1985, Effect of the tripartite leader on synthesis of a non-viral protein in an adenovirus 5 recombinant, Nucl. Acid Res., 13:841.

Berkner K. L., Schaffhausen B., Roberts T. M. & Sharp P. A., 1987, Abundant expression of polyomavirus T antigen and dihydrofolate reductase in an adenovirus recombinant, J. Virol., 61:1213.

Bhat B. M. & Wold W.S.M., 1986, Genetic analysis of mRNA synthesis in adenovirus region E3 at different stages of productive infection by RNA-processing mutants, J. Virol., 60:54.

Borelli E., Hen R. & Chambon P., 1984, Adenovirus2 E1A products repress SV40 enhancer induced stimulation of transcription, Nature, 312:608.

Chinandurai G., 1983, Adenovirus 21p$^+$ locus codes for a 19 kD tumor antigen that plays an essential role in cell transformation, Cell, 33:759.

Davidson D. & Hassel J. A., 1987, Overproduction of polyomavirus middle T antigen in mammalian cells through the use of an adenovirus vector, J. Virol., 61:1226.

Davis A. R., & Huang P. P., 1987, Oral vaccines, European Patent Application EP 0181117.

Davis A. R., Kostek B., Mason B. B., Hsiao C. L., Morin J., Barton J., Dheer S. K., Zandler G. & Huang P. P., 1986, Expression of hepatitis B surface antigen with a recombinant adenovirus, in "Vaccines 86", Cold Spring Harbor Laboratory, Cold Spring Harbor.

Davis A. R., Kostek B., Mason B. B., Hsiao C. L.,.Morin J., Dheer S. K. & Huang P. P., 1985, Expression of hepatitis B surface antigen with a recombinant adenovirus, Proc. Nat. Acad. Sci. USA, 82:7560.

Doerfler W., ed., 1983/1984, "The molecular biology of adenoviruses", Curr. Topics Microbiol. and Immunology, v.109, 110, 111, Springer Verlag, W.Berlin.

Dudding B. A., Top F. H., Scott R. M., Russell P. K. & Buescher E. L., 1972, An analysis of hospitalizations for acute respiratory disease in recruits immunized with adenovirus type 4 and type 7 vaccines, Amer. J. Epidimiol., 95:140.

Fucui Y., Saito I., Shiroke K. & Shimojo H., 1984, Isolation of transforming-defective replication-nondefective early region 1B mutants of adenovirus 12, J.Virol., 44:154.

Ghosh-Choudhury G. & Graham F. L., 1987, Stable transfer of mouse dihydrofolate reductase gene into a deficient cell line using human adenovirus vector, Bioch. Biophys. Res. Commun., 147:964.

Ghosh-Choudhury G., Haj-Ahmad Y. & Graham F. L.,1987, Protein IX, a minor component of the human adenovirus capsid, is essential for the packaging of full-length genomes, The EMBO J.,6,:1733.

Ginsberg H. S., ed." The adenoviruses", Plenum Press, New York.

Gluzman Y., Reichl H. & Solnick D., 1982, Helper-free adenovirus type 5 vectors, in "Eukaryotic viral vectors" Y.Gluzman, ed., Cold Spring Harbor Laboratory, Cold Spring Harbor.

Graham F. L., 1984, Transformation by and oncogenecity of human adenoviruses, in "Adenoviruses", H.S.Ginsberg, ed., Plenum Press, New York.

Graham F. L., Smiley J., Russel W. C., & Nairn R., 1977, Characteristics of a human cell line transformed by DNA from human adenovirus type 5, J. Virol., 36:59.

Green M. & Mackey J., 1977, Are oncogenic adenoviruses associated with human cancer? Analysis of human tumors for adenovirus transforming gene sequences, Cold Spring Harbor Conferences of Cell Proliferation, v.4:1027.

Green M., Wold W. & Mackey J., 1979, Analysis of human tonsil and cancer DNAs and RNAs for sequences of group C human adenoviruses, Proc. Nat. Acad. Sci. USA, 76:6606.

Haj-Ahmad Y. & Graham F. L., 1986, Development of a helper- independent human adenovirus vector and its use in the transfer of the herpes simplex virus thymidine kinase gene, J. Virol., 57:267.

Hay R. T. & McDougall I. M., 1986, Viable viruses with deletions in the left inverted terminal repeat define the adenovirus origin of DNA replication, J. gen. Virol., 67:321.

Jones N. & Shenk T., 1987, Isolation of deletion and substitution mutants of adenovirus type 5, Cell, 13:181.

Jones L., Ristow S., Yilma T. & Moss B., 1986, Accidental human vaccination with vaccinia virus expressing nucleoprotein gene, Nature, 319:543.

Jonson D. C., Ghosh-Choudhury G., Smiley J. R., Fallis L & Graham F. L., 1988, Abundant expression of herpes simplex virus glycoprotein gB using an adenovirus vector, Virology, 164:1.

Ibelgaufts H. 1982, Are human DNA tumor viruses involved in the pathogenesis of human neurogenetic tumors, Neurosurg. Rew., 5:3.

Karlsson S., Humphries R. K., Gluzman Y. & Nienhuis A. W., 1985, Transfer of genes into hematopoetic cells using recombinant DNA viruses, Proc. Nat. Acad. Sci. USA, 82:158.

Katze M. G., Chen Y. T. & Krug R. M., 1981, Nuclear-cytoplasmic transport and VAI RNA-independent translation of influenza viral messenger RNAs in late adenovirus-infected cells, Cell, 37:483.

Kaufman R. J., 1985, Identification of the component necessary for adenovirus translational control and their utilization in cDNA expression vectors, Proc. Nat. Acad. Sci. USA, 82:689.

Klessig D. F., 1984, Adenovirus-SV40 interactions, in "Adenoviruses" H.S.Ginsberg, ed., Plenum Press, New York.

Logan J. & Pilder S., 1985, The use of adenovirus recombinants to study viral gene expression, in "Genetically altered viruses", B.Fields, M.A.Martins & D.Kamely, eds., Cold Spring Harbor Laboratory, Cold Spring Harbor.

Mackey J., Green M., Wold W. & Rigden P., 1979, Analysis of human cancer DNA sequences for DNA of human adenoviruses type 4, J. Nat. Cancer Inst. 62:23.

Mackey J. K., Rigden P. & Green M., 1976, Do highly oncogenic group A human adenoviruses cause human cancer? Analysis of human tumors for adenovirus 12 transforming DNA segments, Proc. Nat. Acad. Sci. USA, 73:4657.

Mansour S. L., Grodzicker T. & Tjian R., 1985, An adenovirus vector system used to express polyomavirus tumor antigenes, Proc. Nat. Acad. Sci. USA, 82:1359.

Meiklejohn G., 1983, Viral respiratory disease at Lowrey Air Force Base in Denver, 1952-1982, J. Infect. Dis., 148:775.

Morin J. E.,Lubeck M. D., Barton J. E., Conley A. J. & Davis A. R., 1987, Recombinant adenovirus induces antibody responce to hepatitis B virus surface antigen in hamsters, Proc. Nat. Acad. Sci. USA, 84:4626.

Naroditsky B. S., Miroshnichenko O .I., Ponomareva T. I. & Tikchonenko T. I., 1983, Construction of the eukaryotic vector using adenovirus type 1, Mol. Genet. Microbiol., Virology, 12:28 (in Russian).

Naroditsky B. S., Miroshnichenko O .I., Ponomareva T. I. & Tikchonenko T. I., 1986, Expression of hepatitis B virus gene inserted into recombinant adenovirus, Mol. Genet. Microbiol., Virology, 1:42 (in Russian).

Roizman B. & Jenkins F. J., 1985, Genetic engineering of novel genomes of large DNA viruses, Science, 229:1208.

Saito I., Oya Y., Yamamoto K., Yuasa T. & Shimojo H., 1985, Construc-

tion of nondefective adenovirus type 5 bearing a 2.8-kilobase
hepatitis B virus DNA near the right end of its genome, J. Virol.,
54:711.

Shenk T. & Williams J., 1984, Genetic analysis of adenoviruses, Curr.
Topics Microbiol. and Immunol., 111:1.

Struk V. I., Tikchonenko T. I. & Sovzova Z. D., 1987, "Infectious
viruses and cancerogenesis", Publ. House "Naukova Dumka", Kiev
(in Russian).

Takafuji E. T., Gaydos J. C., Allan R. G. & Top F. H., 1979, Simultaneous administration of live enteric-coated adenovirus type 4,7 and
21 vaccines: safety and immunochemistry, J. Infect. Dis., 140:48.

Tikchonenko T. I., 1985, Construction of eukaryotic vectors, Biotech-
nol., 2:24 (in Russian).

Tiollais P., Charney P. & Vias G. N., 1981, Biology of hepatitis B
virus, Science, 213:406.

Thummel C., Tjian R. & Grodziker T., 1981, Expression of SV40 large T
antigen under control of Ad promoter, Cell, 23:825.

van der Eb A. J., & Bernards R., 1984, Transformation and oncogenicity
by adenoviruses, Curr. Topics Microbiol. and Immunol., 110:23.

van Doren K., Hanahan D. & Gluzman Y., 1984, Infection of eucaryotic
cells by helper independent recombinant adenoviruses: early
region 1 is not obligatory for integration of viral DNA, J.
Virol., 50:606.

Velcich A. & Ziff E., 1985, Adenovirus E1A proteins repress transcription from SV40 early promoter, Cell, 40:705.

Yamada M., Lewis J. A. & Grodzicker T., 1985, Overproduction of the
protein product of a non-selected foreign gene carried by an
adenovirus vector, Proc. Nat. Acad. Aci. USA, 82:3567.

Young C. S., Shenk T. & Ginsberg H. S., 1984/ The genetic system, in
"Adenoviruses", H.S.Ginsberg, ed., Plenum Press, New York.

# DELTA VIRUS HEPATITIS

Mario Rizzetto

Divisione di Gastroenterologia
Molinette, Cs. Bramante 88, Torino, Italy

## THE VIRUS

The Hepatitis Delta Virus (HVD) occupies a unique niche in human virology. At variance with conventional animal viruses, it cannot replicate autonomously but requires the presence of the Hepatitis B Virus (HBV) to initiate infection. It behaves like a defective interfering agent, in that it ihibits the synthesis of the competent HBV helper once its infection is established. This cooperation is reflected in the HD virion, a 36 nm spherical particle formed by an envelope of HBV-derived surface antigen (HBsAg), which encloses the genome and a specific delta antigen (HD-Ag) (1).

The HDV genome is a small circular single-stranded RNA molecule of which the entire sequence has been decoded (2, 3). The genome has a high prevalence of internal complementary base pairing that under native conditions drives the molecule to assume a rod-like configuration. Genomic and antigenomic RNA could potentially encode for five large polypeptides containing an amino terminal methionine but only a single antigenomic open reading frame (n. 5) appears to encode for polypeptides that possess the immunologic reactivities of HD-antigen; this suggests that HDV may be a negative strand virus (4).

HDV resembles two subgroups of encapsidated RNAs belonging to the microcosmo of the viroids of plants, presently regcognized as the satellite viruses and satellites RNAs (5).

Same as HDV, the RNA satellites and the satellite viruses have small sizes, are encapsidated and capable of protein translation; they are not autonomous but depend for replication on the helper effect of a mature plant virus with which they do not share genomic homologies.

Like the viroids, whose genetic material has been localized primarily in the nuclei of infected cells, in situ hybridization studies have shown that HDV RNA is localized in the hepatocytic nuclei of HDV-infected chimpanzees and woodchucks. Most important, the decoding of the HDV genome has revealed sequences in its RNA that are analogous to short consensus sequences maintained in viroids throughout evolution (6).

The biological dependence of HDV on HBV is also analogous to the dependence of satellites on a concomitant mature plant virus; as HDV derives its capsid from HBV, the satellites obtain the capsid from the helper plant virus. Same as satellites modify the expression of symptoms caused by the helper virus in plants, HDC alters the natural history of HBV infection in man; satellites, however, most often attenuate the disease caused by the competent helper (7), whereas HDV seems to invariably aggravate the natural course of the underlying HBV infection in man.

Properties essential to the survival of HDV are the capacity of adaptation and a high infectious potential. The defective virus is able to exploit helper functions provided not only by HBV but also by the other Hepadnaviruses. It was successfully transmitted to the woodchuck (marmota monax); in this animal the HD virion became coated antigen derived from the woodchuck hepatitis virus (8). In the appropriate host (the carrier of HBV) $HDV$ infections has developed after inoculation of serum diluted up to $10^{-11}$ fold (9); this is the highest infectious titre in transmission of etiologic agents.

The mode of HDV replication remains an enigma. No reverse transcriptase nor other replicases were found in the virion and the size of the RNA does not allow for coding of enzymatic functions. Replication and transcription are not mediated by a RNA dependent/RNA polymerase activity provided by the helper virus. Presently the hypothesis is that HDV subverts the normal replicative mechanism of the hepatocyte to its own advantage in a manner similar to that proposed for the viroids.

Possibly HDV replicates via the rolling-circle mechanism proposed for viroids (10). In this model, genomic and antigenomic RNA would be copied as multimeric plus and minus strands by repeated rounding over the respective input circular strand of opposite polarity. Subsequently, specific cleavage produces unit-length genomic molecules with characteristic end-groups, which are eventually circularized to yield progeny circles, a process reminiscent of RNA splicing. Indirectly evidence that this mechanism may be effective in HDV replication is the finding of circular forms of genomic and antigenomic RNA in livers of chimpanzees infected with HDV, some of which appeared to be twice the length of virion RNA or in form of a double stranded RNA complex (11).

## Epidemiology

The HDV is a global health problem. The occurrence is highest in tropical and subtropical zones, with a prevalence gradient that diminishes in temperature zones in parallel with the decrease of HBV from the equator to the poles. In the Western World HDV is confined within groups that have overt factors predisposing to transmission. The epidemiology of HDV follows therefore two patterns, one of endemic infection in the general HBsAg population and one of sporadic infection with endemic or epidemic clusters within well defined HBsAg-subpopulations.

Inferential evidence implicates the HDV as the cause of outbreaks of fulminant hepatitis that have ravaged the South American Continent since the beginning of the century (12) and markers of HDV have been found in immunoglobulins prepared in the US in 1944 (13).

No predisposing modalities of transmission are detectable in endemic areas indicating that HDV is spread by the inapparent permucosal or percutaneous routes of HBV. Spreading occurs primarily within the household and is facilitated by promiscuity and overcrowding; the incidence is therefore highest in the poor population of the third world.

Clinical studies have confirmed that HDV is a major cause of chronic and aggressive hepatitis leading to liver failure and surveys of severe acute liver illnesses have demonstrated that the defective virus is a major cause of fulminant hepatitis world-wide. The role of coinfection and superinfection in the pathogenesis of fulminant HDV hepatitis varies according to local epidemiologic features; in tropical areas the majority follow superinfection but in the Western World they occur predominantly in drug-addicts who acquire the HDV together with HBV (14).

The role of HDV in hepatocellular carcinoma is debated. Previous reports from Italy (15), US (16) and South Africa (17) have negated relation, but studies from the Middle East (18, 19) have recently pointed to an association between the defective virus and primary hepatocellular carcinoma.

In areas where HDV is non-endemic (North America and Western Europe), Hepatitis D occurs sporadically and is confined to high risk groups; the common predisposing factor is parenteral inoculation of blood and blood derivates.

However, the risk of HDV from tranfusion of blood that passed the HBsAg test is low; in a series of 262 patients with post-transfusional HBsAg-positive hepatitis collected world-wide, only 9 cases (3.4%) were found positive for anti-HD (1).

The risk is instead excessive in intravenous drug-addicts (1). These individuals are the major victims of HDV and the major vehicle of the virus into Northern Europe, where the infection was virtually unknown prior to the diffusion of the drug habit in the seventies.

The risk from blood products is limited to persons exposed to products pooled from multiple donors and is likely to be proportional to the number of donors. HDV markers were consistently negative in HBsAg-positive haemophiliacs given clotting factors produced locally from single or mini pool donors controlled for HBsAg; instead in HBsAg-positive haemophiliacs treated with commercial derivatives prepared from pools formed by several thousands donors, the prevalence of HDV has varied from 27 to 85%.

Unexpectedly, the HDV has not diffused among homosexuals as one could predict from the high rate of parenterally-transmitted agents in these individuals (20, 21, 22). In this population, a strong association with HIV is emphasized by the presence of anti-HIV in all but one of the homosexuals circulating anti-HD. It is unclear whether the HDV is relatively infrequent in homosexuals because it has been only recently introduced in their communities and sufficient time has not elapsed for significant spread, or because rectal intercourse is not an efficient way of transmission.

REFERENCES

1. Rizzetto M., Verme G., Gerin J.L., Purcell R.H.
   Hepatitis delta virus disease.
   In Popper H., Schaffer E. (eds.): Progress in liver Disease,
   Grune and Stratton, New York, VIII:417-413, 1986.

2. Wang K-S, Choo Q-L, Weiner A.J. et al.
   Structure, sequence and expression of the hepatitis delta ( ) viral genome.
   Nature 323:508-513, 1986.

3.  Kos A., Dijkema R., Arnberg A.C. et al.
    The hepatitis delta ( ) virus possesses a circular RNA.
    Nature 323:558-560, 1986.

4.  Weiner A.J., Choo Q-L, Wang K-S et al.
    A single antigenomic open reading frame of the Hepatitis Delta Virus
    encodes the epitope(s) of both Hepatitis Delta Antigen Polypeptides
    p24* and p27*.
    J. Virol. 62:594-599, 1988.

5.  Diener T.O.
    Autonomous and helper-dependent small pathogenic RNAs of plants:
    viroids and satellites.
    Prog. Clin. Biol. Res., Alan R. Liss, New York, vol. 234:3-18, 1987.

6.  Makino s., Chang M.F., Shiek C.K. et al.
    Molecular cloning and sequencing of a human hepatitis delta ( ) virus
    RNA.
    Nature 329:343-346, 1987.

7.  Francki R.I.B.
    Plant virus satellites.
    Ann. Rev. Microbiol. 39:151-174, 1985.

8.  Ponzetto A., Cote P.J., Popper H et al.
    Transmission of the hepatitis B virus-associated agent to the Eastern
    woodchuck.
    Proc. Natl. Acad. Sci. 81:2208-2212, 1984.

9.  Ponzetto A., Hoyer B.H., Popper H et al.
    Titration of infectivity of hepatitis D virus in chimpanzees.
    J. Infect. Dis. 155:72-78, 1987.

10. Branch A.D., Robertson H.D.
    A replication cycle for viroids and other small infectious RNAs.
    Science 223:450-455, 1984.

11. Chen P.J., Kalpana G., Goldberg J. et al.
    Structure and replication of the genome of hepatitis delta virus.
    Proc. Natl. Acad. Sci. 83:8774-8778, 1986.

12. Buitrago B., Hadler S.C., Popper H. et al.
    Epidemiologic aspects of Santa Marta Hepatitis over a 40-year period.
    Hepatology 6 (6):1292-1296, 1986.

13. Ponzetto A., Hoofnagle J.H., Seff L.B.
    Antibody to the hepatitis B virus associated Delta-agent in immune
    serum globulins.
    Gastroenterology 87:1213-1216, 1984.

14. Saracco g., Macagno S., Rosina F., Rizzetto M.
    Serologic markers with fulminant hepatitis in persons positive for
    hepatitis B surface antigen. A worldwide epidemiologic and clinical
    survey.
    Ann. Intern. Med. 108 (3):380-383, 1988.

15. Raimondo G., Craxi A., Longo G et al.
    Delta infection in hepatocellular carcinoma positive for hepatitis B
    surface antigen.
    Ann. Intern. Med. 101:343-344, 1984.

16. Govindarajan S., Hevia F.J., Peters R.L.
    Prevalence of delta antigen/antibody in B-viral-associated
    hepatocellular carcinoma.
    Cancer 53:1692-1694, 1984.

17. Kew M.C., Dusheiko G.M., Hadziyannis S.J., Patterson A.
    Does delta infection play a part in the pathogenesis of hepatitis B
    virus related hepatocellular carcinoma?
    Br. Med. J. 288:1727, 1984.

18. Shobokshi O.A., Serebour F.E.
    Prevalence of delta antigen/antibody in various HBsAg positive patients
    in Saudi Arabia.
    Prog. Clin. Biol. Res. 234:471-475, 1987.

19. Toukan A.U., Abu-el-Rub O.A., Abu-Laban S.AS. et al.
    The epidemiology and clinical outcome of hepatitis D virus (delta)
    infection in Jordan.
    Hepatology 7 (6):1340-1345, 1987.

20. Jacobson I.M., Dienstag J.L., Werner G et al.
    Epidemiology and clinical impact of hepatitis D virus (Delta)
    infection.
    Hepatology 5:188-191, 1984.

21. Hess G., Bienzle U., Slusarczyk J. et al.
    Hepatitis B virus and delta infection in male homosexuals.
    Liver 6 (1):13-16, 1986.

22. Solomon R.E., Kaslow R.A., Phair J.P. et al.
    Human immunodeficiency virus and hepatitis delta virus in homosexuals
    men. A study of four cohorts.
    Ann. Intern. Med. 108 (1):51-54, 1988.

# NATURAL KILLER CELL RESPONSE TO RESPIRATORY

## SYNCYTIAL VIRUS IN THE BALB/C MOUSE MODEL

J.J. Anderson, M. Serin, J. Harrop, S. Amin,
G.L.Toms, and R. Scott
Department of Virology, University of Newcastle
Upon Tyne, The Medical School, Framlington Place
Newcaste Upon Tyne NE2 4HH, United Kingdom

## INTRODUCTION

Infections of the respiratory tract with enveloped single-stranded RNA viruses have become established as common and important causes of infant morbidity. Among these viruses respiratory syncytial (RS) virus is now recognized as the single most common respiratory viral pathogen of infancy (Gardner P.S.,1977). The initial site of virus replication is the respiratory epithelium of the upper nasal passages and the nasopharynx, but in certain instances, the virus spreads to the lower respiratory tract infecting the bronchial and bronchiolar epithelia as well as cells of the pulmonary parenchyma (Aherne et al;1970). RS virus therefore causes a variety of respiratory illnesses, which vary in severity from mild febrile upper respiratory tract disease to severe acute bronchiolitis, bronchitis, pneumonia and croup. The more severe forms of illness however are largely confined to infants less than 12 months of age and regular annual winter epidemics of RS virus infection in this age group have been reported throughout the temperate regions of the world.

The high infant morbidity together with the unique epidemiology of RS virus has led to a dramatic drive to develop a safe effective means of providing prophylaxis. However, to date attempts to produce a vaccine have proved essentially ineffectual. Indeed, in trials initiated in the late 1960s, vaccinees given a formalin inactivated vaccine were found to be predisposed to severe disease when they next encountered the virus (Chin et al.,1969; Kapikian et al.,1969; Kim et al.,1969.).

Recent research which has analysed and compared the neutralizing and glycoprotein-specific serum antibody responses produced by vaccinees in these trials has indicated that disease potentiation may have resulted from a type III hypersensitivity reaction within the lungs of vaccine recipients (Murphy et al., 1986). Studies of natural infection conducted in Buffalo, USA have also suggested that development of severe disease within the 'normal' infant population may be related to development of a type I hypersensitivity response by certain individuals. Correlations have been described between disease severity and acute RS virus specific nasal and serum IgE histamine and leukotriene levels as well as inversely with circulating $CD8^+$ lymphocyte numbers and impaired suppressor function (Welliver et al.,1981; Welliver et al.,1984; Volovitz et al.,1988). Although it is not universally accepted that there may be some pathological involvement in severe RS virus bronchiolitis the above findings as well as the killed vaccine experience have highlighted the need for a better understanding of the role of the immune response in this disease. A detailed understanding of the interaction of the virus or virus subunits with various aspects of the host immune system will be required if a safe effective vaccine is to be developed.

Analysis of the infants' immune response to infection particularly within the lower respiratory tract has been severely restricted due to technical difficulties. However the use of animal models has facilitated analysis of the roles performed by different aspects of the host's immune response in the lung during acute RS virus infection. In this respect the characterization of RS virus infection of Balb/c mice by Dr. G. Taylor and her colleagues has provided an extremely useful model permitting detailed study of the pathogenesis of RS virus infection (Taylor et al., 1984). Research has found that virus specific cytotoxic T lymphocytes (CTL) may be detected in the lungs of RS virus infected Balb/c mice by 5 to 6 days after infection while free virus-specific IgG antibody may not be detected until 10 days post-infection (Taylor et al.,1985; Anderson et al.,1988). The current report compares the kinetics of pulmonary NK cell and interferon responses to A2 strain RS virus infection, with specific CTL and antibody responses in order to assess their potential roles in the recovery process.

THE STUDY

Virus Replication in the Lungs of Balb/c mice was monitored daily during the 12-day period following infection ($1.5 \times 10^5$ PFU/mouse) by assaying the quantity of infectious virus present in individual lung homogenates

212

prepared from groups of three adult mice (mean age 9.8 $\pm$ 1.5 weeks) by plaque assay upon monolayers of Vero cells.

NK cell and CTL activity - Lung and peripheral blood lymphocyte suspensions were prepared by fractionation of homogenates of collagenase digested lung tissue and heparinised blood samples respectively upon discontinuous sodium metrizoate and dextran gradients (density = 1.086 $\pm$ 0.002g/ml). Splenic leucocyte suspensions were prepared from crude spleen cell suspensions following removal of erythrocytes by treatment with tris buffered ammonium chloride solution. Cytotoxic activity in lung and peripheral blood lymphocyte suspensions as well as in splenic leucocyte suspensions was determined at intervals after infection by employing conventional $^{51}$Cr release assays. YAC-1 lymphoma cells were used as target cells in 4 hour NK cell assays while syngeneic Balb/c 3T123 fibroblasts infected with RS virus served as target cells in 8 hour CTL assays.

Local Interferon (IFN) Production was monitored by assaying the level of IFN in tracheo-bronchial lung washes collected from groups of 3 mice between 1 and 12 days after infection. A modified dye uptake assay based upon inhibition of the cytopathic effect produced upon monolayers of murine L929 cells by encephalomyocarditis (EMC) virus was used to establish the IFN content of individual specimens. The IFN content of each specimen was expressed in international units per ml (i.u./ml) relative to a National Institute of Biological Standards and Control (NIBSC) standard mouse alpha-IFN preparation (NIBSC 70/332). The mean lower level of sensitivity of the assay was found to be 5.0i.u. IFN/ml.

Virus-specific antibody production was assayed by indirect membrane immunofluorescent antibody tests (Scott et al., 1976). These tests employed Fc fragment specific goat anti-mouse IgG and IgA fluorecein isothiocyanate conjugates. RS virus specific IgG and IgA titres of sera and pooled concentrated lung lavage samples were determined at intervals throughout the immediate 34 day period following primary experimental infection.

FINDINGS OF THE STUDY

Virus replication - Virus titres in the lungs of experimentally infected mice increased progressively from one day post infection to reach maximum levels (3465 $\pm$ 705 pfu/g tissue) four days after infection and then

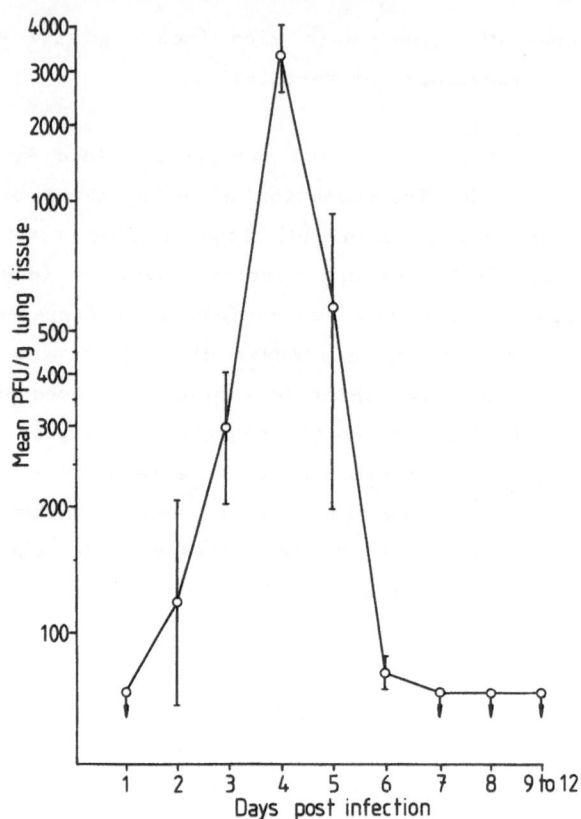

Fig. 1. O: arithmetic mean yield $\pm$ 1 standard
deviation ($\sigma_{n-1}$) from groups of 3 mice

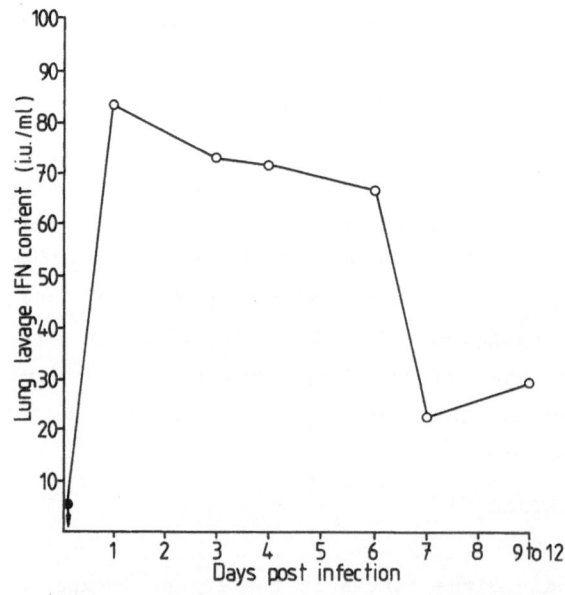

Fig. 2. ●:normal control mice pre-infection (n=3)
:RS virus infected mice (n=3)

declined subsequently until no infectious virus could be recovered from any animal examined after day 7 post-infection (Figure 1).

Innate IFN and NK cell responses to infection

Mean IFN levels in tracheobronchial washes rose rapidly after infection and remained elevated between day 1 and 6 post-infection before declining sharply on day 7 (Figure 2). However; interferon levels in tracheobronchial washes of normal Balb/c mice were consistently found to be below the mean lower level of sensitivity of the bioassay.

Only low basal levels of NK cell activity were found in lung lymphocyte (% lysis = 17.2), splenic leucocyte (% lysis = 2.0) and PBL (% lysis = 7.2) populations collected from normal Balb/c mice. However intransal RS virus infection was found to induce a significant local NK cell response in the lungs of these mice. NK cell activity in lung lymphocyte suspensions was found to rise from pre-infection levels to optimum levels 3 days after infection (% lysis = 42.8) before declining back to basal levels by day 7 (Figure 3b). In control tests "mock infection" of mice intranasally with Hep-2 tissue culture supernatent was shown not to enhance the NK cell activity of pulmonary lymphocytes recovered from mice 3 days after infection, which corresponded with the time of optimal NK cell activity in the lungs of infected animals. Recently, we have found that increasing the initial intranasal virus inoculum given to mice (2.5 fold increase) does not further significantly increase the peak NK cell activity found in lung lymphocytes from mice, but it has been found to enhance the persistence of this response. A plateau of NK cell cytotoxic activity was observed between day 3 and day 5 post infection with decline following over the next 4 days (unpublished observations).

The NK cell response induced by RS virus infection in Balb/c mice appeared to be essentially confined to the lungs of these animals. Intranasal infection did not augment peripheral blood NK cell activity and induced only a slight transient rise in splenic NK cell activity (Figure 3b). Further assessment of the ability of RS virus to activate non-pulmonary populations of NK cells was carried out by comparing the splenic responses induced by intravenous infection with RS virus and influenza virus (FLU A/PR8/34). Although in these cicumstances influenza A virus induced a rapid marked increase of splenic NK cell activity, RS virus again only marginally enhanced splenic NK cell activity (Figure 3a)

A virus specific class I MHC restricted CTL response to infection similar to that described by Taylor and colleages(1985), was found to develop following the early NK cell and IFN responses produced in the lungs of our animals (Figure 4a). Cytotoxic T cell activity was

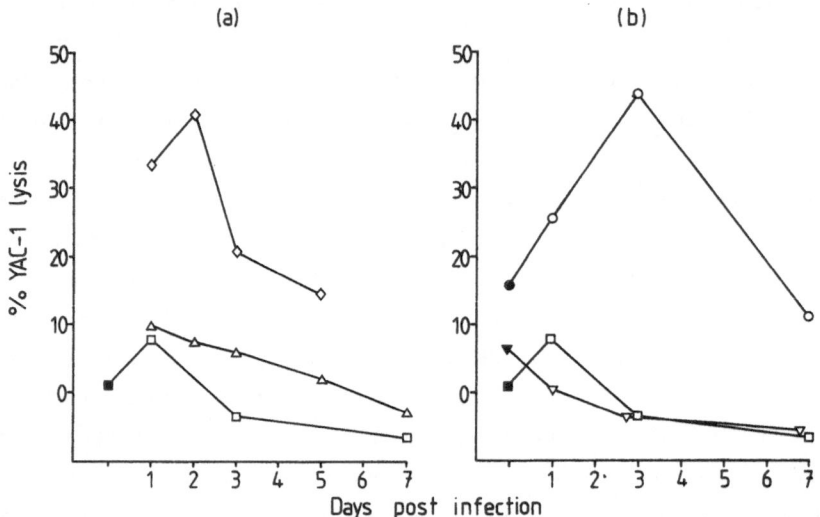

Fig. 3 (a) □ Splenocytes from mice infected intranasally with RS virus
        △ Splenocytes from mice infected intravenously with RS virus,
        ◇ Splenocytes from mice infected intravenously with influenza
          A virus
     (b) ○ Lung 'lymphocytes from mice infected intranasally with
          RS virus
        ▽ PBL from mice infected intranasally with RS virus
        ■ ● ▼ Splenocytes, lung lymphocytes and PBL from normal
          control mice

first detectable in lymphocytes recovered from lung tissue 6 days after infection, one day after the initial decline of virus shedding. Peak CTL activity was recorded on day 7 post-infection (% lysis = 49.0). This response subsequently declined, slowly at first between day 7 and day 9 but then more rapidly to day 12.

A virus-specific serum IgG antibody response was detectable only 3 days after infection (Figure 4b). Although IgG was the first class of immunoglobulin to be found free in lung lavages it was not detected until 10 days after infection. Free RS virus-specific IgA could not be detected in lung lavages until day 24 and not in serum until day 34.

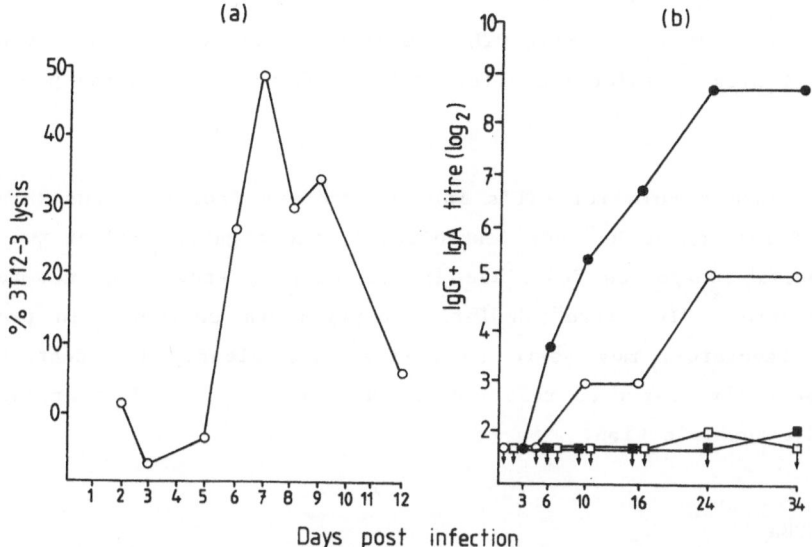

Fig. 4.(a)  CTL lysis determined by lung lymphocytes from RS virus infected mice
    (b)  ● Mean virus specific IgG titre of sera; O virus specific IgG titre of lung lavages
         ■ Mean virus specific IgA titre of sera; □ virus specific IgA titre of lung lavages

COMMENTS

Our research has shown that IFN and NK cell responses are elicited rapidly within the lungs of Balb/c mice infected with $A_2$ strain of RS virus. Both responses preceeding the development of the specific immune responses. Our findings indicate that virus replication within the lungs continued to increase between day 1 and day 4 post-infection and began to decline on day 5 , one day prior to the detection of RS virus specific CTL's in the lung. Decline of virus was also 4 days prior to the first appearance of free virus specific IgG antibody in lung lavages.

Elevated levels of both IFN and NK cells were detectable, in our study, prior to the initial decline of virus in the lung. Previous studies in cotton rats have implicated "NK-like" cells as possible effectors of recovery from infection (Sun et al.,1983; Kumagai et al., 1985). These studies, however, were unable to exclude the involvement of CTL's in the recovery process. Evidence for the involvement of CTLs in recovery has been provided by Dr. Cannon and his colleagues (1987) using the Balb/c mouse model. Adoptive transfer of "in vitro" stimulated RS virus specific CTL's to immuno deficient recipients within the first week of infection effectively resolved an otherwise persistent infection.

Although a role for CTL's in the recovery from RS virus infection in the Balb/c mouse has been established, the significance of the early non-specific responses described in the current study requires further investigation. "In vitro" depletion experiments currently in progress in our laboratory may serve to define more clearly the contribution of those early responses relative to that provided by CTL's in recovery from RS virus infection.

REFERENCES

AHERNE W.A., BIRD T., COURT S.D.M., GARDNER P.S., McQUILLIN J., 1970
Pathological changes in virus infections of the lower respiratory tract in children
J. Clin. Pathol., 23: 7-18

ANDERSON J.J., TOMS G.L., SCOTT R., 1986
Local immune responses in the lungs of Balb/c mice infected with respiratory syncytial virus
Sixth International Congress of Immunology Abstract 1.42.22p51

CANNON M.J., STOTT E.J., TAYLOR G., ASKONAS B.A. 1987
Clearance of persistent respiratory syncytical virus in immunodeficient mice following transfer of primed T-cells
Immunology. 62; 133-138

CHIN J., MAGOFFIN R.L., SHEAPER L.A., SCHIEBLE J.H., LENNETTE E.H., 1969

Field evaluation of respiratory syncytial virus vaccine and tri-valent influenza virus vaccine in a pediatric population
Am.J.Epidemiol., 89: 449-463


GARDNER P.S., 1977
Deaths associated with respiratory viral infections in childhood
Rev. Fr. Mal. Resp., 5: 503-514


KAPIKIAN A.Z., MITCHELL R.H., CHANOCK R.M., SCHVEDOFF R.A., STEWART C.E. 1969
An epidemiologic study of altered clinical reactivity to respiratory syncytial (RS) virus infection in children previously vaccinated with an inactivated RS vaccine
Am. J. Epidemiol., 89: 401-421


KIM H.W. CANCHOLA J.G., BRANDT C.D., PYLES G., CHANOCK R.M., JENSEN K., PARROTT R.H., 1969
Respiratory syncytial virus disease in infants despite prior admini-stration of antigenic inactivated vaccine
Am. J. Epidemiol., 89: 422-434


KUMAGAI T., WOND D.T., OGRA P.L., 1985
Development of cell-mediated cytotoxic activity in the respiratory tract after experimental infection with respiratory syncytial virus
Clin. Exp. Immunol., 61: 351-359


MURPHY B.R., PRINCE G.A., WALSH E.E., KIM H.W., PARROTT R.H., HEMMING V.G., RODRIGUEZ W.J., CHANNOCK R.M. 1986
Dissociation between serum neutralising and glycoprotein antibody res-ponses of infants and children who received inactivated respiratory syncytical virus vaccine
J. Clin. Microbiol., 24: 197-202


SCOTT R., DE LANDAZURI M.O., GARDNER P.S., OWEN J.J.T., 1976
Detection of antibody to respiratory syncytial virus by membrane fluores-cence
Clin. Exp. Immunol., 26: 78-85


SUN C.S., WYDE P.R., KNIGHT V., 1983

Correlation of cytotoxic activity in the lungs to recovery of normal and gamma-irradiated cotton rats from respiratory syncytial virus infection
Am. Rev. Respir. Dis., 128: 668-672


TAYLOR G., STOTT E.J., HAYLE A.J., 1985
Cytotoxic lymphocytes in the lungs of mice infected with respiratory syncytial virus
J. Gen. Virol., 66: 2533-2538


TAYLOR G., STOTT E.J., HUGHES M., COLLINS A.P., 1984
Respiratory syncytial virus infection in mice
Infect. Immun., 43: 649-655


VOLOVITZ B., FADEN H., OGRA P.L.
Release of leukotriene C4 in respiratory tract during acute viral infection
J. Pediatr., 112: 218-222


WELLIVER R.C., KAUL T.N., SUN M., OGRA P.L., 1984
Defective regulation of immune responses in respiratory syncytial virus infection
J. Immunol., 133: 1925-1930


WELLIVER R.C., WONG D.T., SUN M., MIDDLETON T., VAUGHAN R.S., OGRA P.L., 1981
The development of respiratory syncytial virus-specific IgE and the release of histamine in nasopharyngeal secretions after infection
New. Engl. J. Med., 305: 841-846

# HUMORAL RESPONSE TO INACTIVATED POLIOVACCINE IN ANTI-HIV POSITIVE INFANTS

Maria Bardare, Maria Barbi*,  Anna Plebani, Gabriele Ferraris,
and Alessandro Remo Zanetti*

*1st Paediatric Clinic and Virology DPT
University of Milan, Italy
Via Commenda 9 and via Pascal 38  Milan, Italy

## INTRODUCTION

The growing number of paediatric immune deficiency virus ( HIV )
infections has raised some problems about the appropriate policies for
vaccinations. Some aspects must especially be pointed out: 1) the possible
severe side-effects as well as disseminated infection caused by live vaccines
in immunodeficient subjects; 2) the excessive stimulation of T Lymphocytes
by both live and inactivated vaccines, with the possibility to accelerate
the course of HIV infection; 3) the capacity of HIV-infected children to
respond properly to immunisation; 4) the course of vaccine-preventable-
diseases in HIV-infected children.

As for the first point, it is difficult to assess the immunodeficiency
of the HIV-infected children and, moreover, it is not easy to distinguish
in the first 15 months of life between healthy and affected seropositive
infants. The danger of live vaccines seems prooved only for BCG[1,2] and
small-pox vaccine[3],while there are some reports on the safety of measles
and oral polio vaccine[4]. Anyway, the possibility of spreading the infection
to the immunodeficient cohabiting people must be considered[5].

The possibility that activation of CD4 cells by multiple immunisations
could result in the passage from the state of asymptomatic to that of
symptomatic infection has been frequently envisaged[6], but convincing
evidence is still lacking.

The response to vaccination is inversely correlated to the stage of
HIV infection, being porer in full-blown AIDS[7]. It has been reported a scarce
primary as well as secondary response to pneumococcal vaccine and tetanus
toxoid[8,9] but, on the other hand, the majority of HIV-infected children
respond to measles vaccine and to polio vaccine[4]; the diphteria sero=
conversion seems unsatisfactory[10].

In considering the risk-benefit ratio one must take into account the
often unfavourable course in HIV-infected patients of some diseases, such
as pneumococcal pneumonia, measles and tuberculosis[11].

The scarce and often conflicting experiences justify the different
suggestions on the vaccination practice: while there is general agreement

in avoiding BCG ( and small-pox too )[4,7,11,12,13], the use of the other live vaccines is controversial. The recommendation of employing live vaccines in every case of HIV infection [12,13] is balanced by the recommendation of employing live vaccines in asymptomatic but not in symptomatic subjects[4].

A good alternative to oral polio vaccine (OPV) is thought to be the inactivated polio vaccine (IPV), which gives comparable antibody titres.

Furthermore, pnuemococcal and Hemophilus vaccines are advisable after 2 yrs of age, influentiae vaccine after 6 mo and DTP in any case at the usual schedule time[4].

The existing controversies have prompted us to vaccinate with IPV some infants born to HIV seropositive mothers, monitoring the antibody response and the behaviour of HIV infection.

## MATERIALS AND METHODS

Twenty five infants, age range 5-14 mo (mean 8,9 mo) born to HIV positive mothers, entered the trial. All the patients were anti HIV seropositive both by EIA (Abbott Lab.) and by Western Blot (Sorin) assays. Three doses of inactivated polio vaccine containing respectively 40-8-32 DU of poliovirus serotypes 1,2,3 (IPV - Institut Merieux France) were administred one month apart.

Serum samples were collected in any case at the time of the first dose and one month later after the second administration; in 8 cases further samples were collected at different times after the end of vaccine cycle.

In each sample the titres of neutralizing poliovirus antibodies were evaluated by serial dilutions against 100 $DC_{50}$ of polio strains 1,2 and 3, incubated 1 h. at room temperature and then with Vero cells for 5 days at 37°C.

The response to the vaccine was considered positive when it involved fourfold increase of the antibody titres. Specific antibodies to the core and envelope (env) viral proteins were tested by both Western Blot and a competitive immune assay using recombinant antigens (ABBOTT Lab.).

HIV-antigen was detected by a solid phase, sandwich-type enzyme immune assay (ABBOTT Lab.) using human and rabbit polyclonal anti-HIV IgG as capture and probe antibodies, respectively.

Goat antibody to rabbit IgG conjugated with horse radish peroxidase was used to identify a positive reaction.

## RESULTS

Fifteen out of 25 children responded to 3 polio strains, 10 after 2 doses and 5 after 3 doses. Four of these 15 cases were HIV infected, as showed by WB, and in all of them p24 antigenemia was negative. Five children, all negative for antigenemia, responded only to 2 strains, after 2 administrations: one of them was positive by WB.

One infant, positive both for antigenemia and WB, yielded positive results to only 1 polio strain, after 3 doses of vaccine.

Four cases gave no response even with 3 doses of vaccine : all these infants were symptomatic and immunodeficient.

A booster dose was administred in 3 cases, resulting in an optimal response in two (from 1:4 to 1:256), unchanged in the third.

Data on the infected children are summarized in Table 1.

Table 1.  Main features of infected children and antibody response

| Case | sex | age | class | HIV-Ag | antibody response to poliovirus type | | |
|------|-----|-----|-------|--------|------|------|------|
|      |     |     |       |        | 1 | 2 | 3 |
| M.P.  | M  | 11 m | P2-D1  | +   | – | – | – |
| F.B.  | F  | 13 m | P1-B   | +   | – | – | + |
| A.B.  | M  | 14 m | P1-A   | –   | + | + | + |
| S.B.  | F  | 7 m  | P2-C   | –   | + | + | + |
| A.S.  | F  | 13 m | P2-F   | –   | + | + | + |
| I.P.  | M  | 6 m  | P1-C   | –   | + | + | + |
| V.P.  | M  | 7 m  | P1-B   | –   | + | + | – |
| A.B.  | M  | 14 m | P1-B   | ±   | + | + | + |
| S.L.  | M. | 6 m  | P2-D1  | +   | – | – | – |
| J.P.  | F  | 5 m  | P2-BD1 | +   | – | – | – |
| G.DR. | M  | 7 m  | P2-D1  | –   | – | – | – |

CONCLUSIONS

In agreement with other experiences, we have noticed an inverse relationship between stage of disease and immune response to vaccine, which was
absent in the most compromised children, who were for the majority antigen-
and WB-positive. Four HIV-infected children responded to 3 strains and 1
to 2 strains, but in two cases 3 doses have been necessary to elicit a
response.

In no case an acceleration of the course of disease was caused by IPV.

It can be concluded that IPV is safe and works, provided it is employed
in asymptomatic HIV-infected infants or in the very early stages of the
disease.

REFERENCES

1.  Disseminated Mycobacterium bovis infection from BCG vaccination of a
        patient with acquired immunodeficiency syndrome, MMWR, 34 : 227 (1986)
2.  J. B. Nousbaum,M. Garre,J. M. Boles, Deux manifestations inhabituelles
        d'un infection par le virus LAV-HIV : BCGite et varicelle pulmonaire,
        Rev. Pneumol. Clin., 42 : 310 (1986)
3.  R. R. Redfield,D. C. Wright,T. S. Jones,C. Brown,D. S. Burke,Disseminated
        vaccinia in a military recruit with human immunodeficiency virus,
        N. Engl. J. Med,, 316 : 673 (1987)
4.  CDC, Immunisation of children infected with human T-lymphotropic virus
        type III/lymphadenopaty-associated virus, MMWR, 35 : 595 (1986)
5.  CDC, Paralitic poliomyelitis-United States 1982 and 1983, MMWR, 33 :
        635 (1984)
6.  D. Zagury,J. Bernard,R. Leonard,R. Cheynier,M. Feldman,P. S. Sarin, R.
        C. Gallo, Long-term cultures of HTLV III-infected T cells : a model
        of cytopathology of T-cell depletion in AIDS, Science, 231 : 850
        (1986)
7.  N. A. Halsey, D. A. Henderson, HIV infection and immunisation against

    other agents, N. Engl. J. Med., 316 : 683 (1987)

8.  A. J. Amman,G. Schiffman,D. Abrams,P. Volberding,J. Ziegler,M. Conant, B-cell immunodeficiency in acquired immune deficiency syndrome, Jama, 251 : 1447 (1984)

9.  L. J. Bernstein,H. D. Ochs,R. J. Wedgwood,A. Rubinstein, Defective humoral immunity in pediatric acquired immune deficiency syndrome, J. Pediatr., 107 : 352 (1985)

10.  W. Borkowsky,C. J. Steele,S. Grubman,T. Moore,P. La Russa,K. Krasinski, Antibody response to bacterial toxoids in children infected with human immunodeficiency virus, J. Pediat., 110 : 563 (1987)

11.  C. F. Von Reyn,C. J. Clement,J. M. Mann, Human immunodeficiency virus infection and routine childhood immunisation, Lancet, II : 669 (1987)

12.  A. G. M. Campbell, Immunisation for the immunosuppressed child, Arch. Dis. Child., 63 : 113 (1988)

13.  Joint WHO - Unicef statement on immunisation and AIDS, Wkly Epidem. Rec., 62 : 53 (1987)

DIFFERENT EXPRESSION OF HLA CLASS I ANTIGENS IN LIVER OF CHILDREN WITH

CHRONIC HEPATITIS B, EVALUATED BY IMMUNOHISTOCHEMICAL METHOD

Calzia R., Marazzi M.G.*, Campelli A.*, Piscopo R.*,
Dorati M., Puppo F., Borgiani L.**, and  Canepa M.**

Dept. of Internal Medicine and 2nd Clinic Infectious Disease*
University of Genoa; Dept. of Pathology, Galliera Hospital
Genova**, Genova  Italy

SUMMARY

Studies on the quantitative expression of the Major Histocompatibility
Complex (MCH) in hepatocytes chronically infected by Hepatitis  B  Virus
(HBV) report  that an increased expression  of these antigens  could  be
related to a good immunological response.
In  the  present  work we analyze the expression of the  MCH  antigens  in
cryostatic  sections of liver biopsies taken from subjects (19  children)
with various forms of HBsAg positive  chronic hepatitis.
A high  expression  of HLA class I antigens  and  a  high  degree  of
hepatocyte  necrosis  was  evident in Chronic Active Hepatitis  (CAH)  and
Chronic  Lobular  Hepatitis  (CLH).   On  the  contrary,   subjects  with
histological diagnosis of Chronic Persistent Hepatitis (CPH) showed  a low
expression  of such antigens.  There was however,  the difference that  in
subjects  with high hepatic cytolysis and high expression of HLA  class  I
antigens,   serum HBV-DNA was  clearly present in almost all the cases with
CAH, but not detectable in all  cases with CLH.
The expression of HLA class II antigens and of Beta2 microglobulin was the
same in all 19 cases. All  cases with HBV-DNA positivity with high class I
antigen  expression  had active hepatitis which seems to suggest that  all
attemps  at viral clearance on the part of the immune system have been  in
vain.  We hope our  paper  will be  an additional parameter for evaluating
the course of hepatitis during Interferon treatment.

INTRODUCTION

The  inability to obtain the viral clearance which occurs in  chronic
hepatitis  B is probably due to immunological defects which,  even  today,
are not clearly understood.  Recently some Authors have undertaken studies
on  the quantitative expression of HLA class I and II antigens and Beta  2
microglobulin in hepatocytes chronically infected by HBV.
In  1982 Montaño et al. (5),  by means of  an immunofluorescence  study,

225

found in those patients with markers of high viral replication a weaker expression of HLA class I antigens than in those with a low viral replication.

In 1985 Thomas H.C. and Pignatelli M., by an immunofluorescence method (9), observed a high HLA-I expression on hepatocytes of a Chimpanzee affected by acute experimental Hepatitis B. During Interferon therapy they noticed in biopsies of subjects affected by chronic hepatitis B, a pattern similar to that which appeared in experimental hepatitis in the Chimpanzee.

De Vos et al, in 1986, described HLA class I antigen expression on hepatocyte membranes using electron transmission microscope (3).

In another work in 1986 Negafuchi and Scheuer (6) with an immunohistochemical technique, studied Beta 2 microglobulin expression on the hepatocyte membranes of subjects affected by acute and chronic hepatitis B and other forms of liver disease.

The quantitative evaluation of this antigen was related to that of HLA class I antigens.

The interesting interpretations suggested by these authors on this immunological aspect of host response versus HBV, led us to study the possibility of there being an immunity impairment in children affected by chronic hepatitis B.

The aim of the present work was to study, by an immunohistochemical technique in liver biopsies of children affected by chronic HBsAg positive hepatitis, the quantitative expression of Major Histocompatibility Complex (MHC) products (HLA-I and-II antigens, and B2-microglobulin) and its possible correlation to viral replication, (evaluated in liver biopsies and in serum) and to the clinical state of the subjects.

Moreover this work was performed as a preliminary study to treatment with interferon (IFN) on a selected group of children.

PATIENTS AND METHODS

The case-list includes 19 children, 3 to 14 years old, affected by chronic HBsAg hepatitis, and 2 children, as controls, affected by chronic HBsAg negative hepatitis.

The histological diagnosis taken from the liver biopsies of these subjects were: Chronic Active Hepatitis (CAH) 7 cases, Chronic Persistent Hepatitis (CPH) 3 cases, Chronic Lobular Hepatitis (CLH) and Chronic Minimal Hepatitis (CMH), respectively 6 and 3 cases.

All the subjects had been evaluated not only with the clinical and biohumoral parameters, but also taking into consideration the serological markers of viral replication: HBeAg/HBeAb (RIA Abbott) and HBV-DNA (spot Hybridization technique using a P32 Radiolabelled Probe).

Of the 2 HBsAg negative children, evaluated as controls, one was affected by hepatosplenomegaly following surgical treatment of Wilm's tumor and the other by hepatocellular carcinoma.

Liver biopsies were performed by the Menghini method.

Every liver specimen was both cryostatated and paraffin embedded: on the cryostatic section we evaluated the expression of the MHC antigens; on the paraffin embedded samples we studied the histology of the liver disease and the viral antigens expression.

The peroxidase anti-peroxidase (PAP) method was employed to detect the viral antigens: HBsAg and HBcAg. An indirect method was performed for detecting Hepatitis Delta Virus (HDV).

5 u sections, fixed in formalin and paraffin embedded were deparaffinated, transferred in 1% $H_2O$ in methanol for 30 m' to remove the endogenous peroxidase, and then were digested by an enzyme-solution composed of 0.05% trypsine with 0.05% Ca chlorure at pH 7,8, at 37° C for 10 minutes. Sections were washed in PBS and covered with Swine's non immune serum at 1:20 dilution, for ten minutes. We used the following primary antibodies: anti-HBs (Behring) 1:100 diluted, anti-HBc (Dakopatts) 1:500 diluted; anti-HDAg (gift from Dr. M. Rizzetto) 1: 3000 diluted. Staining was performed by a diaminobenzidina solution (DAB) prepared just before use. Counterstaining was performed with haematoxylin. Liver biopsies for the evaluation of MCH products by an immunohistochemical method, to be treated with monoclonal antibodies to HLA class I and II and B2 microglobulin, were frozen in liquid nitrogen and stored at -70° C. 5 $\mu$ sections were air dried overnight, then fixed in aceton for 10' at room temperature and fixed again in chloroform for 10'.

Sections were dried again, rehydrated in TBS and than the APAAP procedure was employed. Sections were left in normal rabbit serum 1: 50 diluted (Dakopatts) for 30' at the following dilutions:

anti-HLA class I W6/32 (Dr. Ferrone-Dakopatts) 1: 100; anti-HLA class II Q5/13 (Dr. Ferrone-Dakopatts) 1: 10; anti-Beta 2 microglobulin NAMB I 1: 200. Sections were covered with rabbit anti-mouse immunoglobulins (Dakopatts) 1: 25 diluted for 30 m'.

Then were incubated with APAAP complex (Dakopatts) 1: 25 diluted for 30 m'. Substrate for staining included: New Fuchsin to 5%, Naphtol AS-B1 phosphate with Levamisol 1 M (400 ul).

The slides were evaluated, in double blind, by a Zeiss, light microscope with built-in camera to take photomicrographs. The film used was Kodak Ektachrome professional ISO 160 ISO.


RESULTS

All the sections evaluated for HLA class I antigens showed a clearly positive staining on sinusoidal living cells, on mononuclear cells in the portal and periportal inflammatory infiltrates, and on bile-duct epithelial cells.

A positive staining was also evident on the hepatocyte membranes in sections of subjects displaying some amounts of HLA class I antigens.

The positive staining appeared only on wedge shaped groups of the lobule cells like a slight thin line, or in a granular fashion along the sides of the liver cells (Fig. 1 and Fig. 2).

When the sections were treated with monoclonal antibodies to HLA class II antigens, a positive staining on sinusoidal cells (Kupffer and endothelial cells) was seen. A high positivity was also seen on inflammatory portal infiltrates and on intralobular mononuclear cells (Fig. 3).

When the monoclonal antibody to B2-microglobulin was used, staining appeared on the whole membrane of each hepatocyte in a "honeycomb" pattern. Obviously, the mononuclear inflammatory and sinusoidal cells were positive, just as when the antibodies to both class I and class II HLA antigens were used (Fig. 4).

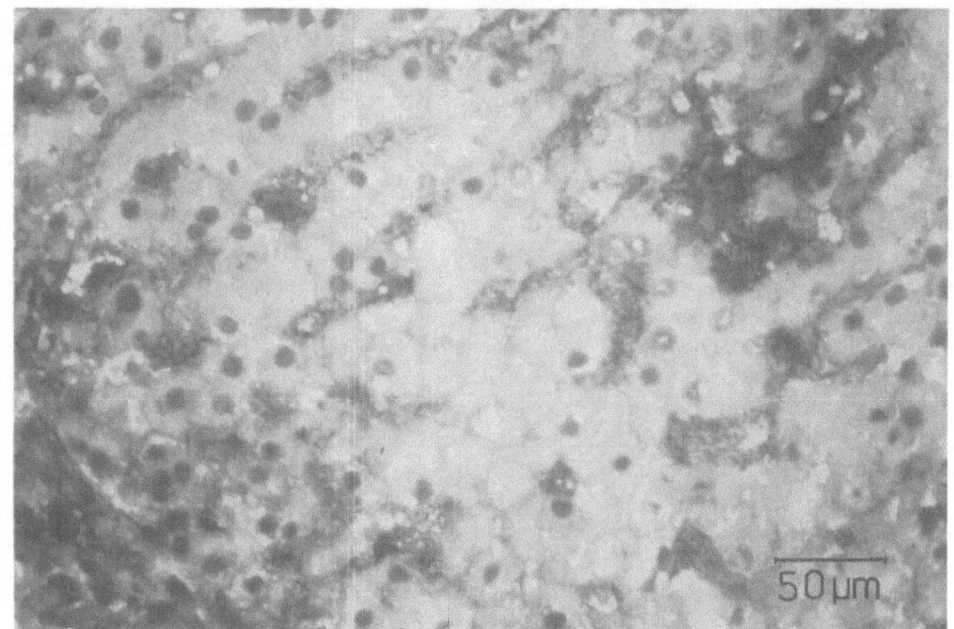

Fig. 1. A part of hepatic lobule, near an infiltrated portal space (bottom left) with hepatocyte membranes displaying HLA class I products in a high percentage of cells.

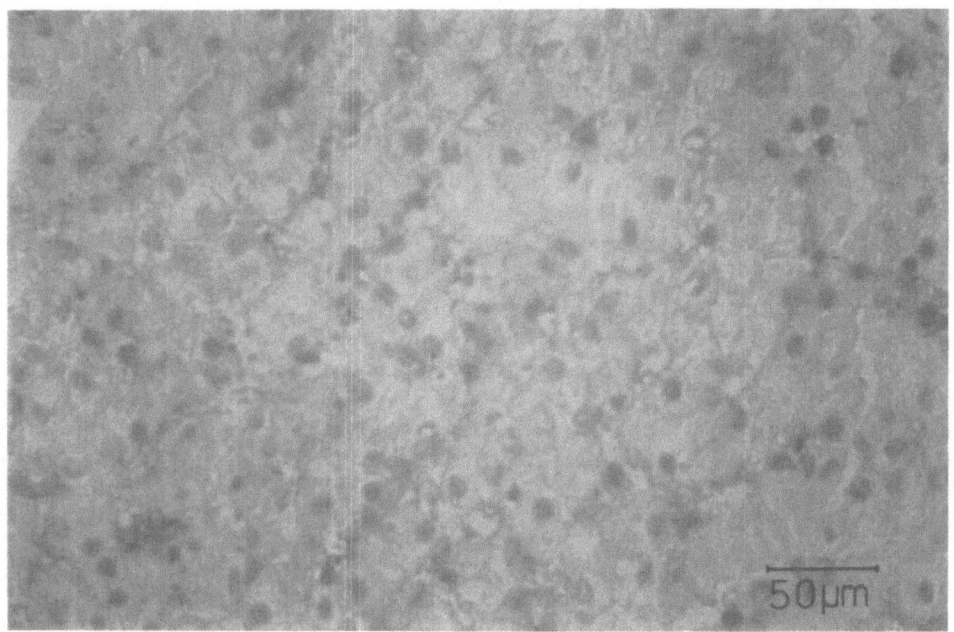

Fig. 2. A case without staining for HLA-I on the hepatocyte membranes. The visible staining refers to endothelial or/and Kupffer cells, and infiltrating cells.

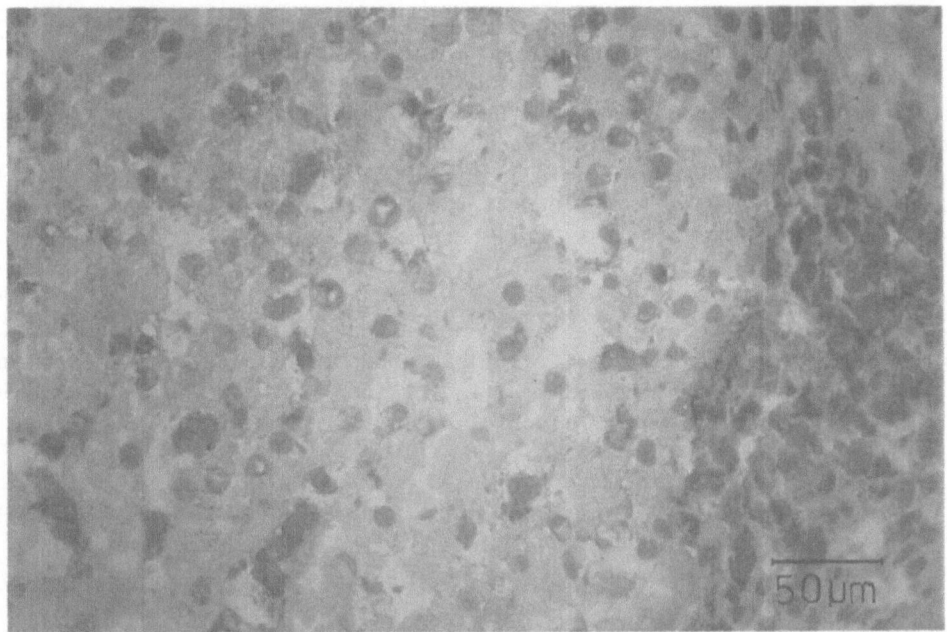

Fig. 3. A section of liver tissue processed and stained for HLA class II antigens. The hepatocytes are negative. The only stained cells are sinusoidal and endothelial cells, Kupffer cells and immunocytes infiltrating into the portal tract (right) and the hepatic lobule.

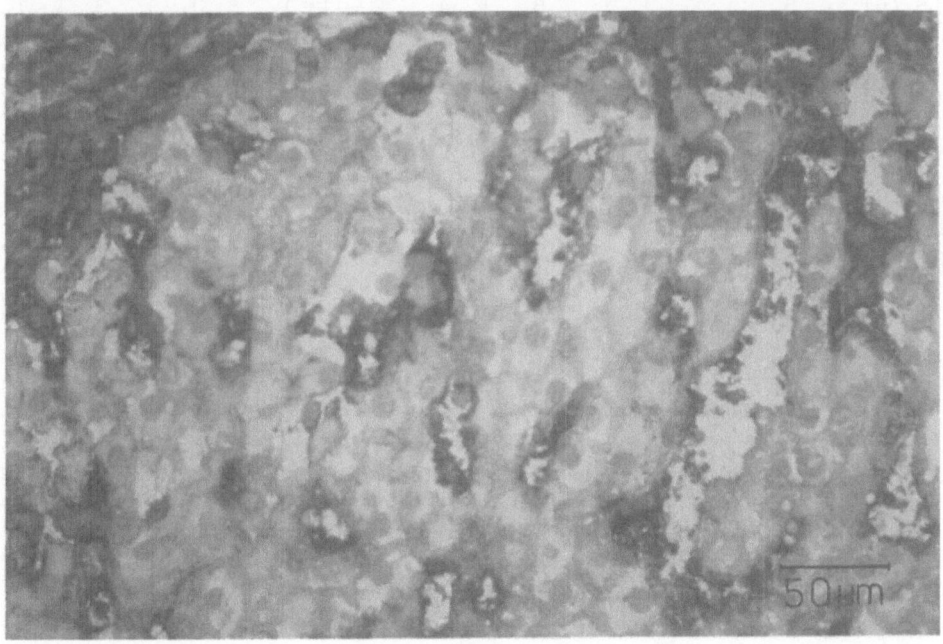

Fig. 4. With its positive staining the whole membrane of each hepatocyte clearly expresses Beta 2-microglobulin. A "honeycomb" pattern is the result. The endothelial cells of the sinusoids, Kupffer cells and infiltrating cells are also heavily stained.

Because of the homogeneity of results obtained with antibodies to both HLA class II and B2 microglobulin, we evaluated only the expression of HLA class I antigens on hepatocyte membranes and the distribution of intralobular positive liver cells. Nevertheless we did not take into consideration the expression of HLA class I antigens on infiltrating cells and sinusoidal cells. As is well known, according to Montaño et al (1982), Thomas H.C. and Pignatelli M. (1985), Nagafuchi Y. and Scheuer P. (1986), Van der Oord J.J. and Desmet V.J. (1984), in this type of study it is of no interest to evaluate HLA I expression in inflammatory infiltrates that are always positive. When more than 20% of the cells on field were stained, the expression of HLA antigens was considered high, and below 20% the expression was considered mean and/or low. We correlated these data with histological and virological pictures of the subjects. We found correlation between the severity of the histological picture and HLA class I antigens expression in the liver (Tab. 1).

TABLE 1

EXPRESSION OF HLA CLASS I ANTIGENS ON LIVER CELLS OF 19 CHILDREN WITH CHRONIC HBV CHRONIC HBV HEPATITIS

| | LIVER | ! | SERUM | | ! | LIVER | | |
|---|---|---|---|---|---|---|---|---|
| | HLA Class I ! high low ! | HBV-DNA ! pos neg ! | HBeAg ! pos neg ! | HBsAg ! ++ + ! | HBcAg ! +++ + ! | HDAg |
| CAH (7) | 6 1 ! | 5 2 ! | 4 3 ! | 3 4 ! | 3 1 ! | 1 |
| CLH (6) | 5 1 ! | 1 5 ! | 4 2 ! | 1 5 ! | 3 3 ! | 1 |
| | ! | | ! | | ! | ! | | ! |
| CPH (3) | — 3 ! | 3 — ! | 3 — ! | 2 1 ! | 3 — ! | — |
| CMH (3) | — 3 ! | 1 2 ! | 1 2 ! | 1 2 ! | 2 1 ! | — |

The table summarize the results of the expression of the HLA products on the hepatoccyte membranes and compares them with the markers of active viral replication.

High HLA class I display was seen in patients with Chronic Active Hepatitis (CAH) and Chronic Lobular Hepatitis (CLH). The differences between the two groups were not only histological but also consisted in a different behaviour of the replicative viral markers. The subjects with CAH showed a high level of viral replication (HBV-DNA in serum and high percentage of focal core in the liver).
However, most cases with CLH had no detectable HBV-DNA in serum, and HBcAg in the liver was present only in half of them.
Low HLA-I expression was seen in all cases of Chronic Minimal Hepatitis (CMH) and Chronic Persistent Hepatitis (CPH). Four of these subjects exhibited markers of active viral replication (HBV-DNA+++, HbeAg+ and a strong expression of focal HBcAg). It must be pointed out that some of these patients had previously undergone immunosuppressive treatment when the disease had entered particularly severe phases.
Worthy of note are some peculiar aspects of the viral antigens expression, in a child affected by CPH. Membranous HBsAg was evident in almost all the hepatocytes, and cytoplasmic HBcAg was seen in more than 50% (Fig. 5).

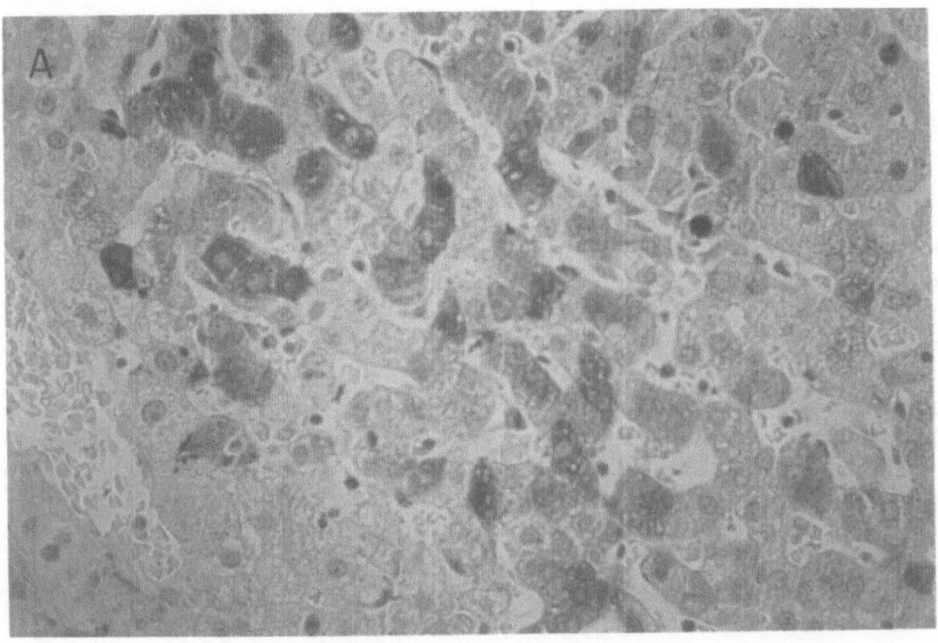

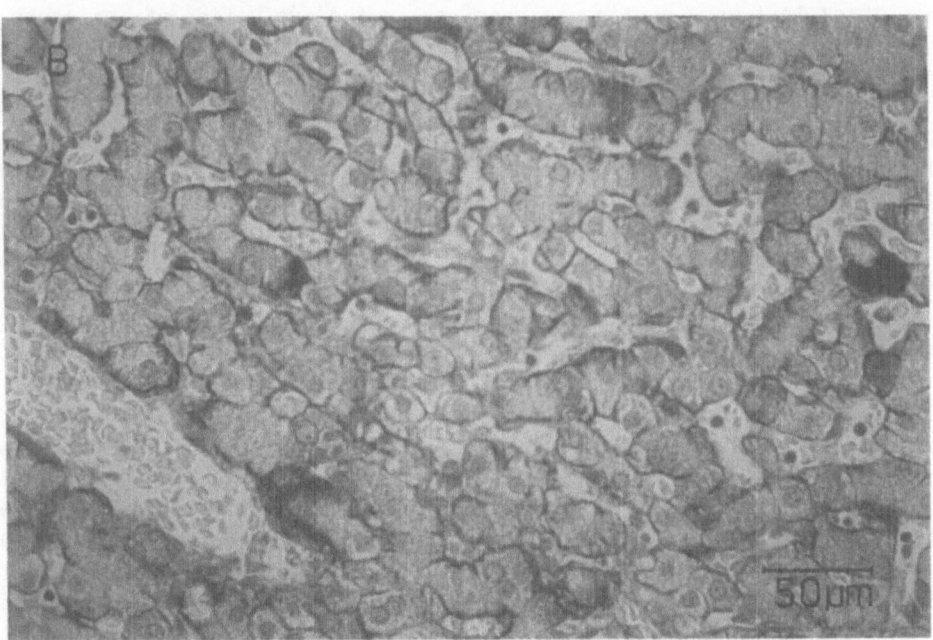

Fig. 5. The photomicrographs show two sections of the same lobular area taken from a single liver sample. The cuts were made very near to each other. A: Cytoplasmic spotty nuclear HBcAg. B: membranous HBsAg in almost all hepatocytes.

As in the other cases with CPH, a low HLA class I antigen display was found.

DISCUSSION

In our patients, high HLA class I antigens expression was present only in those biopsies with histological pictures of active disease with necrosis (CAH e CLH).
This type of cellular response could be the result of a series of attempts on the part of host cell-mediated immunity, to obtain viral clearance.
We tried to evaluate if there were any differences between these two groups. In the case of CAH, the attempts of the host immune response seem to be ineffective because of the persistence of a severe liver injury associated with high levels of HBV replication. In fact most cases of CAH (5/7) showed detectable HBV-DNA and HBeAg in serum, and focal HBcAg in a great number of hepatocytes.
The pattern of high HLA class I antigens display in CLH could reflect viral clearance which is imminent. Almost all these patients did have undetectable HBV-DNA in serum and half of them had no HBcAg in liver.
The follow-up could make these results and their prognostic significance clearer.
Low expression (in some cases less than 1%) on liver cells was seen in all the biospies with histological pictures of CPH and CMH (and in only 2 cases of CAH). In the case of CMH this pattern of HLA class I antigen expression seems to be related to a stable remission of hepatitis.
We think that more caution is required in the evaluation of the cases of CPH because of the persistence of active viral replication in all the subjects (HBV-DNA and HBeAg in serum and HBcAg in liver).
One of the subjects of this group, the child M.M., is worthy of particular mention. He was affected by CPH and presented a peculiar immunohistochemical picture, which differed from the usual one found in children. It differed in that the prevaling cytoplasmic HbcAg, was evident in 50% of hepatocytes with an associated expression of spotty nuclear HBcAg, and membranous HBsAg, displayed by 90% of liver cells (Fig.5A, 5B). The low expression of HLA class I antigens in this subject may indicate an impaired immune response versus HBV, perhaps caused by defective Interferon production. This pattern could indicate a negative outcome of hepatitis.
Low expression of HLA class I antigens could be a results of predisposing genetic factors, as previously suggested by Montaño et al. (1982), or of a defective association of HLA class I antigens with those of HBV on the hepatocyte membranes, thus impairing T-cells lysis function. We would like to examine another case regarding a child observed during recovery after HDV superinfection. The histological pattern was that of CLH. Nevertheless, serological levels of HBV replication (HBV-DNA and HBeAg) and the expression of both HBV and HDV antigens in liver, were completely negative.
There was a high expression of HLA class I antigens on the hepatocyte membranes. This aspect reflects the efficiency of the immune response in obtaining viral clearance. Most cases of chronic hepatitis in childhood show focal HBcAg and cytoplasmic HBsAg and the expression of the latter is less evident than that of HBcAg.

In their studies the Authors compared the HLA class I antigens display with viral antigens expression in subjects with HBV chronic hepatitis.

The increased of histocompatibility antigens expression on the hepatocytes was associated with decreased viral marker expression on the liver cells (Cytoplasmic HBsAg and absence of HBcAg). On the other hand a low HLA class I antigens expression seems to correspond to a membranous HBsAg and a focal HBcAg. It is not clear whether this different expression of the MHC products really has any effect on the efficacy of the cytotoxic mechanism in the lysis of infected hepatocytes. As can be seen from Tab. 1 our case-list of children did not provide us with any evidence of correlation between MHC and viral markers in liver (HBcAg and HBsAg).

For instance a case of a child affected by CPH with high expression of cytoplasmic HBsAg, had no markers of viral replication (HBV-DNA negative, anti HBe+, HBcAg negative) associated to a low display of HLA class I antigens. This subject could be an example of a tendency of chronic hepatitis towards recovery.

Neither the data regarding HLA class II antigens nor those relative to the B2 microglobulin can be discussed because of the homogeneity of our results. We found no expression of HLA class II antigens on the hepatocye membranes, and this is in agreement with Pignatelli et al. (1986) and disagreement with Van den Oord et al. (1986).

This disagreement could be a result of the fact that we used a light instead of a trasmission electronic microscope (TEM). It was with the latter that Van den Oord observed HLA class II antigens in small groups of hepatocytes near the portal spaces.

We obtained a homogeneous pattern of expression of Beta 2- microglobulin that was similar to that obtained by Nagafuki et Scheuer (1986). But it should be pointed out that we used a monoclonal (NAMB I) and they used a polyclonal antibody.

For this reason it seems that, for a preliminary evaluation of these subjects to be treated with interferon (INF), the expression of Beta 2-microglobulin on liver cells is not in itself sufficient. As has already emerged in previous studies, when the Beta 2-microglobulin is displayed on liver cell membrane, it is nonconvalently associated with HLA class I molecules.

It is just possible that there is greater synthesis of the Beta 2-microglobulin, than of the HLA I glicoprotein.

Beta 2-microglobulin is detectable in serum after an adequate stimulus, but to date this does not appear to be the case for HLA class I antigens. This indirectly confirms the above hypothesis.


CONCLUSIONS

Our observations support the hypothesis that there is a genetic predisposition to a chronic outcome of HBV infection, both in the case of a patient with a chronic disease, and in "healthy" carriers. One of the intenstions of the study is to introduce another parameter for the evaluation of the effects of interferon treatment, which today is also used for children with chronic hepatitis B.

The adoption of this parameter should make it possible to observe HLA class I antigens on liver cell membranes after a period of from 6 to 12

months of INF treatment to evaluate an INF induced increase of HLA class I display on HBV infected hepatocytes.
The subjects having a low expression of HLA class I before the INF treatment will be of primary importance in confirming these hypothesis.


REFERENCES

1.  G. Ballardini, F.B. Bianchi, R. Mirakian, M. Fallani, E. Pisi, G.F. Bottazzo, 1987, HLA-A,B,C, HLA-D/DR and HLA-D/DQ expression on unfixed liver biopsy sections from patients with chronic liver disease.
    Clin. Exp. Immunol. 70, 35-46.
2.  C. Chia-Ming, S. Wei-Chue, K. Ruey-Wen, L. Yun-Fan, 1987, HLA class I antigen display on hepatocyte membrane in chronic hepatitis B virus infection: its role in the pathogenesis of chronic type B hepatitis.
    Hepatology 7, 6: 1311-1316.
3.  R. De Vos, C. De Wolf-Peters, J. Van den Oord and V. Desmet, 1985, Ultrastructural immunocytochemical demonstration of HLA class I antigens in human pathological liver tissue.
    Hepatology 5, 6: 1071-1075.
4.  M. Igarashi, L. Imberti, M. Maio, M. Tsujisaki, F. Perosa, K. Sakaguchi, P.G. Natali, S. Ferrone, 1986, Preparation characterization, and utilization of monoclonal antibodies to the gene products of the HLA-D region with special enphasis on those to polymorphic determinants. In "HLA class II antigens" pagg. 224-248 B.G. Solheim, E. Moller, S. Ferrone eds. Springer-Verlag - Berlin.
5.  L. Montano, G.C. Miescher, A.H. Goodali, K.H. Wiedmann, G. Janossi, H.C. Thomas, 1982, Hepatitis B virus and HLA antigen display in the liver during chronic hepatitis B virus infection.
    Hepatology 2, 5: 557-561.
6.  Y. Nagafuchi, P. Scheuer, 1986, Expression of Beta 2-microglobulin on hepatocytes in acute and chronic type B hepatitis.
    Hepatology 6, 1: 20-23.
7.  R.G. Paul, S.T. Roodman, D.A. Paul, R.P. Perrillo, 1987, Elevated HLA class I antigen expression on peripheral blood mononuclear cells of HbsAg carriers with coexistent human immunodeficiency virus infection.
    Hepatology 7, 6: 1326-1328.
8.  M. Pignatelli, J. Waters, D. Brown, A. Lever, S. Iwarson, Z. Shaff, R. Gerety, H.C. Thomas, 1986, HLA class I antigens on the hepatocyte membrane during recovery from acute hepatitis B virus infection and during Interferon therapy in chronic hepatitis B virus infection.
    Hepatology 6, 3: 349-353.
9.  H.C. Thoms, M. Pignatelli, 1985, Is modulation of HLA display by Interferon important in preventing the development of the chronic carrier state of hepatitis B virus in adults?
    Gastroenterol. Clin. Biol. 9: 287-289.
10. J.J. Van den Oord, V.J. Desmet, 1984, Distribution patterns of major histocompatibility antigens in normal and diseased liver tissue.
    Leber, Magen, Darm 14: 244-254.

11.  J.J. Van den Oord, R. De Vos, V.J. Desmet, 1986, In situ
     distribution of major histocompatibility complex products and viral
     antigens in chronic hepatitis B virus infection:  evidence that HBc-
     containing hepatocytes may express HLA-DR antigens.
     Hepatology 6, 5: 981-986.

# SPECIFIC LYMPHOCYTE PROLIFERATIVE RESPONSE IN VITRO

## AFTER cDNA HBsAg IMMUNIZATION

A. Degrassi, E. Mariani, P. Roda, R. Miniero, M. Capelli,
M.C. Honorati, A. Astaldi and A. Facchini

Istituto di Clinica Medica e Gastroenterologia,
Laboratorio Centralizzato, Universita' di Bologna,
Istituti Ortopedici Rizzoli, Bologna, Smith Kline &
French, Milano,Italy

## INTRODUCTION

The protective mechanism of hepatitis B vaccination using surface antigen is the induction of neutralizing antibodies. Hepatitis B surface Antigen (HBsAg) has proven to be well tolerated and effective in many clinical trials [1,2].

The limited availability of serum derived from chronic carriers of HBsAg for the production of plasma derived human hepatitis B vaccines together with the fear that these preparations may vehiculate transmittable pathogens prevented the large diffusion of these vaccines.

The recent possibility to produce HBsAg in yeast by recombinant DNA (cDNA) technique offers an alternative to plasma derived vaccines [3]. Several clinical studies have been conducted on cDNA HBsAg vaccines and have demonstrated that immunogenicity of yeast derived vaccines is similar to that of plasmas derived, with no difference in the specificity and avidity of antibody synthetised [4].

On the contrary only few studies have been conducted on the cellular response in vitro in recipients of cDNA derived vaccines. Studies on subjects injected with plasma purified vaccines demonstrated that the induction of anti-HBs production in vitro is dependent by T lymphocytes and antigen presenting cells [5] but a proliferative response could not be easily obtained in vitro using peripheral blood lymphocytes as effectors. HBsAg induced proliferation was obtained only when cells from acute hepatitis patients were used and even in this case the response was low [6].

In this study we have evaluated the proliferative response in vitro of lymphocytes obtained from healthy subjects immunized with two different cDNA yeast derived hepatitis B vaccines: Engerix-B (SK&F) containing 20 μg HBsAg and Gen-H-B-VAX (MSD) with 10 μg of antigen.

Our preliminary data show a seroconversion rate of 100 % after the second boost and a logaritmic increase of serum antibody titres after the third boost. The cellular proliferative response together with the effect of recombinant interleukin 2 (rIL-2) on pre-activated cells has been assessed in vaccinated subjects for one month after the third boost.

Only at the second week after the third boost a proliferative response in vitro to HBsAg was detectable in vitro. The latter could be increased by the addition of rIL-2. Also after the third boost a spontaneous and mitogen induced anti HBs production in vitro was found, although with a broad variability among subjects.


MATERIALS AND METHODS

## Patients and Protocol

Two groups of ten healthy subjects from hospital personnel and medical students were selected; mean age±SD: 33 ± 1.3 (first group: 5 male - 5 female; second group: 4 male - 6 female)

Subjects were negative for all Hepatitis B Virus markers (HBsAg, anti-HBs, HBeAg, anti-HBe, anti-HBc).

Two different recombinant HBsAg vaccines were used: the first group was injected with Engerix-B (SKF) containing 20 µg of antigen while the second group was vaccinated with Gen-H-B-Vax (MSD) containing 10 µg of antigen. Vaccination schedule provided four injection I.M. in deltoid on months 0, 1, 6, and 12. Blood samples were drawn on weeks 0, 4, 8, 24, 25, 26, 27 and 28. Cells from one non vaccinated healthy subject negative for HBV markers was used as control.

## Serological Assay

A commercially available radioimmunoassay (AB-AUK-3 Sorin, Italy) was employed to measure the level of anti-HBs antibody in serum samples at months 0, 1, 2, 6, and 7 and in supernatants of in vitro cultures. The assay was sensitive to 0.1 ng of HBsAb. Routine biochemical and hematological parameters were assessed for safety evaluation of vaccinees.

## Antigen

HBsAg used for the stimulation in cultures was prepared from the plasma of previously infected donors (kindly provided by Dr. Giacosa, SORIN SpA, Saluggia, Vc, Italy).

## Lymphocyte Proliferative Assay

Peripheral Blood Mononuclear Cells (PBMC) were isolated by Ficoll-Hypaque gradient centrifugation, washed twice and resuspended in RPMI 1640 medium supplemented with 25 mM Hepes buffer, 10% heat inactivated fetal calf serum, 4mM glutamine, penicillin (100 IU/ml) and streptomycin (100 µg/ml). PBMC ($1 \times 10^5$/well) were stimulated with different doses of antigen (1, 0.5, 0.1, 0.05, 0.01 µg/ml) in a 6 - day antigen blastogenic response. As source of Antigen Presenting Cells (APC), $5 \times 10^5$ autologous irradiated (2000 RAD) PBMC were added to each well.

A mitogen induced proliferative response was also performed as positive control of the assay: Concanavalin-A (ConA) was used at a final concentration of 13 µg/ml and negative control cultures were performed by incubating cells in medium without any stimulation. Cultures were pulsed with $^3$H-thymidine (0.4 µCi/well)(specific activity 25 Ci/mmMOL) (Amersham, U.K.) after six days of culture and harvested on filters (Skatron, Norway) 18h later. Filter discs were placed in a β counter to determine the incorporation of $^3$H-thymidine into DNA. Stimulation indices were calculated as ratio between $^3$H-thymidine incorporation of antigen stimulated cells and that of cells incubated in medium alone.
On the second day the effect of IL-2 on HBsAg proliferative response was assessed by adding the lymphokine (10 IU/ml final concentration) to other cultures performed as outlined above (rIL2 was kindly provided by Dr. A. Galazka, Glaxo I.M.B., S.A., Geneva, Switzerland).

## Anti-HBs production in vitro

PBMC ($1x10^6$/ml) in 1 ml medium were placed in 12x75 mm polystyrene test tubes (Falcon) and two different policlonal B cell activation pathways were tested: Pokeweed Mitogen (PWM), a B-$T_h$ cell mitogen at a final concentration of 1/200 and anti-CD3 monoclonal antibody from hybridoma supernatant with tested mitogenic activity on $T_h$ cells. After cells were incubated with mitogens at 37°C for 4 days they were added with 1 ml of fresh medium for an additional 5 days culture. Supernatant was then collected to assay for anti-HBs antibody production. Spontaneous production was also tested as outlined above in absence of any mitogenic stimulus in culture medium.

## RESULTS

### Serological data

The Geometric Mean Titres (GMT) of antibody levels at different times of immunization are shown in Fig. 1.

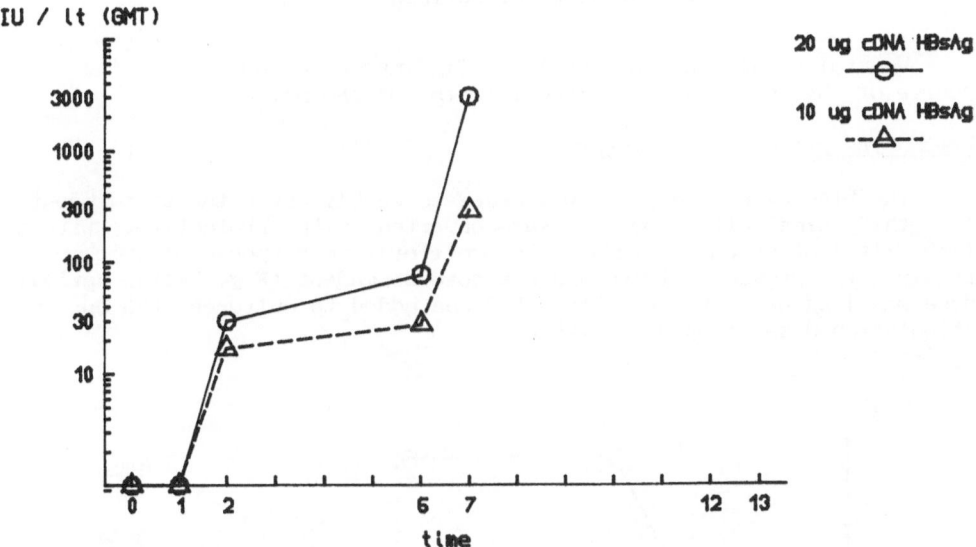

Fig. 1.    Kinetics of serum Anti HBs Antibody titres in subjects vaccinated with different doses of antigen.

After the first dose of vaccine two out of twelve patients seroconverted while one month after the second dose of vaccine the seroconversion rate was 100 % with a GMT of 30 IU/lt for Engerix-B group, and 17 IU/lt for Gen H-B-Vax vaccined subjects respectively; moreover 10/10 (100 %) subjects vaccinated with 20 µg (Engerix-B) of antigen presented a serum titre of 10 IU/lt, generally considered as the protective level, while this title was present only in 5/10 (50 %) patients vaccinated with 10 µg of antigen (Gen H-B-Vax). The GMT of anti HBs levels immediately before the third boost was respectively 71 for Engerix-B and 39 for Gen H-B-Vax group.
One month after the third boost the GMT was 3027 IU/lt. for 20 µg vaccinated people and only 288 IU/lt. for 10 µg.
When anti HBs serum levels were evaluated in females and males

regardless to the vaccine used: the GMT was always higher in the female group when compared to the male ones (Fig. 2).

**M O N T H S**

| vaccine | 0 | | 1 | | 2 | | 6 | | 7 | |
|---|---|---|---|---|---|---|---|---|---|---|
| | 20 µg | 10 µg | 20 µg | 10 µg | 20 µg | 10 µg | 20 µg | 10 µg | 20 µg | 10 µg |
| female | 0 | 0 | 0 | 0 | 35 | 20 | 80 | 26 | 3298 | 332 |
| male | 0 | 0 | 0 | 0 | 27 | 11 | 45 | 37 | 2778 | 187 |

Fig. 2. Serum anti HBs titres in male and female subjects vaccinated with different doses of antigen.

Hematological and biochemical findings were normal during the course of the immunization protocol (data not shown).

Lymphocyte proliferative assay

The blastogenic response was assessed weekly after the third boost. At that time all subjects seroconverted with protective titres (≥10IU/lt.) of serum anti HBs. The proliferative response of lymphocyte required the presence of APC and was dose dependent (Fig. 3);the optimal dose was 1 µg HBs although when rIL-2 was added to cultures, the optimal stimulation dose dropped to 0.05 µg HBs.

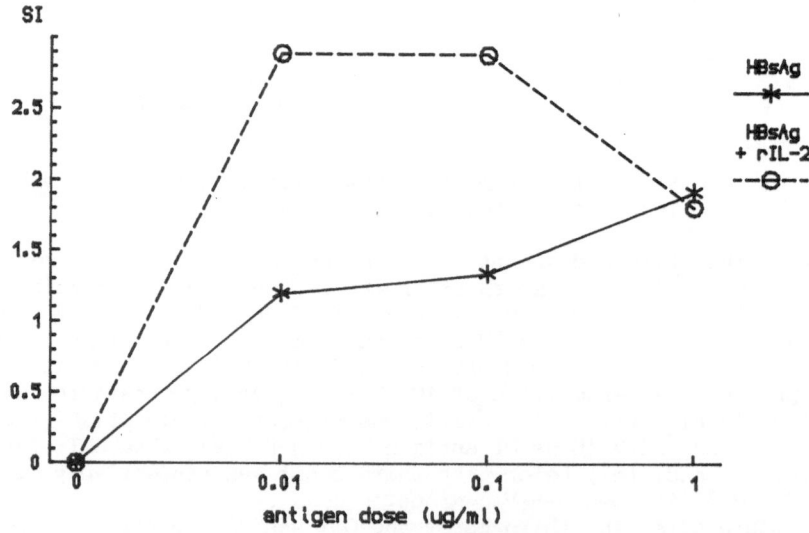

Fig. 3. Dose-response curve of lymphocytes proliferation at different invitro attivation conditions.

When the kinetic of proliferative response was analyzed weekly, no response could be observed 1 week after the third immunization.The responsive cells appeared only on the second week and were not detectable on the third and fourth week after the third boost.

When proliferative data from the two groups were compared, the group who received 20 μg presented a higher proliferative response (Fig.4).

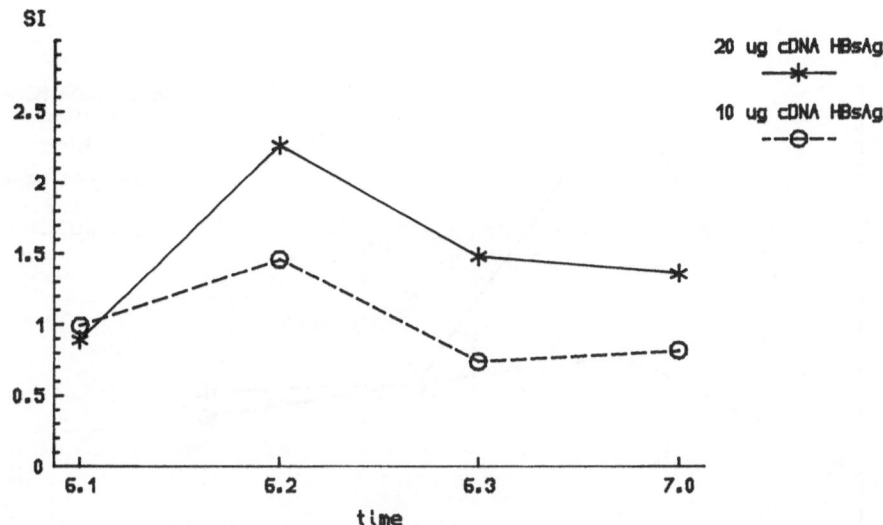

Fig. 4.     Kinetics of antigen induced lymphocyte proliferative assay. cells were obtained from subjects vaccinated with different doses of antigen.

Negative control did not show any proliferation after stimulation with the antigen at any time during the experiments (data not shown).

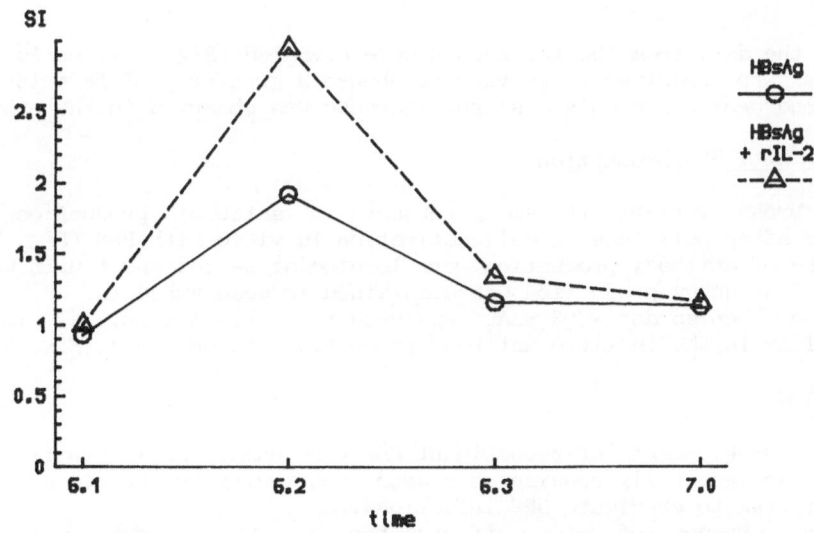

Fig. 5.     Effect of recombinant IL-2 addition to culture stimulated with HBsAg.

Effect of IL-2 on Ag dep. prol. resp.

The effect of IL-2 on HBsAg induced lymphocyte proliferation is represented in fig. 5: a significant increase of lymphocyte proliferation was observed when rIL-2 was added to the cultures on the second day of incubation and this was particularly evident two weeks after the third dose of vaccine.

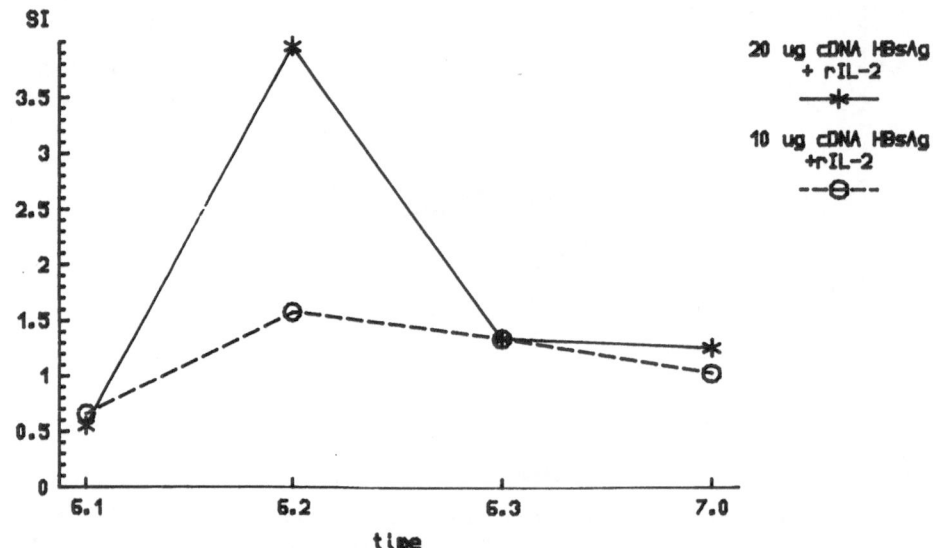

Fig. 6.    Lymphocyte proliferative assay after incubation of cells with HBsAg + IL-2: cells were obtained from subjects vaccinated with different doses of Antigen.

If the data from the two groups were compared (Fig. 6), cells from subjects who received 20 µg vaccine showed a greater proliferation on the second week while only a slight increase was observed in the others.

In vitro anti HBs production

A broad variability among subjects in antibody production was observed after policlonal B cell activation in vitro with PWM (Fig. 7). The peaks of antibody production were located at second and fourth week, while a continuous production was maintained in some subjects.
Also when an anti-CD3 MoAb was used for stimulation, a similar variability in the in vitro antibody production was observed (Fig. 8).

DISCUSSION

The development of recombinant HBsAg as human anti hepatitis B vaccine is generally considered a good opportunity to perform a large scale program to eradicate HBV infections.
The absence of side effects together with biochemical and hematological normal findings support the use of rHBsAg when used for vaccination.

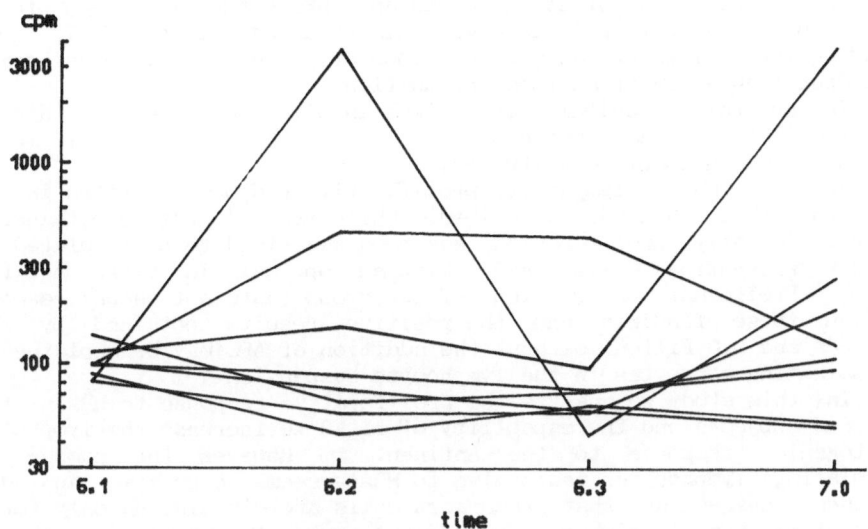

Fig. 7.  Kinetics of Anti HBs production after stimulation in  vitro
with PWM (eight subjects).
Results are expressed as cpm from the Anti HBs  RIA.

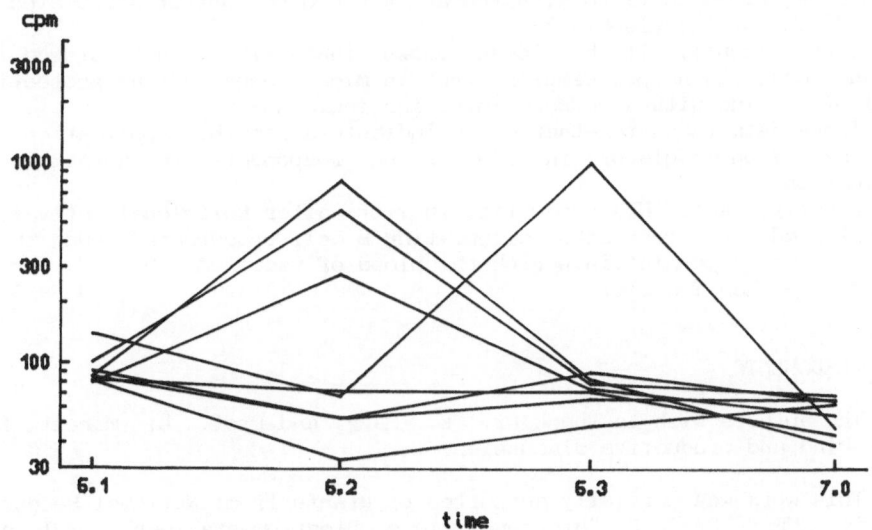

Fig. 8.  Kinetics of Anti HBs production after stimulation in  vitro
with anti CD3 MoAb (eight subjects).
Results are expressed as cpm from the Anti HBs  RIA.

The analysis of seroconversion rates and serum anti-HBs titres after the third dose of vaccine showes that both were immunogenic. The difference in serum antibody titres seems likely to be due to a dose dependent response.

The observation of an increased anti HBs serum level in female group could be referred to a sex related upregulation of Ir genes encoding for antigen processing functions or, more likely, to an hormone dependent regulation of antibody production.

The humoral and cellular mechanisms involved in protection against infection after rHBsAg vaccination as well as the immunological memory has not yet been completely studied.

Antigen induced lymphocyte proliferative response in vitro in HBs vaccined subjects has been described, while it could not be obtained by others. Recently proliferation has been described to be regulated by Antigen Presenting Cells (APC) with an optimal in vitro APC/PBMC ratio[7]. Preliminary observations of our group (data not shown) seems to confirm these findings and the positive results obtained by M.W. Steward[8] and L.G.Filion[9] without the addition of APC may be explained by contaminating monocytes in the lymphocyte suspensions.

In this study we analyze the proliferative response to HBsAg from vaccined subjects and the capability of rIL-2 to increase the lymphocyte blastogenic response to the antigen. However the number of circulating lymphocytes responsive to HBsAg seems to be time and dose dependent suggesting that precursors cells are circulating only for a short time but in higher number in the subjects vaccined with the highest dose of antigen. The further disappearance of proliferative response could indicate the activation of a suppressive pathway with regulatory functions or eventually the compartimentalization of precursors in lymphoyd organs.

Furthermore the effect of rIL-2 on the lymphocyte proliferative response after incubation with the antigen, indicates that circulating precursors primed in vitro with the antigen can be further stimulated by the addition of interleukin-2.

Very recently it has been shown that rIL-2 administered in conjunction with herpes simplex virus in mice, enhanced the protective level of vaccine with a reduced infection incidence[10].

These data taken together could indicate a possible application of rIL2 as immunomodulator in low or no responders to hepatitis B vaccination.

Finally, anti HBs production in vitro after policlonal activation demonstrated the presence of circulating B cell precursors, and these cells seems to persist in peripheral blood of vaccined subjects for a longer time than T cells.


ACKNOWLEDGMENTS

The authors wish to thank Dr. E. Maggi and Prof. L. Moretta for their kind and productive discussion.

This work was partially supported by grants from: National Research Council CNR - Project "Biotecnologie e Biostrumentazione", MPI and Istituti Ortopedici Rizzoli.

REFERENCES

1. W. Szmuness, C. E. Stevens, E. J. Harley, E. A. Zang, W. R. Oleszko, D. C. William, R. Sadowsky, J. M. Morrison, and A. Kellner, Hepatitis B vaccine. Demonstration of efficacy in a

controlled clinical trial in a high-risk population in the United States, <u>N. Engl. J. Med.</u> 303: 833 (1980).

2. W. Szmuness, C. E. Stevens, E. A. Zang, E. J. Harley, and A. Keffner, A controlled clinical trial of the efficacy of the hepatitis B vaccine (Heptavax B): a final report, <u>Hepatology</u> 1: 377 (1981)

3. W. J. McAleer, E. B. Buynac, R. Z. Maigetter, D. E. Wampler, W. J. Miller, and M. R. Hilleman, Human hepatitis B vaccine from recombinant yeast, <u>Nature</u> 307: 178 (1984)

4. W. Jilg, B. Lorbeer, M. Shmidt, B. Wilske, G Zoulek, and F. Deinhardt, Clinical evaluation of a recombinant hepatitis B vaccine, <u>Lancet</u> ii: 1174 (1984)

5. T. R. Cupps, J. L. Grein, R. H. Purcell, P. K. Goldsmith, and A. S. Fauci, In vitro antigen-induced antibody response hepatitis B surface antigen in man, <u>J. Clin Invest.</u> 74: 1204 (1984)

6. G. Fattovich, A. Alberti, C. Crivellaro, P. Pontisso, F. Noventa, and G. Realdi, Cellular immunity to the hepatitis B virion in acute hepatitis type B, <u>Clin. Exp. Immunol.</u> 53: 645 (1983)

7. U. Hellstrom, S. Sylvan, and P. Lundberg, Regulatory functions of T- and accessory-cells for hepatitis B surface antigen induced specific antibody production and proliferation of human peripheral blood lymphocytes, in vitro, <u>J. Clin. Lab. Immunol.</u> 16: 173 (1985)

8. M. W. Steward, B. M. Sisley, C. Stanley, S. E. Brown , and C. R. Howard, Immunity to hepatitis B: analysis of antibody and cellular responses in recipients of a plasma-derived vaccine using synthetic peptides mimicking S and pre-S regions, <u>Clin. Exp. Immunol.</u> 71: 19 (1988)

9. L. G. Filion, R. Saginur, and N. Szczerbak, Humoral and cellular immune responses by normal individuals to hepatitis B surface antigen vaccination, <u>Clin. Exp. Immunol.</u> 71: 405 (1988)

10. A. Weinberg, and T. C. Merigan, Recombinant interleukin-2 as an adjuvant for vaccined-induced protection, <u>J. Immunol.</u> 140: 294 (1988)

# INFLUENZA VIRUS INFECTION REDUCES CELLULAR IMMUNITY FUNCTIONS

Del Gobbo V., Balestra E., Marini S., Villani N. and Calió R.

Department of Experimental Medicine
II University of Rome

## INTRODUCTION

In recent years, detailed information has been accumulated[1,2,3], that cellular, as well as humoral, immune responses play an important role in the clearing of influenza virus and in the recovery phase of infection. On the other hand, influenza virus, such as many viruses, during the acute stages of infection is able to inhibit cellular immune functions as shown by Reiss[4], and Rodgers[5]. Little is known about this mechanism of immunosuppression. Previous our studies[6] described a strong alteration of thymocytes and reticulo-epithelial cells in mice infected with PR8 influenza virus.
The aim of the present investigation was to analysed the impairment of natural and cellular immune function in PR8 influenza virus infected mice and to identified the mechanism involved. Our data indicate that PR8 virus infection results in decrease of T-lymphocyte blastogenic response, IL-2 production and natural killer cell activity by stimulating T-cell suppressor subpopulation.

## MATERIALS AND METHODS

### Mice

Balb/C mice, four to five week old provided by Charles River were used in these studies.

### Virus

A/PR/8/34 Influenza virus ($H_1N_1$) was growth in the allantoic cavity of 10-day old embryonated eggs; after 48 hours, the allantoic fluid was harvested. Virus was stored at -70° C.

### Virus inoculum

Four-five week old Balb/C mice were inoculated intranasally

(i.n.) with 50 ul of PR8 virus 12 or 6 haemagglutinating units (HAU)/ml under light ether anesthesia.

## Preparation of spleen cell suspensions

At various time intervals control and infected mice were sacrificed by cervical dislocation and the spleens were aseptically removed. Cell suspensions were obtained from individual spleens by passage through 10 ml sterile syringe and by filtering through Nytex mesh. Cells were washed twice and resuspended in RPMI 1640 medium containing 10% heat-inactivated fetal calf serum, supplemented with L-glutamine (2mM/ml), 100 units/ml of penicillin and 100 µg/ml of streptomicin. The viability, as determined by trypan blue exclusion, were >95%.

## Concanavalin A induced proliferation

Individual spleen cells from normal and virus infected mice were suspended in complete culture medium and cultured in microtiter plates with Concanavalin A (Con A) for three days. Finally each well was pulsed with 1 µCi of tritiated thymidine for 18 hr.
After incubation, the cells were collected by a semiautomatic cell harvester and the radioactivity was measured as counts per minute (cpm) in a $\beta$ counter. The results, expressed as stimulation ratio (SR), were obtained by dividing the mean cpm incorporated in triplicate wells containing mitogen, by mean cpm incorporated in triplicate wells not containing mitogen.

## IL-2 production and titration

The assay was based on the method of Gillis[7]. Briefly murine IL-2 production was induced by stimulating individual spleen cells from infected and control mice with ConA and phorbol myristic acetate (PMA). After 48 hr incubation at 37°C, the supernatants were collected and 2-fold serially diluited in medium; the samples of each diluition were added into 96 well microtiter plates , containing IL-2 depedent cytotoxic T-lymphocyte line (CTLL) cell suspension. After 20 hr incubation, all microcultures received [3]H-thymidine and, 4 hr later, cells were harvested by a semiautomatic cell harvester. IL-2 activity measured as [3]H-thymidine incorporation by IL-2 depedent CTLL clone, was expressed as the reciprocal of supernatant dilution producing 50% maximal [3]H-TdR incorporation with reference to a standardized IL-2 preparation.

## Nk cytotoxicity assay

The assay was based on the method of Herberman[8]. Briefly YAC-1 cells,used as target cells, were labeled with [51]Cr and placed in 96 well microtiter plates with individual effector spleen cells. At least two effector to target cell ratio were used (100:1 and 50:1). Tests were run in quadruplicate and spontaneous release never exceed 10% of total release. After 4hr incubation at 37°C the specific [51]Cr release was calculated as follow: specific $^{51}$Cr release % = (mean counts in the presence of effector cells - mean spontaneous release counts)/ (mean total release counts - mean spontaneous release counts) x 100.

## T-cells enrichment

T-cells were enriched by the direct "panning" method[9]. Briefly, polystyrene petri dishes were coated with Ab specific for mouse Ig. Spleen cells from infected and control mice were added and allowed to adhere to prepared plates for 90 min. at 4°C. Non-adherent cells were harvested and resuspended in culture complete medium. T-cell recovery, determined by FACS, was ranging from 90 to 95% and their viability were > 95%.

## T-suppressor cell depletion

Spleen lymphocytes from infected and control mice were suspended in RPMI 1640 medium containing 1:20 dilution of Lyt 2 anti-serum and incubated at 4°C for 45 min. After incubation the cells were washed, and incubated in culture medium, containing 1:10 dilution of rabbit complement (C') for 1 hr at 37°C. Spleen cells incubated with either medium alone or with 1:10 dilution of rabbit complement were used as control. These cell suspension (analyzed by FACS) contained less than 0,6 % of Ts subpopulation and cell viability were > 90%.

## RESULTS

### PR8 inhibits lymphocyte blastogenic response to Concanavalin A

Initial experiments were performed to analyse whether the morfological and functional alteration of thymus, observed in PR8 infected mice, results in impairment of T cell proliferative response to Con A. Fig.1 shows that spleen cells from Balb/C mice given PR8 failed to proliferate when cultured in vitro with Con A. A lower $^3$H-thymidine incorporation was detected in groups of mice inoculated with 12 HAU/ml. Spleen cells from mice given UV-inactivated PR8 virus respond normally (data not shown).

### PR8 inhibits interleukin-2 production

In order to investigate whether the inhibition of DNA synthesis can be attribuited to the lack of IL-2 production, spleen cells from PR8 infected mice were tested for IL-2 production at same intervals (Fig.2). The results confirm that PR8 virus infection significantly reduced the levels of IL-2 production by spleen cells and this decrease is dose-dependent. In fact mice infected with 6 HAU/ml show a significant decrease of IL-2 production at later period and this effect is shorter. When UV-inactivated PR8 virus was used normal levels of IL-2 production were obtained (data not shown).

## Modulation of NK cell activity

The data from the first set of experiments indicate that
influenza virus infection results in an impairment of T
helper cells function. Therefore we examined whether the
lack of IL-2 production may modulate natural immune
response. At same time intervals spleen cells from control
and infected mice were assayed for NK cell activity (Fig.
3). The results confirm this hypothesis; in fact, after a
transient increase, NK cell activity was significantly
reduced in virus infected mice and the minimum [51]Cr release
appears to be correlated with the lower IL-2 production in
both groups of infected mice. These results point out that
influenza virus infection causes a impairment of natural and
cellular functions, through a reduced IL-2 production by T
helper cells.

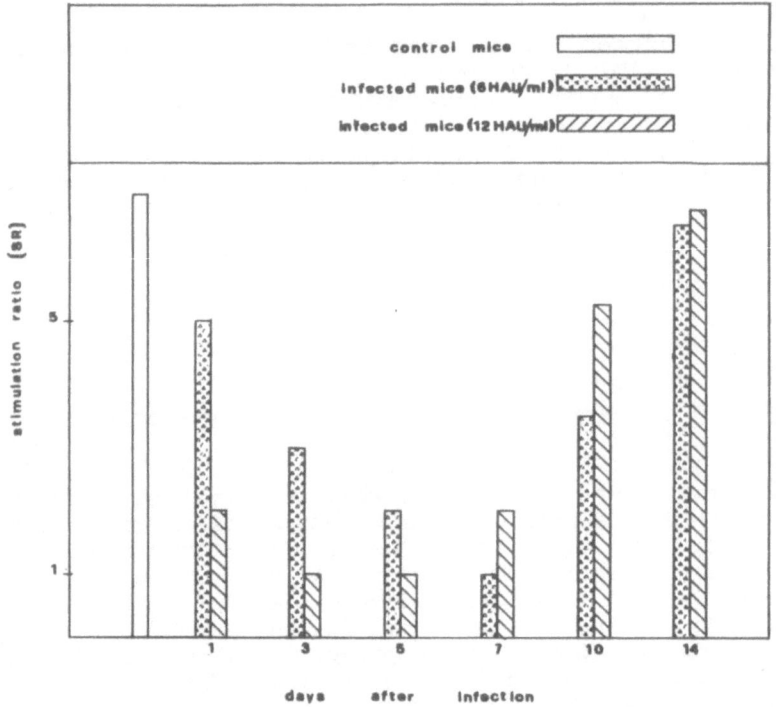

FIG. 1. The figure illustrates the blastogenic responses to
Concanavalin A in spleen cells from control and infected
mice. A stimulation ratio (SR) for each set of cultures is
calculated from mean cpm incorporated in triplicate cultures
containing mitogen, divided by the mean cpm incorporated in
triplicate cultures not containing mitogen.

## T-cell enrichment not modifies the levels of IL-2 production

In previous experiments we observed that , during the course
of influenza virus infection the spleen weight was markedly
reduced and that T helper/cytotoxic-suppressor cell ratio

was significantly alterated. In order to examined whether the deficit in IL-2 production can be due to T helper cell reduced number, we enrhic T cells by direct panning tecnique (Fig.4,B). The results showed that T cell enrhicment was inable to restore normal levels of IL-2 production by spleen cells from infected mice.

## T-cell suppressor depletion is able to restore normal IL-2 activity

Recent data from literature has been demonstrated[10] that PR8 influenza virus inoculum was able to give an increase of T-suppressor cells (Ts). That is in keeping with our FACS observation, wich showed an increased number of Ts cell subpopulation in spleen of PR8 infected mice. Therefore we

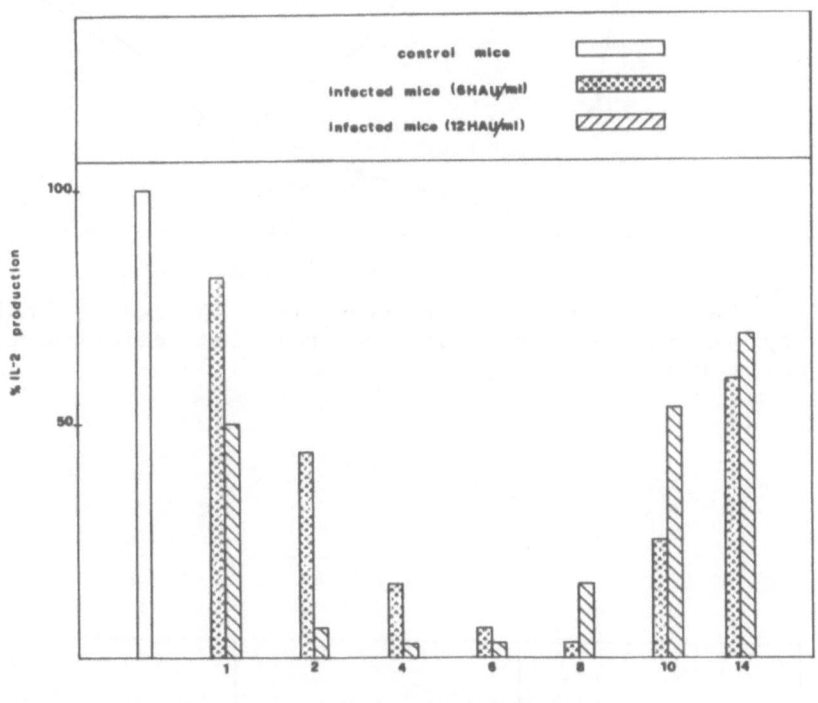

FIG. 2. The figure illustrates the production of interleukin 2 (IL-2) in control and infected mice. IL-2 activity, measured as $^3$H-TdR incorporation by the IL-2 dependent CTLL clone, is expressed in percent respect to maximal $^3$H-TdR incorporation of standardized IL-2 preparation.

considered the possibility that the IL-2 deficit may be due to PR8-activated Ts cells.
In order to analyse this finding, Ts cells were removed by anti-Ly2 antiserum + complement. Fig. 4,C shows IL-2 production after Ts depletion. The results demonstrate that T suppressor cells are responsable for defective IL-2 production. In fact, spleen cells from PR8 infected mice, Ts free, were able to produce normal levels of IL-2.

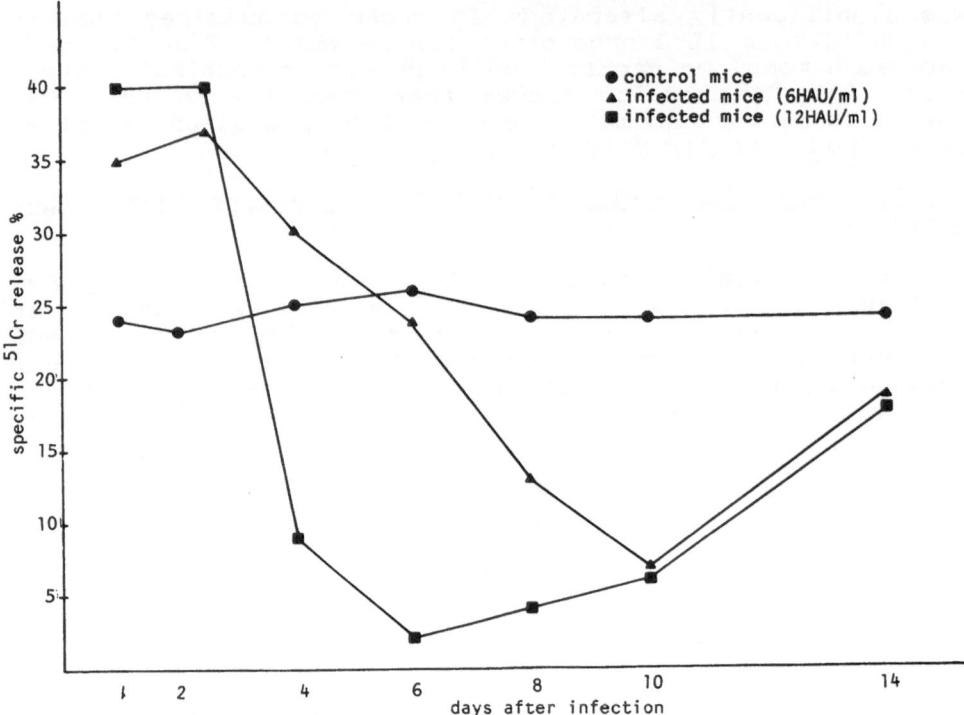

FIG. 3. The figure illustrates the NK cell cytolitic
activity in control and infected mice at effector-target
cell ratio 100:1.

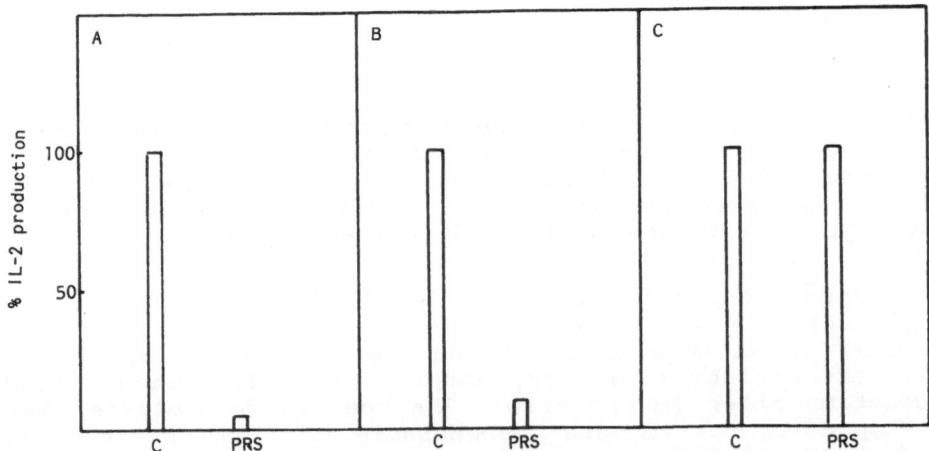

FIG. 4. The figure illustrates the IL-2 production by
splenocytes (A) from control and infected mice (12 HAU/ml)
at 6 day postinfection, (B) from control and infected mice

after T-cell enrichment, (C) from control and infected mice
after Ts removal.


DISCUSSION

The present study analyses cellular and natural reactivity
in mice given PR8 influenza virus. The results show that
this infection results in an inhibition of in vitro T-cell
proliferative response to Con A and NK cell activity by
affecting the production of IL-2. This impairment is
dose-dependent and related to virus replication. In fact,
mice inoculated with UV-inactivated virus ( which completely
destroyed viral infectivity) show normally natural and
cellular reactivity (data not shown).
The deficit of IL-2 production is not be due to T helper
cell reduced number, observed in spleen of infected mice. In
fact, T cell enrhicment does not reduce the defective IL-2
production.
Another working hypothesis is whether the IL-2 deficit can
dependent by alterated IL-1 production; our data indicated
that adherent cells were not involved in this phenomena
(data not shown).
The results, described in this report, point out that this T
helper cell impairment is due to PR8-activated suppressor
cells, which inhibit the IL-2 production. Such results are
in good agreement with Ghovais[11,12]observation, that T
suppressor cells can modulate IL-2 production. The mechanism
of action of PR8-induced T suppressor cells must further
investigate.
Several report have indicate[13,14,15]that other viruses
inhibit IL-2 production. This depressed IL-2 production may
be due to an inhibitory molecules released by Ts, or by
direct cell to cell contact.
Further studies are required on the T cell subset from
infected mice in order to define the Ts role in influenza
virus infection.


REFERENCES

1)   Ada G.L., Leung K.N., Ertl H.C.: An analysis of effector
     T cell generation and function in mice exposed to
     Influenza A or Sendai viruses. Immunol.Rev., 58, 5,
     (1981).

2)   Daisy J.A., Tolpin M.D., Quinnan G.V., Rook A.H., Murphy
     B.R., Mitthal K., Clemenis M.L., Mullinix M.G., Kiley
     S.C., Ennis F.A.: Influenza A infection in human
     volunteers, pag.443 in The replication of negative
     strand virus, Bishop, D.M.L., Compans, R.W. Ed.,
     Elsevier, North-Holland, (1981).

3)   Ennis A.: Some newly recognized aspects of resistence
     against and recovery from influenza. Archives of
     Virology, 73, 207, (1982).

4)   Reiss C.S., Schulman J.L.: Cellular Immune Responses of
     Mice to Influenza Virus Infection., Cellular Immunology,
     56, 502, (1980).

5) Rodgers B., Mins C.A.: Interaction of Influenza Virus with mouse macrofages. Infect. Immun., 31, 751, (1981).

6) Garaci E., Calió R., Djaczenko W.: Effect of Influenza virus PR8 infection on thymus in intact and adrenolectomized mice. Experentia, 30, 358, (1974).

7) Gillis S., Ferm M.M. Ou W., Smith KA.: T-cell growth factor: parameters of production and a quantitative microassay for activity. J. Immunol., 120, 2027, (1978).

8) Herberman RB. and Ortaldo J.R.: Natural killer cells: their role in defenses against disease. Science, 24, 214, (1981).

9) Wysocki L.J. and Sato V.L.: "Panning" for lymphocytes: A method for cell selection. Proc. Natl. Acad. Sci. USA, 75, 2844, (1978).

10) Hurwitz J.L. and Hackett C.J.: Influenza-specific suppression: contribution of major viral proteins to the generation and function of T suppressor cells. J.Immunology, 135, 2134, (1985).

11) Chouaib S., Chatenoud L., Klatzmann D. and Fradelizi D.: The mechanisms of inhibition of human IL-2 production. J.Immunology, 132, 1851, (1984).

12) Chouaib S. and Fradelizi D.: The mechanism of inhibition of human IL-2 production. J.Immunology, 129, 2463, (1982).

13) Wainberg M.A., Spira B., Boushira M. and Margolese R.G.: Inhibition by human T-lymphotropic virus (HTLV-I) of T-lymphocyte mitogenesis: failure of exogenous T-cell growth factor to restore responsiveness to lectin. Immunology, 54, 1, (1985).

14) Blackett S. and Mims C.A.: Studies of depressed interleukin-2 production by spleen cells from mice following infection with cytomegalovirus. Arch. Virol. 99, 1, (1988)

15) Colonna Romano G., Dieli F., Abrignani S., Salerno A. and Colizzi V.: Inhibition of lymphocyte mitogenesis in mice infected with Newcastle disease virus: viral interference with the interleukin system. Immunology, 57, 373, (1986).

# CHARACTERIZATION OF ANTI LIVER KIDNEY MICROSOMAL ANTIBODY ASSOCIATED WITH CHRONIC HDV INFECTION BY IMMUNOBLOTTING

M. Lenzi, M.Fusconi, L.Selleri, A.Caselli, F.Cassani, A. Craxí* and F.B. Bianchi

Istituto di Clinica Medica II University of Bologna and Clinica Medica R and Pneumologia*, University of Palermo

Cytoplasmic antibodies directed against antigen(s) present in the membranes of the endoplasmic reticulum have been reported in a proportion of patients with chronic HBV-HDV infection (Crivelli et al.,1981). While the immunofluorescence (IFL) pattern associated with LKMδ antibody is quite similar to that of the liver kidney microsomal antibody (LKM) described by Rizzetto et al. (1973) and associated with an autoimmune type of chronic active hepatitis (Thomas, 1980), the target antigen(s) was supposed to be different on the basis of blocking experiments. The target antigen of LKM antibody has been recently partially identified as a 50kD protein located in the membranes of the smooth endoplasmic reticulum (Alvarez et al., 1985), while no information is at present available on the nature of LKMδ associated antigen.

The aim of the present study was to evaluate the prevalence of LKMδ antibodies in a series of patients with HDV infection and to characterize the target antigen using an immuno "western" blotting tecnique (IB).

## PATIENTS ANS MATERIALS

Sera from 173 HBsAg positive patients with chronic delta superinfection were studied by an indirect IFL test performed according to standard procedures (Roitt and Doniach, 1969 ) using 5μ cryostat section of normal human liver (surgical specimen) and rat liver and kidney as substrates. Sera were screened at 1:20 dilution and the positive ones were titred by doubling dilutions to end point. Immunoblotting was performed according to Fusconi et al. (1988) using a rat liver subcellular fraction enriched in microsomes as source of antigen. Ten sera positive for antimitochondrial antibody from patients with primary biliary cirrhosis, 10 from patients with LKM positive CALD, 10 from patients with autoimmune CALD positive for antinuclear and/or anti-smooth muscle antibodies and 10 from healthy blood donors were used as controls.

## RESULTS

Sixteen out of 173 sera (9.2%) gave a bright cytoplasmic fluorescence on human liver and were scored as LKMδ positive with a median titre of 1:80 ( 1:20- 1:160). All of them were also positive on rat liver and 6 reacted with rat kidney too, giving an IFL pattern similar to that of the LKM of the autoimmune type, the only difference being the positivity of kidney medullar tubules and a less sharp demarcation of the fluorescence positivity between the portal space and liver parenchima. Titres of the sera giving renal positivity were significantly higher than those reacting with human and rat liver only ($p < 0.02$: Wilcoxon rank sum test). Results of IB experiments showed that 5 aut of 16 LKM$\delta$ positive sera (31%) gave a positive reaction with a peptide of 54kD molecular weight which was unreactive with LKM positive and control sera. The titres of the sera reacting in IB were not significantly higher than those negative in IB. In particular, no reaction was obtained with the 50kd microsomal protein which is at present considered the target of the autoimmune LKM type.

Moreover, none of the LKM$\delta$ positive sera gave a precipitin reaction when tested in a counterimmunoelectrophoresis test against a rat liver subcellular fraction enriched in smooth endoplasmic reticulum as obtained with LKM antibody (Lenzi et al.,1988).

DISCUSSION

It is confirmed that in sera of patients with HDV infection an autoantibody is found which reacts with cytoplasmic structures of the liver; its prevalence is low and comparable to that reported in literature (9.2%). In our experience no difference could be demonstrated between human and rat liver substrates in terms of titre or pattern of reactivity. In a small number of cases (6) this antibody reacted also with rat kidney giving a pattern superimposable to that given by LKM antibody found in an autoimmune subgroup of CALD. Hovever, in spite of a similar IFL pattern no cross reactivity could be demonstrated between the two antibodies, both by counterimmunoelectrophoresis and IB. Five (31%) of the LKM$\delta$ positive sera reacted with a peptide of 54kD molecular weight present on the membrane of rat liver endoplasmic reticulum. On the basis of these preliminary results we are trying to evaluate the reactivity of the same sera on human substrate and to isolate the antigen in order to define its chemical nature and functional properties.

REFERENCES

Alvarez, F., Bernard, O., Homberg, J.C. and Kreibich, G., 1985, Anti-liver-kidney microsome antibody recognizes a 50000 molecular weight protein of endoplasmic reticulum. J. Exp. Med. 161:1231.
Crivelli, O., Lavarini, C., Chiaberge, E., Amoroso, A., Farci, P., Negro, F.and Rizzetto, M.,1983, Microsomal autoantibodies in chronic infection with the HBsAg associated delta agent. Clin. Exp Immunol. 54:232.
Fusconi, M., Ghadiminejad, I., Bianchi, F.B., Baum H., Bottazzo G.F.and Pisi, E., 1988, Heterogeneity of antimitochondrial antibodies with M2-M4 pattern by immunofluorescence as assessed by western immunoblotting and enzyme linked immunosorbent assay. Gut. 29:440
Lenzi,M., Fusconi, M., Selleri, L., Caselli A., Cassani F., Bianchi, F.B. and Pisi, E., 1988, Counterimmunoelectrophoresis (CIE) for the detection of anti-liver-kidney microsome (LKM) antibodies in the sera of patients with chronic liver disease. J.Immunol. Methods in press.
Rizzetto, M., Swana, G. and Doniach, D., 1973, Microsomal antibodies in active chronic hepatitis and other disorders. Clin .Exp. Immunol. 15:331.
Roitt, I.N. and Doniach D., 1969, Mammalian Autoimmune Serology. World Health Organisation, Geneve.
Thomas, H.C., 1980, Introduction. Springer Semin. Immunopathol. 3:281.

# EXTRACELLULAR HSV 2 INHIBITION OF LYMPHOCYTE RESPONSE

## IN HEALTHY SUBJECTS

G. Maietta*, P. Grima°, L. Tagliaferro**,
and V. Pellegrino*

\* Immunology Department, Pignatelli Institute, Lecce – Italy
\*\* Virology Department, Pignatelli Institute, Lecce – Italy
° Microbiology Department, University of Lecce – Italy

## INTRODUTION

The relation between herpes simplex virus and immune system are the aim of some studies in these last years. The phenomenon of herpes recurrences has induced some scientists to look for an imbalance in the immune response to the virus. Some data have shown an alterations in the motility of polimorfonuclear cells,[1] an alteration in the ability of lymphocyte to respond to HSV in vitro and to produce lymphokine,[2-8] an alteration in natural killer citotoxicity [9].

The hypothesis that HSV itself is able to inhibit the immune response induced Wainberg et al.[10] to study the effects of HSV on lymphocyte blastogenesis. They reported that HSV was able to inhibit lectin–driven mitogenesis of peripheral blood lymphocyte in patients with HSV recurrences. This phenomenon was non–dependent of virus infectivity because it was demontrable too if the virus was UV treated before the experiments.

Following these evidences we decided to evaluate the herpes simplex virus (HSV) ability to prevent lectin–driven mitogenesis of coincubated human peripheral blood lymphocytes independently of virus infectivity in healthy subjects, so trying to explain the immunosoppressive role of HSV.

## MATERIALS AND METHODS

Healthy subjects – The study consisted of 20 subjects (age 20–40 years) without clinical features of herpes simplex labialis or genitalis infections, all HSV seronegative (Elisa – Clinical Sciences/Medical System).

Virus source – The experiments were performed with HSV type 2 (HSV 2). The virus was isolated from genitalis herpes vescicle ant typed by direct immunofluorescence using monoclonal antibodies (Pathfinder–Kallestad). The virus was grown and quantified in Vero cells cultures (Flow). The culture medium was used as virus source and previously exposed to UV light (OSRAM lamp 15 W) at a distance of 60 cm for 45 minutes to abolish virus infectivity.

Before UV treatment, the presence of HSV in culture medium was documented by the appearance of cytolytic plaque when incubated for 24 hours with a monolayer of Vero cells, not previously infected. The HSV 2 amount was quantified and expressed as Cytolytic Plaque Forming Units (CPFU). After UV exposure the culture medium was inable to induce cytolytic plaque if incubated with Vero cells.

Lectin-driven mitogenesis - Mononuclear cells (mc) were separated from whole blood of healthy subjects using Ficoll-Isopaque (Lymphocyte Separation Medium - Flow) centrifugation, as reported by Boyum [11] and employed at a concentration of $2x10^6$/ml in RPMI 1640 Medium (Flow), supplemented with 10% fetal calf serum (Flow). Phytohemoagglutinin (PHA M - Difco) was used at a concentration of 1:30. Extracellular HSV 2 was used at different concentration so that HSV/lymphocyte ratio was 0.1, 0.5, 1.0, 3.0 and 6.0 .

Mononuclear cells, PHA and HSV 2 were cultured in the same time in sterile microplate (Nunc). $^3$H-thymidine (Amersham) was added to the wells (0.5 microCi/well) after 56 hours of a 72 hours incubation period at 37° C in humidified atmosphere of 5% $CO_2$. After 72 hours, mononuclear cells were collected by means of a cell harvester (Flow) and the proliferation rate was extimated as measurement of incorporated isotope with a beta counter and expressed as Stimulation Index (S.I.) ($\frac{\text{cpm stimulated mc}}{\text{cpm unstimulated mc}}$).

Culture medium of HSV 2 not infected Vero cells was used as control at the same concentration of extracellular HSV 2.

RESULTS

The results of lectin-driven mitogenesis after coincubation between mononuclear cells and extracellular HSV 2 are reported in table 1.

TABLE 1 - LECTIN-DRIVEN MITOGENESIS (S.I.).

| HSV/LY = 6 | | HSV/LY = 3 | | HSV/LY = 1 | | HSV/LY = 0.5 | | HSV/LY = 0.1 | |
|---|---|---|---|---|---|---|---|---|---|
| (−) | (+) | (−) | (+) | (−) | (+) | (−) | (+) | (−) | (+) |
| 25.8 | 9.7 | 27.3 | 15.1 | 28.8 | 22.6 | 27.4 | 22.7 | 28.4 | 24.2 |
| 7.2 | 5.8 | 7.4 | 6.4 | 8.0 | 7.5 | 7.5 | 7.5 | 8.5 | 6.0 |
| 14.7 | 8.4 | 15.3 | 12.9 | 18.4 | 15.8 | 16.5 | 16.1 | 18.3 | 17.2 |
| 18.7 | 8.7 | 20.5 | 13.9 | 20.9 | 18.9 | 20.9 | 19.6 | 23.2 | 19.6 |
| 20.0 | 6.3 | 23.4 | 17.7 | 30.8 | 21.1 | 33.0 | 24.3 | 32.8 | 28.5 |
| 30.5 | 9.9 | 49.8 | 33.0 | 56.5 | 39.6 | 55.6 | 46.6 | 56.5 | 48.8 |
| 6.1 | 1.2 | 15.0 | 3.3 | 15.7 | 8.7 | 15.3 | 10.2 | 15.5 | 12.3 |
| 24.4 | 12.1 | 37.2 | 18.0 | 38.0 | 29.5 | 39.0 | 35.6 | 39.7 | 38.2 |
| 24.6 | 9.0 | 37.7 | 20.3 | 33.9 | 27.9 | 36.9 | 27.3 | 36.7 | 28.2 |
| 59.5 | 11.0 | 60.9 | 37.4 | 61.5 | 53.2 | 74.4 | 53.4 | 74.6 | 54.2 |
| 10.0 | 2.5 | 14.0 | 2.7 | 16.4 | 9.4 | 15.2 | 8.8 | 15.6 | 9.7 |
| 22.7 | 13.4 | 24.6 | 18.0 | 24.8 | 20.2 | 24.9 | 20.5 | 25.2 | 22.4 |
| 43.7 | 23.8 | 44.9 | 30.5 | 45.2 | 33.8 | 45.1 | 38.9 | 45.8 | 39.7 |
| 14.0 | 8.7 | 15.2 | 9.5 | 14.8 | 9.9 | 15.3 | 11.8 | 15.8 | 12.3 |
| 27.9 | 10.6 | 28.5 | 19.4 | 28.5 | 22.3 | 29.2 | 26.8 | 29.0 | 26.6 |
| 51.5 | 38.1 | 51.8 | 42.4 | 52.3 | 45.7 | 52.7 | 46.5 | 52.6 | 47.8 |
| 12.3 | 9.0 | 13.1 | 9.7 | 13.5 | 10.3 | 13.4 | 11.8 | 13.8 | 12.1 |
| 15.6 | 10.8 | 16.3 | 13.4 | 16.5 | 14.5 | 17.2 | 14.9 | 17.0 | 15.6 |
| 38.7 | 15.6 | 39.0 | 25.2 | 39.2 | 29.8 | 39.2 | 34.3 | 39.5 | 36.0 |
| 29.9 | 18.5 | 30.6 | 22.6 | 30.8 | 25.3 | 31.2 | 25.8 | 31.0 | 26.1 |

(−) Absence of HSV 2 in culture medium.
(+) Presence of HSV 2 in culture medium.

258

The proliferation rate is expressed as stimulation index (S.I.).
A decrease of stimulation index is shown in all experiments ant the percentage of inhibitiob of lymphocyte mitogenesis increase with the amount
of extracellular HSV 2 present in the culture, as reported in table 2
and figure 1.

TABLE 2 – PERCENTAGE OF INHIBITION OF LIMPHOCYTE MITOGENESIS

| HSV/LYMPHOCYTE RATIO | % |
|---|---|
| 0.1 | 15.70 + 8.6 |
| 0.5 | 16.81 + 10.4 |
| 1.0 | 22.13 + 10.2 |
| 3.0 | 35.42 + 18.3 |
| 6.0 | 51.63 + 18.6 |

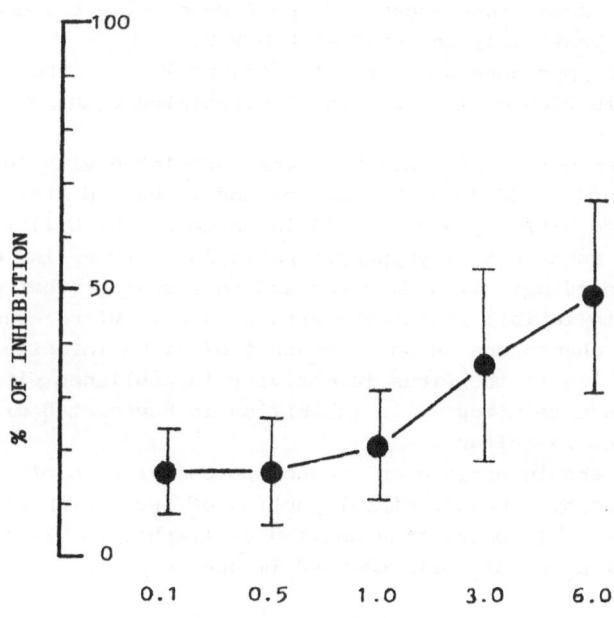

HSV/LYMPHOCYTE RATIO

FIGURE 1 – PERCENTAGE OF INHIBITION OF LYMPHOCYTE MITOGENESIS.

# DISCUSSION

Little is known about immune response to HSV recurrent infections. Some data have shown an alteration at different levels of the immune response.

Plager-Marshall S. et al. in 1978 and Wainberg M. et al. in 1985[12,10] reported that HSV inhibits the lymphocyte blastogenesis in vitro in patients with HSV recurrences, independetly of its infectivity.

Following these data, we studied the effects of culture medium of HSV 2 infected Vero cells (extracellular HSV 2) on lymphocyte blastogenesis after stimulation with phytohemoagglutinin (PHA) in healthy subjects, using UV treated virus.

The lymphocytes were obtained from periferal blood of healthy subjects (20-40 years), seronegative for HSV.

The virus, HSV 2, was isolated from an herpes genitalis vescicle, typed with monoclonal antibodies in direct immunofluorescence and grown in Vero cells culture. The medium was collected and used as virus source (extracellular HSV 2),[13] after exposure to UV light to abolish its infectivity. The presence of HSV 2 in the culture medium was documented by the appearance of cytolytic plaque when an aliquot of medium was incubated for 24 hours with a monolayer of Vero cells. This phenomenon was abolished after UV treatment. The HSV 2 amount was quantified and expressed as Cytolytic Plaque Forming Units (CPFU).

Lymphocyte from healthy subjects were cultured with aliquots of extracellular HSV 2 at different concentrations and with phytohemoagglutinin. After 56 hours $^3$H-thymidine was added in the culture medium and at the 72nd hour the cells were harvested and the proliferation rate was estimated as amount of isotope incorporated from lymphocyte and expressed as Stimulation Index.

As control, the same experiences were performed using culture medium from Vero cells not previously infected with HSV 2.

In all cases the presence of extracellular HSV 2 in lymphocyte cultures induced an inhibition of mitogenesis if correlated with control cultures (table 1).

Moreover the percentage of inhibition was correlated with the amount of extracellular HSV 2 added in the culture and it varied from 15.70 ± 8.6% (when we used an HSV/lymphocyte ratio in culture like 0.1) to 51.63 ± 18.6% (when we used an HSV:lymphocyte ratio in culture like 6.0).

These data, accordingly with Wainberg and co., suggest that extracellular HSV 2 is able to inhibit lymphocyte mitogenesis in vitro even in healthy subjects. This phenomenon is not dependent of virus infectivity, because it is observed too if the virus infectivity is abolished with UV treatment. Last, the lymphocyte mitogenesis inhibition is correlated with the amount of virus present in culture.

These findings should explain an immunosoppressive role of HSV related to the interference by some structural portion of the virion with some membrane protein of cell presenting antigen or lymphocyte implicated in the process of activation of cell mediated immune response.

# REFERENCES

1. Rabson AR, Whiting DA, Anderson R., Glover A, Koornhof HJ, <u>J.Infect. Dis.</u> 1977; 135: 113–6.
2. El Araby II, Chernesky MA, Rawls WE, Dent PB, <u>Clin.Immunol. Immunopathol.</u> 1978; 9: 253–63.
3. Kirchner H, Schwenteck M, Northoff H, Schöpf E, <u>Clin. Immunol. Immuno-pathol.</u> 1978; 11: 267–74.
4. Corey L, Reeves WC, Holmes KK, <u>N. Engl. J. Med.</u> 1978; 299: 986–91.
5. O'Reilly RJ, Chibbaro A, Anger E, Lopez C, <u>J. Immunol.</u> 1977; 118: 1095 – 102.
6. Rasmussen LE, Jordan GW, Stevens DA, Merigan TC, <u>J. Immunol.</u> 1974; 112: 728–36.
7. Shillitoe EJ, Wilton JMA, Lehner T, <u>Infect. Immun.</u> 1977; 18: 130–7.
8. Rattray MC, Peterman GM, Altman LC, Corey L, Holmes KK, <u>Infect. Immun.</u> 1980; 30: 110–6.
9. Thong YH, Vincent MN, Hensen SA, Fuccillo DA, Rola-Pleszczynski M, Bellanti JA, <u>Infect. Immun.</u> 1975; 12: 76–80.
10. Wainberg MA, Portnoy JD, Clecner B, Hubschman S, Lagacé-Simard J, Rabinovic N, Remer Z, Mendelson J, <u>J. Infect. Dis.</u> 1985; 152: 441–7.
11. Boyum A, <u>Tissue Antigens</u> 1974; 4: 269.
12. Plaeger-Marshall S, Smith JW, <u>J. Infect. Dis.</u> 1978; 138: 506–11.
13. Welling-Wester S, Huitema BA, Wilterdink JB, <u>J. Immunol. Methods.</u> 1985; 76: 239–46.

# EXPRESSION OF HERPES SIMPLEX VIRUS 1 GLYCOPROTEIN B IN HUMAN CELLS AND PROTECTION OF MICE AGAINST LETHAL HERPES SIMPLEX VIRUS 1 INFECTION

Roberto Manservigi[+], Massimo Negrini[+], Rita Gualandri[+], Gabriele Milanesi[*] and Giuseppe Barbanti-Brodano[+]

[+]Institute of Microbiology, [*]Institute of Istology and Embriology, University of Ferrara, I-44100, Ferrara

## INTRODUCTION

Expression of cloned viral and cellular genes in eukaryotic cells is relevant to production of vaccines (Melnick, 1986; Moss and Flexner, 1987) and gene therapy (Anderson, 1984). To this purpose numerous eukaryotic vectors have been developed. Since the level of expression of exogenous genes is directly related to the degree of gene amplification (Rigby, 1983) episomal eukaryotic vectors have been constructed where a viral origin of replication directs vector amplification in suitable cells (DiMaio, 1987; Milanesi et al., 1984). However, only two eukaryotic vectors (Milanesi et al., 1984; Sugden et al., 1985) are specific for human cells and they allow expression of genes endowed with their own regulatory sequences. We have therefore devised a viral vector (pBK-1) that replicates and persists episomally in human cells; The structure and general characteristics of this vector have been described in detail (Milanesi et al., 1984). It contains DNA sequences from the plasmid pML, a deletion derivative of pBR322, as well as the origin of replication and the early region of BK virus (BKV), a human papovavirus that efficently replicates in human cells.

The herpes simplex type 1 (HSV-1) genome encodes five major glycoproteins, gB1, gC1, gD1, gE1 and gH1, wich are found in the viral envelope as well as on the surface of infected cells (Spear, 1985). Glycoprotein gB1 is one of the more abundant viral glycoproteins and is essential for entry of the virus into cells. Studies with temperature-sensitive (ts) mutants of gB1 indicate that this glycoprotein is involved in a membrane fusion activity wich is required for penetration (Little et al., 1981).

The essential role or roles carried out by gB may also be important in the replicative cycles of other herpesviruses because gB homologs have been described for a number of human and animal herpesviruses (Gong et al., 1987; Emini et al., 1987; Snowden et al., 1985). There is good evidence that gB, as well as other glycoproteins, plays an important role in the host immune response. Neutralizing antibodies are produced against gB in infected animals and polyclonal antisera to gB mediate immune cytolysis of HSV-infected cells (Spear, 1985).In addition, purified gB can stimulate human memory T lymphocytes (Torseth et al., 1987) and helper T cells induced by purified gB can protect mice against HSV-1 infection (Chan et al., 1985). However, there is also evidence that gB is not a major target for cytotoxic T lymphocytes in mice (Rosenthal et al., 1987).

The gB1 gene has been recently cloned in Escherichia Coli (Bzick et al., 1984) and expressed in yeast (Nozaki et al., 1985), in cells from syrian hamster (Arsenakis et al., 1986) and chinese hamster (Pachl et al., 1987) as well as in simian cells either transiently (Ahmed Ali et al., 1987) or through a vaccinia virus vector (Cantin et al., 1987). Due to the importance of gB1 in HSV-1 biology and phatogenicity, and because man is the natural host for HSV-1 infection, we have cloned the gB1 gene into pBK-1 and detected its expression in transfected human cells.

We have chosen to produce sufficient amounts of gB1 to aid in elucidating the immune response of the HSV-infected host and to test as a subunit vaccine. In this paper we report the establishment of 293 cells expressing cell-associated forms of gB1 and the use of these lines to immunize mice against primary virus infection.


MATERIALS AND METHODS

Plasmid pBK-1 includes sequences from plasmid pML and from the papovavirus BK. Plasmid pJYMMT(E) (Hsiung et al, 1984) consists of a 4 Kb EcoRI fragment containing the mouse metallothionein-I gene end its promoter inserted in pBR322. pSV2-neo (Southern and Berg, 1982) contains sequences from the bacterial gene for aminoglycosyde-3'-fosfotransferase, conferring to mammalian cells resistance to the aminoglycoside antibiotic G418.

293 cells are human embryonic kidney cells transformed by the early region of adenovirus type 5 (Graham et al., 1977). They were maintained in Dulbecco's modified Eagle's medium (DMEM) supplemented with 10% fetal bovine serum. Subconfluent cells were supplied with fresh medium 4 h before transfection. Recombinant plasmid DNA (10 µg for about $10^7$ cells) was transfected by the calcium phosphate coprecipitation technique (Graham and van der Eb, 1973). G418 resistant transformants were selected in DMEM containing 400 µg/ml of the antibiotic. Resulting cell colonies were individually cloned and expanded into cell lines.

For immunoprecipitation experiments cells were maintained in methionine-free medium for 7 h prior to labeling. In induction experiments, during the incubation period without methionine the medium was supplemented with one of the following inducers: CdCl (2 µM), ZnCl (100 µM) or dexamethasone (10 µM). Cells were then labeled for 4 h in methionine-free medium containing 50 µCi/ml of $^{35}$S-methionine (specific activity 1,000 to 2,000 Ci/mmol).

Cell extracts were prepared by lysing cells in 10 mM

sodium phosphate buffer, pH 7.4, containing 1% Nonidet P40, 0.5% sodium deoxycolate and protease inhibitors.
For immunoprecipitation, extraxts were mixed with monoclonal antibody I-144, directed against gB1, and held on ice for 1 h. Formalin-fixed Staphylococcus aureus was added to precipitate the immune complexes. Analysis of immunoprecipitates was performed by electrophoresis in 8.5% SDS-polyacrylamide gels. Gels were then subjected to fluorography, dried and exposed to an X-ray film.
ELISA tests for quantitative analysis of gB1 was performed by a capture method as described previously (Corallini et al., 1987).

RESULTS

The vector pBK-BamG (15.3 Kb), containing the entire coding region of HSV-1 gB gene, was constructed by inserting

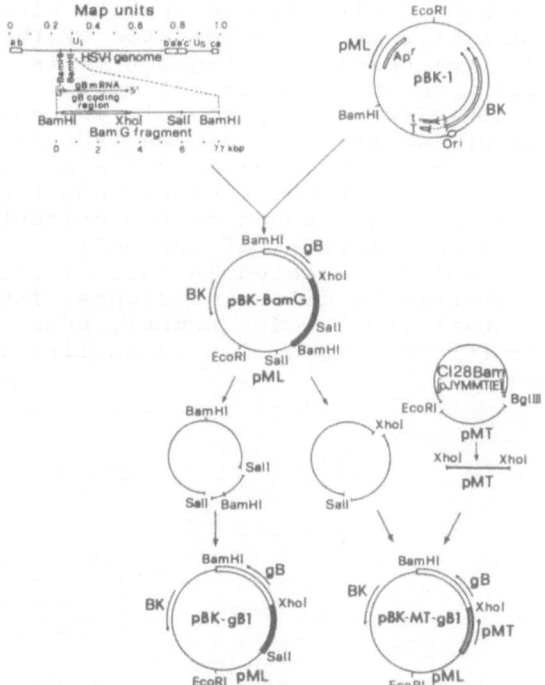

FIG. 1. In the hupper left corner the HSV-1 genome is represented schematically. pBK-BamG was constructed by inserting the 7.7 Kb BamHI G fragment from the HSV-1 genome into the unique BamHI site of the vector pBK-1. A 1,9 Kb fragment was removed from pBK-BamG by digestion with SalI, followed by circularization to obtain pBK-gB1. To construct pBK-MT-gB1, the promoter of the mouse metallothionein-I gene (pMT) was excised from plasmid pJYMM(E) by digestion with EcoRI and BglI1. Endogenous gB1 promoter sequences were excised from pBK-BamG by digestion with XhoI and SalI. Black boxes mark gB1 promoter and nonessential sequences and white boxes indicate gB1 coding sequences.

the BamHI G fragment from HSV-1 DNA into the unique BamHI
site of the vector pBK-1 (Fig. 1). Sequences from pBK-BamG
were employed to construct two further vectors that were
tested for the expression of gB1 in human cells.
The vector pBK-gB1 (13.4 K) was prepared by digesting
pBK-BamG DNA with SalI. In the second recombinant vector,
pBK-MT-gB1 (12.7 Kb), the gB1 gene was linked to the
promoter region of the mouse metallothionein-I gene.
    We transfected pBK-gB1 into human adenovirus
5-transformed 293 cells that produce transcriptional
activators for several eukaryotic viral and cellular genes.
The vector pSV2-neo, was cotransfected with pBK-gB1 at a
ratio of 1:10 to 1:30 and stable transformants were selected
in medium containing G418. Seven out of twenty five clones,
analysed for gB1 production by immunoprecipitation with
monoclonal antibody I-144, were found to express gB1. Fig.
2A shows immunoprecipitates from two of the positive clones,
4B6 and 8B3. Two bands of 120 and 110 kilodaltons,
corresponding in size to the mature form of gB1 and to its
precursor (pgB1), were clearly visible in both clones. Two
proteins of identical size were precipitated by the same
monoclonal antibody from HSV-1 infected 293 cells.
    Similar results were obtained in 293 cells
cotransfected with pBK-MT-gB1 and pSV2-neo. The data on
immunoprecipitation of two positive clones, 3C4 and 3C5, are
presented in Fig. 2B. gB1 production in 3C4 and 3C5 cell
lines was increased 1.3 to 2.5 folds after induction with
cadmium or zinc (Fig. 3). The amount of recombinant gB1 was
evaluated by ELISA on cell lysates of 293 cell
transformants. The results presented in Table 1 show that
gB1 production is variable in different clones. The quantity
produced correlates with vector copy number, suggesting that
the critical factor for gB1 expression is amplification of
the recombinant DNA.

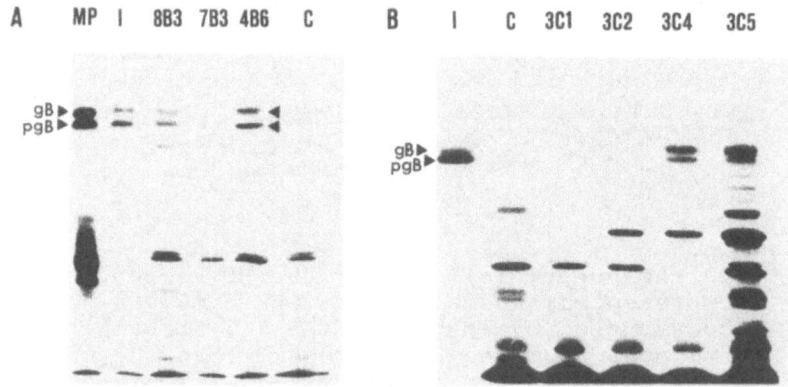

    FIG. 2. Radioimmunoprecipitation by monoclonal antibody
            I-144 of gB1 expressed in selected clones
            obtained by transfection of 293 cells with
            pBK-gB1 (A) and pBK-MT-gB1
            (B). Immunoprecipitates from control cultures
            include uninfected 293 cells (C) and 293 cells
            infected by HSV-1 strain MP (I). The precursor
            (pgB) and mature (gB) forms of gB1 are
            indicated by closed arrows.

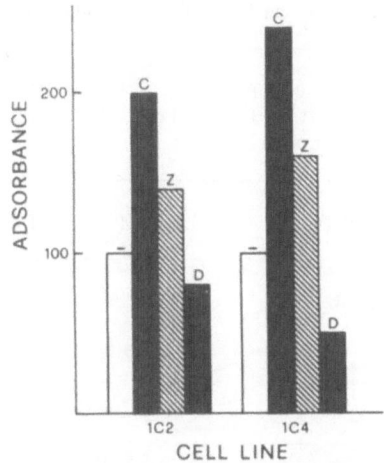

FIG. 3. Regulation of gB1 expression by cadmium (C),
zinc (Z) and dexamethasone (D) in 293 cell 3C4
and 3C5 clones transformed by pBK-MT-gB1.
Adsorbance is in arbitrary units calculated
from densitometric scanning of
immunoprecipitated bands, taking as 100 the
value of the untreated control (-).

Mice immunized with cell lysates of 293 cell clone 3C4
produced neutralizing antibody to HSV-1 strain F and, with a
lower titer, to HSV-2 strain G. They were protected from
paralysis and death after a lethal challange with HSV-1. No
antibody production and protection were detected in mice
immunized with the concentrated culture fluid of 293 cell
clone 3C4 or with cell lysate of normal 293 cells (Table 2).

Table 1. Quantitative determinantion of gB1
produced by 293 cell transformants

| Cells | $OD_{492}/300$ µg protein | Vector Copy Number |
|---|---|---|
| Normal 293 | 0 | ND |
| HSV-1 F infected 293 | >2.00 | ND |
| 293 transformants: | | |
| 4B6 | 1.14 | 20 |
| 8B3 | 0.45 | 7 |
| 3C4 | 1.31 | 20 |
| 3C5 | >2.00 | 28 |

Cell lysates  were analysed  by ELISA as
described under Materials and Methods. Vector copy
number was determined by densitometric analysis of
hybridization bands in comparison with known
amounts of control DNA. ND, not done.

Table 2. Immunogenicity of recombinant gB1 in mice

| Antigen | In vitro neutralization titer($\log_2$) | | HSV-1 challenge Number of mice |
|---|---|---|---|
| | HSV-1 | HSV-2 | Dead/Normal |
| 293 Clone 3C4 cell lysate | 6.6 | 4.4 | 0/10 |
| Normal 293 cell lysate | <3 | <3 | 10/0 |

Cell lysates were prepared as for immunoprecipitation (see Materials and Methods) and clarified by centrifugation at 15,000 g for 30 min. Groups of 10 two-month-old female Balb/C mice were immunized with cell lysates. Each animal received 12 mg of protein. For group 1 this amount is equivalent to 157 $OD_{492}$ (ELISA) of recombinant gB-1. The antigen was mixed with an equal volume of complete Freund adjuvant and administered intraperitoneally. A second immunization was repeated after 20 days and animals were bled 15 days after booster. Neutralization titers of sera were determined as 100% plaque reduction end-points on Vero cells infected by HSV-1 strains F and by HSV-2 strain G. Mice were challenged 30 days after booster by inoculation of the HSV-1 encephalitogenic strain 13 (500 $LD_{50}$) into the rear right foot pad.


DISCUSSION

Transformation of human cells with pBK-1/gB1 recombinants led to the isolation of stable clones that expressed structurally and functionally normal gB1. Since reactions leading to gB1 glycosylation and processing are carried out by cellular factors and enzymes (Campadelli-Fiume and Serafini-Cessi, 1985, the expression in human cells would ensure that gB1 antigenic determinants are identical to those of the virion glycoprotein produced during natural infection in humans.
The level of expression was different in various clones, but some transformants produced gB1 as efficently as in HSV-1 lytic infection. The amount of gB1 produced was related to the properties of the vector that replicated to about twenty copies per cell with consequent amplification of the inserted gB1 gene. We succeed, in obtaining constitutive expression of gB1 in 293 cells, where endogenous adenovirus trans-activating factors may substitute for HSV-1 early proteins in the induction of the transfected gene for gB1, suggesting that herpesvirus and adenovirus early proteins may act through a set of common intermediate cellular factors in the induction of cellular and viral genes.

gB1 is immunogenic, eliciting both humoral and
cell-mediated immune responses (Norrild et al;, 1979; Dix
and Mills, 1985) and anti-gB antibodies are protective
against HSV infection in mice (Balachandran et al., 1982;
Rector et al., 1982; Kino et al., 1985). In these
experiments, mice immunized with recombinant gB1 produced
HSV-1 and HSV-2 neutralizing antibodies and were protected
from a lethal challange with HSV-1, suggesting a possible
development for preparation of a subunit vaccine. Cross
reaction of gB1 with equivalent glycoproteins of HSV-2
(Norrild, 1980) and varicella-zoster virus (Edson et al.,
1985; Kitamura et al., 1986) could be exploited to confer
protection against these infections. The system could also
provide diagnostic material useful to detect and monitor
HSV-1 infection by immunological tests.

This work was supported by the Italian National
Research Council, Special Projects "Oncology", "Control of
Infectious Diseases", "Genetic Engineering and Molecular
Basis of Hereditary Diseases", by North Atlantic Treaty
Organization research grant B6-170 and by Associazione
Italiana per la Ricerca sul Cancro (A.I.R.C.). We thank A.
Bevilacqua and M. Bonazzi for excellent technical assistance
and for preparing the manuscript.

REFERENCES

AHMED ALI, M., BUTCHER, M., and GHOSH, H.P., 1987,
    Expression and nuclear envelope localization of
    biologically active fusion glycoprotein gB of herpes
    simplex virus in mammalian cells using cloned DNA. Proc.
    Natl. Acad. Sci. USA, 84:5675.
ANDERSON, W.F., 1984, Prospects for human gene therapy.
    Science, 226:401.
ARSENAKIS, M., HUBENTHAL-VOSS, J., CAMPADELLI-FIUME, G.,
    PEREIRA, L., and ROIZMAN, B., 1986, Construction and
    properties of a cell line constitutively expressing the
    herpes simplex virus glycopretein B  dependent on
    functional  4 protein synthesis, J. Virol., 60:674.
BALACHANDRAN, N., BACCHETTI, S., and RAWLS, W.E., 1982,
    Protection against lethal challenge of BALB/c mice by
    passive transfer of monoclonal antibodies to five
    glycoproteins of herpes simplex virus type 2.
    Infect. Immun.,37:1132.
BZIK, D.J., FOX, B.A., DeLUCA, N.A., and PERSON, S., 1984,
    Nucleotide sequence specifying the glycoprotein gene, gB,
    of herpes simplex virus type 1. Virology, 133:301.
CAMPADELLI-FIUME, G., and SERAFINI-CESSI, F., 1985,
    Processing of the oligosaccharide chains of herpes
    simplex virus type 1 glycoproteins. In:
    "The Herpesviruses", Vol. 3, B. Roizman, ed., Plenum
    Press, New York.
CANTIN, E.M., EBERLE, R., BALDICK, J.L., MOSS, B.,
    WILLEY, D.E., NOTKINS, A.L., and OPENSHAW, H., 1987,
    Expression of herpes simplex virus 1 glycoprotein B by a
    recombinant vaccinia virus and protection of mice against
    lethal herpes simplex virus 1 infection. Proc. Natl.
    Acad. Sci. USA, 84:5908.

CORALLINI, A., PAGNANI, M., VIADANA, P., SILINI, E., MOTTES, M., MILANESI, G., GERNA, G., VETTOR, R., TRAPELLA, G., SILVANI, V., GAIST, G., and BARBANTI-BRODANO, G., 1987, Association of BK virus with human brain tumors and tumors of pancreatic islets. Int. J. Cancer, 39:60.

DiMAIO, D., TREISMAN, R., and MANIATIS, T., 1982, Bovine papillomavirus vector that propagates as a plasmid in both mouse and bacterial cells. Proc. Natl. Acad. Sci. USA, 79:4030.

GRAHAM, F.L., and van der EB, A.J., 1973, A new technique for the assay of infectivity of huamn adenovirus 5 DNA. Virology, 52:456.

GRAHAM, F.L., SMILEY, J., RUSSEL, W.C., and NAIRN, R., 1977, Characteristics of a human cell line transformed by DNA from adenovirus type 5. J. Gen. Virol., 36:59.

HSlUNG, N., FITTS, R., WILSON, S., MILNE, A., and HAMER, D., 1984, Efficient production of hepatitis B surface antigen using a bovine papilloma virus-metallothionein vector. J. Mol. Appl. Genet., 2:497.

KINO, Y., ETO, T., OHTOMO, N., HAYASHI, Y., YAMAMOTO, M., and MORI, R., 1985, Passive immunization of mice with monoclonal antibodies to glycoprotein gB of herpes simplex virus. Microbiol. Immunol., 29:143.

KITAMURA, K., NAMAZUE, J., CAMPO-VERA, H., OGINO, T., and YAMANISHI, K., 1986, Induction of neutralizing antibody against varicella-zoster virus (VZV) by VZV gp3 and cross-reactivity between VZV gp3 and herpes simplex viruses gB. Virology, 149: 74.

LITTLE, S.P., JOFRE, J.T., COURTNEY, R.J., and SCHAFFER, P.A., 1981, A virion-associated glycoprotein essential for infectivity of herpes simplex virus type 1. Virology, 115:149.

MELNICK, J.L., 1986, Virus vaccines: 1986 update. Prog. Med. Virol., 33:134.

MOSS, B., AND FLEXNER, C., 1987, Vaccinia virus expression vectors. Annu. Rev. Immunol., 5:305.

MILANESI, G., BARBANTI-BRODANO, G., NEGRINI, M., LEE, D., CORALLINI, A., CAPUTO, A., GROSSI, M.P., and RICCIARDI, R.P., 1984, BK virus-plasmid expression vector that persists episomally in human cells and shuttles into Escherichia coli. Mol. Cell. Biol., 4:1551.

NORRILD, B., 1980, Immunochemistry of herpes simplex virus glycoproteins. Curr. Top. Microbiol. Immunol., 90:67.

NORRILD, B., SHORE, S.L., and NAHMIAS, A.J., 1979, Herpes simplex virus glycoproteins: participation of individual herpes simplex virus type 1 glycoprotein antigens in immunocytolysis and their correlation with previously identified glycopolypeptides. J. Virol., 32:741.

NOZAKI, C., MAKIZUMI, K., KINO, Y., NAKATAKE, H., ETO, T., MIZUNO, K., HAMADA, F., and OHTOMO, N., 1985, Expression of herpes simplex virus glycoprotein B gene in yeast. Virus Research, 4:107.

PACHL, C., BURKE, R.L., STUVE, L.L., SANCHEZ-PESCADOR, L., VAN NEST, G., MASIARZ, F., and DINA, D., 1987, Expression of cell-associated and secreted forms of herpes simplex virus type 1 glycoprotein gB in mammalian cells. J. Virol., 61:315.

RECTOR, J.T., LAUSCH, R.N., and OAKES, J.E., 1982, Use of monoclonal antibodies for analysis of antibody-dependent immunity to ocular herpes simplex virus type 1 infection. Infect. Immun., 38:168.

RIGBY, P.W.J., 1983, Cloning vectors derived from animal viruses. J. Gen. Virol., 64:255.

SOUTHERN, P.J., and BERG, P., 1982, Transformation of mammalian cells to antibiotic resistance with a bacterial gene under control of the SV40 early region promoter. J. Mol. Appl. Genet., 1:327.

SUGDEN, B., MARSH, K., AND YATES, J., 1985, A vector that replicates as a plasmid and can be efficiently selected in B-limphoblasts transformed by Epstein-Barr virus. Mol. Cell. Biol., 5:410.

# HEPATITIS DELTA VIRUS AND AUTOANTIBODIES TO EPITHELIAL CELLS

Fabio Miserocchi, Daniela Zauli and Emilio Pisi

Istituto di Clinica Medica II, University of Bologna

Bologna, Italy

## INTRODUCTION

It is well established that chronic liver disease (CLD) due to hepatitis delta virus (HDV) is associated with a variety of autoantibodies (Crivelli et al., 1983; Zauli et al., 1984; Zauli et al., 1986). The type of autoantibody which appears to segregate more closely with such condition is certainly the basal cell layer antibody (BCLA). BCLA decorates in immunofluorescence the basal cells of squamous epithelia, such as rat fore stomach which is currently used to detect the antibody in human sera (Lenkei et al., 1983).

BCLA is detectable in approximately 50% of patients with CLD due to HDV and only in a minority of patients with other types of CLD, including hepatitis B virus CLD. It is absent in other conditions and in healthy controls (Zauli et al., 1984).

In 1985 Lenkei et al. reported the close association of BCLA with an autoantibody reacting with epithelial cells of human thymus (TECA). It was not clear, however, whether the two antibodies are distinct or share the same antigenic specificity.

To answer the question and to better define the antigenic specificity of the two antibodies, we have now performed absorption experiments and immunofluorescence studies with monoclonal antibodies, which will be reported in the present paper.

## MATERIAL AND METHODS

Sixty-eight sera from patients with CLD (chronic persistent and chronic active hepatitis with and without cirrhosis) were studied. All cases were HDV and/or anti-HDV positive. They were obviously also HBsAg and anti-HBc positive. Most of them were anti-HBe and only two were HBeAg positive.

BCLA was determined by an indirect immunofluorescence technique on rat forestomach using serum initially diluted 1:40. TECA was detected by an identical technique on human thymus (obtained from children undergoing cardiac surgery) using serum initially diluted 1:20. A FITC-conjugated rabbit anti-human immunoglobulin (Wellcome) and anti-human IgG (Behring-werke AG) antisera were used as second antibodies.

The following monoclonal antibodies were used on the same tissues by an indirect immunofluorescence technique: AE1 (anti-acid keratins, a kind gift of T.-T. Sun, New York), anti-keratin A, B, C and anti-panfilament (Ortho). A FITC-conjugated anti-mouse immunoglobulin was used as second antibody.

In double immunofluorescence studies with human and monoclonal anti-bodies a TRITC-conjugated anti-mouse immunoglobulin was used.

For absorption experiments rat forestomach and human thymus were ho-mogenized in phosphate buffered saline, centrifuged at 500 g for 30 minu-tes and used to absorb three BCLA and TECA positive sera. Sera were left overnight at 4°C, centrifuged as above and the supernatants tested on rat forestomach and human thymus by an indirect immunofluorescence method.

Statistical analysis was performed by the Wilcoxon rank sum test with Yates' correction.

RESULTS

Of the 68 sera 36 (53%) were positive for BCLA. Twenty-six (72%) of the 36 were also positive for TECA. All but one TECA positive sera were BCLA positive. Titres of BCLA in TECA positive sera were significantly higher than those of TECA negative ones ($p < 0.01$).

Absorption of sera with rat forestomach and human thymus extracts completely abolished reactivities on both tissues.

The reactivities of the various monoclonal antibodies can be summa-rized as follows: AE1 and anti-keratin B antibodies gave the same pattern as spontaneous sera on both rat forestomach and human thymus; anti-keratin A did the same, but also stained the suprabasal layer of rat forestomach, whereas anti-keratin C stained only the latter; antipanfilament antibody was completely negative on both tissues.

In double immunofluorescence the reactivity of AE1 and anti-keratin B antibodies was superimposable on that of BCLA and TECA.

DISCUSSION

The close correlation between BCLA and TECA described by Lenkei et al. (1985) is confirmed by the present data. In our experience 72% of BCLA positive and only one BCLA negative sera reacted with epithelial cells of human htymus. This would indicate that epithelial cells of the basal layer of rat forestomach and of human thymus express antigens recognized by BCLA, although in the latter tissue they are probably represented less. This suggestion comes from the observation that only high-titre BCLA react with human thymus. These conclusions are also supported by the results of the absorption experiments, which demonstrated that the reactivity of both an-tibodies is abolished by the incubation with the two substrates.

The second problem was to identify the target antigen(s) of the anti-bodies. In our hands keratins seem the best candidate, as two anti-keratin monoclonal antibodies display an immunofluorescence pattern superimposable on that of the spontaneous antibodies. The negative results obtained by Lenkei et al. (1985) with a monoclonal antibody to cytokeratin are proba-bly due to the fact that the types of keratin to which the antibody was directed are not expressed by epithelial cells of the basal layer and of the thymus. At least 19 different keratins have been identified, which are differently expressed by the different epithelial tissues. This is confirmed by the negative results obtained by the use of the anti-keratin C monoclonal antibody. Thus the discrepancy may be related to the different antigenic specificity of the anti-keratin antibodies employed.

In conclusion, we suggest that the autoantibodies associated with chronic liver disease due to HDV are directed to keratins or alternatively to antigens which are co-expressed with keratins. The use of other techniques might clarify the problem. Preliminary immunoblot experiments with keratins extracted from rat forestomach have, however, given, at least in our hands, unconclusive results.

## REFERENCES

Crivelli, O., Lavarini, C., Chiaberge, E., Amoroso, A., Farci, P., Negro, F., and Rizzetto, M., 1983, Microsomal autoantibodies in chronic infection with the HBsAg associated delta agent. Cli. exp. Immunol., 54:468.

Lenkei, R., Biberfeld, G., Magnius, L.O., Fagraeus, A., and Biberfeld, P., 1985, Autoantibodies to the basal cells of squamous epithelium react with thymic epithelial cells. Clin. Immunol. Immunopathol., 34:11.

Lenkei, R., Biberfeld, G., Buligescu, L., Tovaru, S., Biberfeld, P., Hilborn, L., and Magnius, L., 1983, Autoantibodies against the basal cell layer: an association with chronic hepatitis B. Clin. Immunol. Immunopathol., 26:436.

Zauli, D., Fusconi, M., Crespi, C., Bianchi, F.B., Craxi, A., and Pisi, E., 1984, Close association between basal cell layer antibodies and hepatitis B virus-associated chronic delta infection. Hepatology, 4:1103.

Zauli, D., Crespi, C., Bianchi, F.B., Craxi, A., and Pisi, E., 1986, Auto immunity in chronic liver disease caused by hepatitis delta virus. J. Clin. Pathol., 39:897.

# IMMUNODEFICIENCY ASSOCIATED WITH MUMPS VIRUS.

## STUDY ON A FOUR-PEOPLE FAMILY

F. Squadrini, F. Barboni*, F. Taparelli. B. De Rienzo, E. Bertrandi*, and G. Lami

Istituto di Clinica delle Malattie Infettive e Tropicali, Università di Modena
* Lab. Analisi Cliniche, Ospedale Prov. Spec. Malpighi, Bologna

## INTRODUCTION

The monoclonal antibodies technique (KOHLER and MILTSEIN, 1975) allowed to identify a great number of human T-lymphocyte sub-populations, showing different characteristics. Such method revealed that the most important immunological anomalies are connected with one or both those cellular T populations.

Then, a high number of TS has been reported in some viral infections (E.B.V., C.M.V.) (ROUTHIER et al., 1980).

Taking into account these new data, we decided to study a four-people family in which two members (mother and son) did not show a C.F. antibody seroconversion towards the epidemic parotitis v., after a clinically typical disease in both, while the son showed a meningoencephalitic complication.

## Methods

All patients underwent the following immunological tests: serum immunoglobulins determination, study of lymphocytes subpopulations (rosette E, EAC) monoclonal OKT4, OKT8 antibodies for TH and TS subpopulations (NIH long term) and OKT4 antisera for OKT8 inducers for suppressors both diluted 1 : 100. "In vitro" blastization test through PHA; con.A., PWM with incorporation of TH3 thymidine.

## Results

As far as the rosette test is concerned, only the patient with meningoencephalitis showed an increase of T and a decrease of lymphocytes B. The blastization test showed a decrease of the response only in mother and son. Examining the sub-population, we observed an increase of T8 with an OKT4/ OKT8 ratio of 0.6 for the son and 1 for the mother (physiological 1.5), while father and daughter were normal.

Table 1

| daughter | IgG | IgA | IgM | Rosette % E-ART | RAC |
|---|---|---|---|---|---|
| patient | 615 | 65 | 57 | 87 | 12 |
| daughter | 631 | 40 | 86 | 77 | 18 |
| mother | 886 | 88 | 89 | 77 | 24 |
| father | | | | 73 | 24 |
| controls | 700–1700 mg% | 70–350 mg% | 70–210 mg% | 78:3 | 20:3 |

The table show a great decrease of all immunoglobulins, most evident in the two children, regarding IgA.

Table 2

Reactivity with Monoclonal AB

| | T4 | T8 | OKT4 / OKT8 |
|---|---|---|---|
| p. | 40 | 60 | 0.6 |
| d. | 60 | 30 | 2 |
| m. | 50 | 50 | 1 |
| f. | n.d. | n.d. | n.d. |
| c. | 60 | 40 | 1.5 |

## Conclusion

In a four-people family, two members (mother and son) were afflicted with parotitis (mother) and parotitis with meningoencephalitis (son). Three members showed an evident immunoglobulinic deficit, particularly regarding IgA in the two children. The blastization test showed a deficit in the "in vitro" response to mitogens only in the two patients.

In the two patients (mother and son) we could report, by making use of the monoclonal antibody technique, an immunological hyperactivity connected with T-suppressor.

REFERENCES

1. Kohler C., Milstein C.
   Continuous culture of fused cells secreting antibody of predefined specificity.
   Nature 1975, 236, 495-497.

2. Routhier G., Epstein O., Janossy G., Thomas H.C., Sherlock S.
   Effects of cyclosporin a on suppressor and inducer T lymphocytes in primary biliary cirrhosis.
   Lancet, 6, 1223, 1980.

# CELL-MEDIATED IMMUNITY IN INFECTIOUS MONONUCLEOSIS

G. Verucchi, L. Attard, M. Maldini* G. Gelati, A. Boschi,
E. Beltrandi* G. Fasulo, A. Moroni° R. Mancini* and F. Chiodo

Infectious Diseases Institute - University of Bologna
° Microbiology Institute - University of Bologna
* Central Dept. of Clinical Pathology - M. Malpighi Hospital
Bologna - Italy

## INTRODUCTION

Infectious Mononucleosis (I.M.) is an acute viral illness caused by Epstein-Barr virus (EBV), the only Herpes virus of humans that is lymphotropic (Sugden, 1982). It has been recognized all over the world, but the clinical disorder is rarely reported in developing countries.

Primary infections with EBV occur commonly during childhood but usually remain asymptomatic (Fleisher et al. 1979) whereas 50% of infected young adults (15-25 years) develop a clinical syndrome (Crawford and Edwards, 1988). However, EBV antibodies are detectable in 90% of the adult population (Vella, Rocchi, 1988).

Oropharyngeal excretion of EBV, linked to infection transmission, occurs in approximately 80% of patients with acute I.M. and continues for months thereafter.

Following an incubation period of 30-50 days, in children this is shorter (10-14 days), a prodromal stage of 3-5 days including mild symptoms such as headache, malaise and fatigue is frequent (Niederman, 1977).

In the manifested phase of the disease, clinical features of I.M. are extremely variable in severity; in more than 80% of patients they include fever, exudative tonsillitis or pharyngitis and cervical lymphoadenopathy. Generalized adenopathy may also develop during the course of the illness. Moderate enlargement of the spleen develops in about 50% of the cases (Niederman, 1977; Crawford and Edwards, 1988).

Frequently, liver function tests result abnormal for several weeks, but jaundice is uncommon (4-5%) and hepatomegaly is detectable in only 15-25% of cases.

During the early phases of I.M. a faint rash on the trunk and proximal extremities can appear. This eruption is either maculopapular and frequent in the patient treated with ampicillin, or erythematous and rarer in all the others. (Crawford and Edwards, 1988).

Bilateral supro-orbital edema may also be a transient but typical finding early in the course of the disease.

Fewer than 1% of patients with I.M. have complaints concerning the nervous system. These manifestations range from aseptic meningitis and encephalitis to Guillain-Barrè syndrome but most patients recover completely (Baker et al., 1983; Vella, Rocchi, 1988).

During the acute phase, the diagnosis is confirmed by the detection of specific IgM and IgG antibodies to viral capsid antigen (VCA). Only after 2-3 months are found the antibodies to nuclear antigen (EBNA) (Fleisher et al., 1979; Sumaya, 1986; Marklund,et al. 1986).

A small number of patients do not recover completely and develop "Chronic Mononucleosis Syndrome", difficult to define and characterized by fatigue, general malaise, slight temperature, myalgia, limphoadenopathy,headache, nausea, depression, lack of concentration and decrease in work ability.In this pathological picture generally an anomalous persistence of the titre of the antibodies to precocious antigens (EA) and the absence of EBNA antibodies is revealed. (Tobi et al., 1982; Horwitz, et al. 1985;Jones, 1986; 1986; Kormaroff, 1987; Straus, 1988; Vella, Rocchi, 1988).

A typical feature of I.M. recognized around the end of the first or early in the second week of illness is a sharp increase of lymphocytes,including atypical cells. In fact, after entering the organism, the EBV attaches itself to specific receptors (EBVR) present on the B lymphocyte membrane, thereby infecting them. As a consequence, a T lymphocyte activation stimulated by infected B lymphocytes occurs, causing the onset of atypical elements with cheracteristic monocytoid morphology. (Pattengale, et al.1974; Tosato et al., 1979, Crawford et al., 1981, Sugden, 1982; Svedmyr, et al. 1984). From a functional point of view, this behaviour represents a tendency towards the self-limitation of the disease.

By employing instruments like the flow cytofluorimeter it has been possible, in addition to improving diagnosis, to identify particular cellular expressions which better define the immunitary response in I.M.


MATERIALS AND METHODS

Fifty three patients, ages ranging from 4 to 28 years, 29 males and 24 females, affected by acute I.M. were examined and a 1-6 months follow-up for 28 of them was made.

Diagnosis was made by means of a symptomatological picture characterized by angina, lymphoadenopathy, hepatomegaly, splenomegaly, fever, rash, together with typical lymphocytosis, positivity of the Paul-Bunnell reaction and detection of specific EBV antibodies.

VCA antibodies of IgM and IgG classes were detected by indirect immunofluorescense technique, EBNA antibodies by employing anti-complement immunofluorescence.

In evaluating the lymphocyte subsets, an EDTA-treated peripheral blood sample was incubated for 15 minutes at room temperature with fluorescent monoclonal antibodies (Becton-Dickinson) employing a double labelling technique, and, after lysogenic solution supplementation, for 10 minutes at room temperature, the specimen was centrifugated and washed in PBS. Afterwards, this preparation was analyzed on a cytofluorimeter FACSCAN.

Lymphocyte functional activity by means of aspecific mitogens (PHA,PWM, ConA) was evaluated.

As a casual cross-section of healthy population, 16 people were tested as a comparison.


RESULTS

The acute phase of the disease in our patients was characterized by a considerable increase of CD8 lymphocyte per cent value and by a CD4 per cent decrease, thereby causing CD4/CD8 cells ratio inversion.

These alterations have resulted expressive as regards healthy population (Fig. 1). After data analysis of absolute values, a CD8 increase was noted whereas CD4 remained in the normality (Fig. 2).

By the use of double labelling immunofluorescence, a lymphocyte subset co-expressing both CD3 and HLA-DR antigens as surface markers was revealed during the acute stage (Fig. 3A). After further probing investigations, it was remarked that most of these "activated" cells belong to CD8 phenotype subset (Fig. 3B).

In the acute phase a reduced "in vitro" lymphocyte response to aspecific mitogens was observed (Fig. 4).

During the follow-up of our patients, a progressive normalization of the above-mentioned immunological alteration was revealed. In particular, the CD4/CD8 ratio inversion noted in 83% of the patients in acute phase, was seen in 32.1% after 1 month and in 11.5% after 3 months; also the lymphocyte subset espressing double receptors showed a gradual decrease from 90.6%, in the acute phase, to 46,4% after 1 month, reaching 15.4% after 3 months. No alteration was observed after 6 months following acute phase (Tab. 1).

An observation of specific EBV antibodies has shown that 6 months, following the acute phase of I.M., 19.1% of the patients had not yet developed the antibodies to EBNA, whereas high titres of anti-VCA IgG (1:256) persisted (Tab. 1). These were 4 patients, presenting a low percentage of HLA-DR/CD3 double receptors, who after 1 month following the acute phase revealed a CD4/CD8 ratio > 1 and disappearance of double receptors cells.

From a clinical point of view, the disease in these patients has had a severe course and, during the convalescent stage, an abnormal predisposition to be affected by respiratory infections and persistent state of fatigue even 6 months after the acute phase, has been verified.

Tab.1: Serological and immunological findings during I.M.: specific antibodies behaviour, HLA-DR/CD3 double receptors presence and CD4/CD8 ratio inversion.

|  |  | Acute phase | 1 month | 3 months | 6 months |
|---|---|---|---|---|---|
| No pts |  | 53 | 28 | 26 | 21 |
| Anti-VCA | IgM | 100 % | 39.3% | – | – |
|  | IgG | 100 % | 100 % | 100 % | 100 % |
| Anti-EBNA |  | – | 14.3% | 46.1% | 80.9% |
| HLA-DR/CD3 |  | 90.6% | 46.4% | 15.4% | – |
| CD4/CD8 < 1 |  | 83.0% | 32.1% | 11.5% | – |

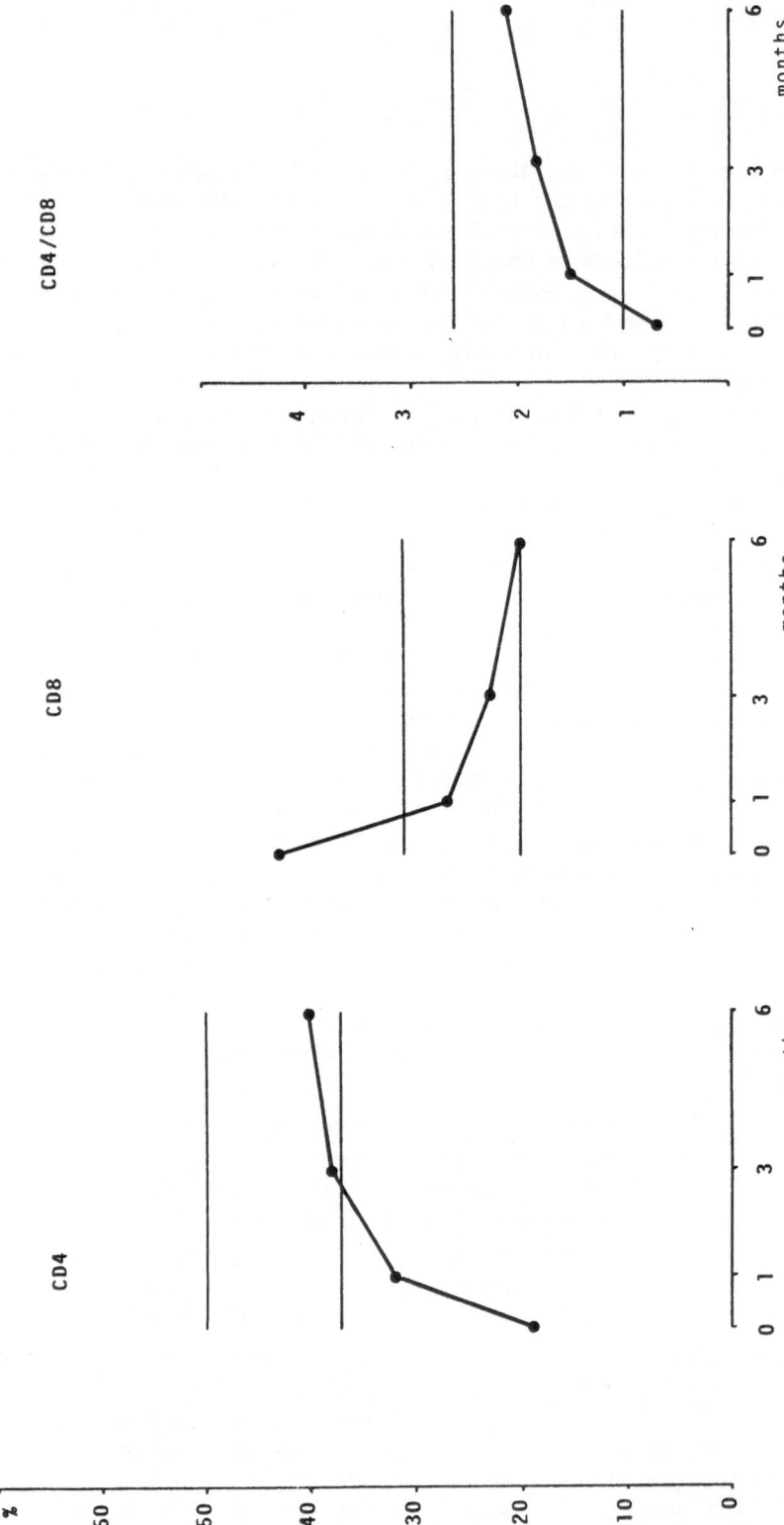

Fig. 1. Percentages of the CD4 and CD8 lymphocyte subsets in patients with I.M.

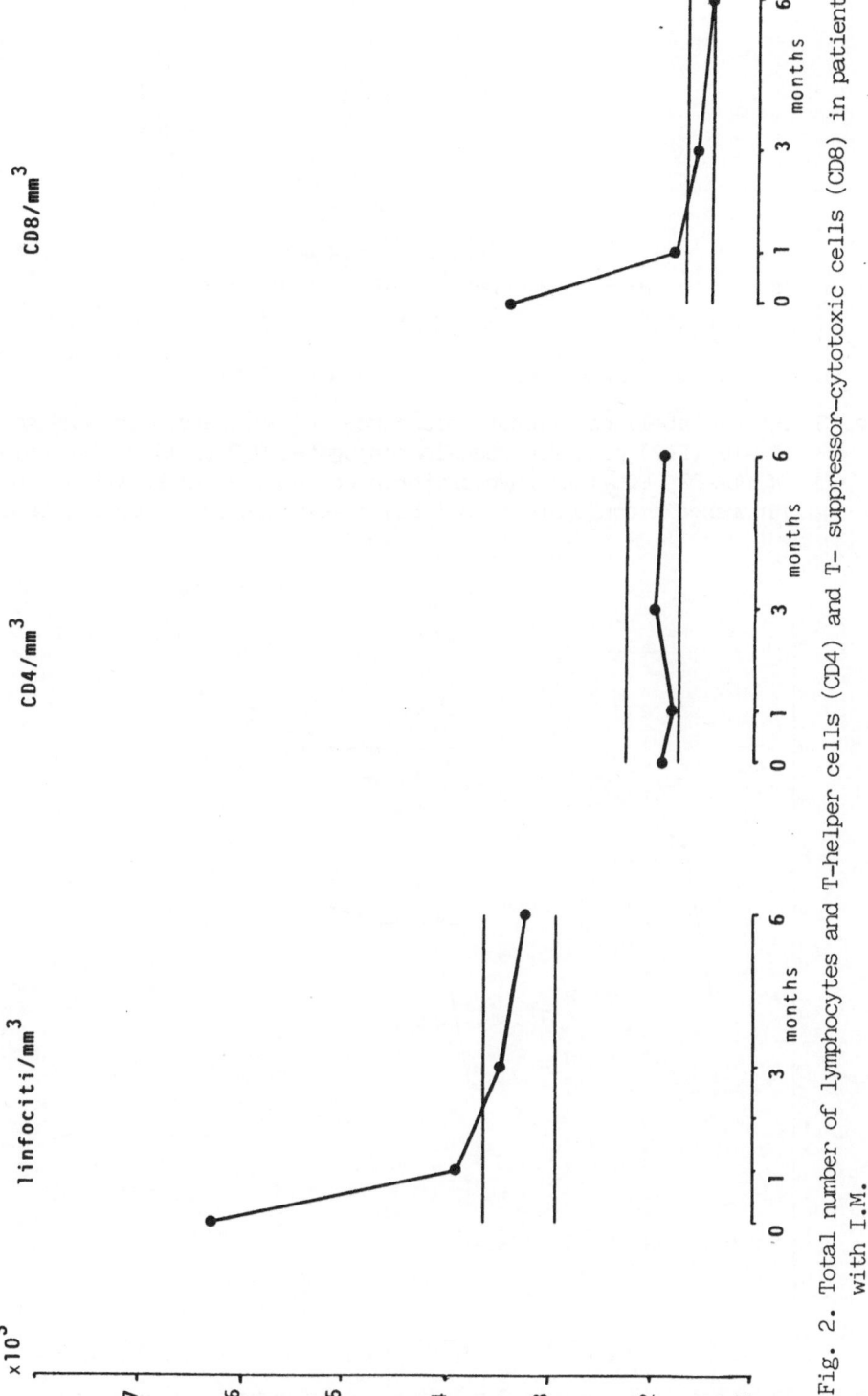

Fig. 2. Total number of lymphocytes and T-helper cells (CD4) and T- suppressor-cytotoxic cells (CD8) in patients with I.M.

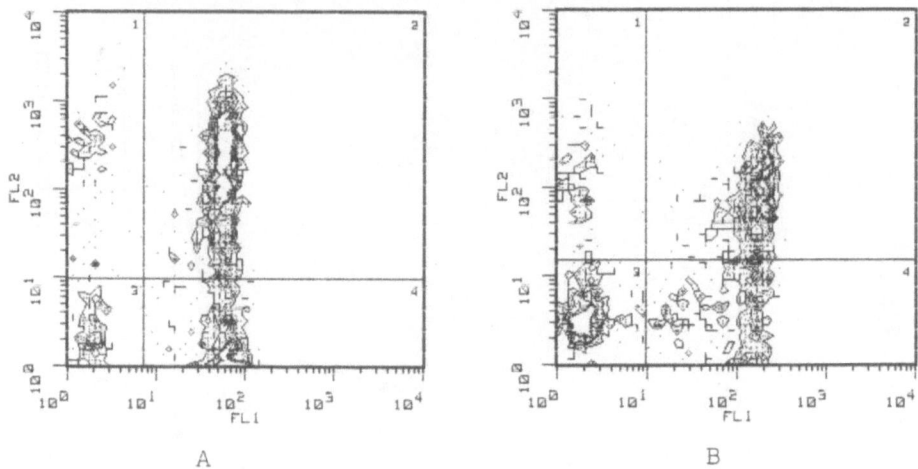

Fig. 3. Double labelling immunofluorescence: A: phycoerythrin conjugated
   HLA-DR (FL2) plus fluorescein conjugated CD3 (FL1) co-expression
   of HLA-DR, CD3+. B: phycoerythrin conjugated HLA-DR (FL2) plus
   fluorescein conjugated CD8 (FL1) co-expression of HLA-DR, CD8.

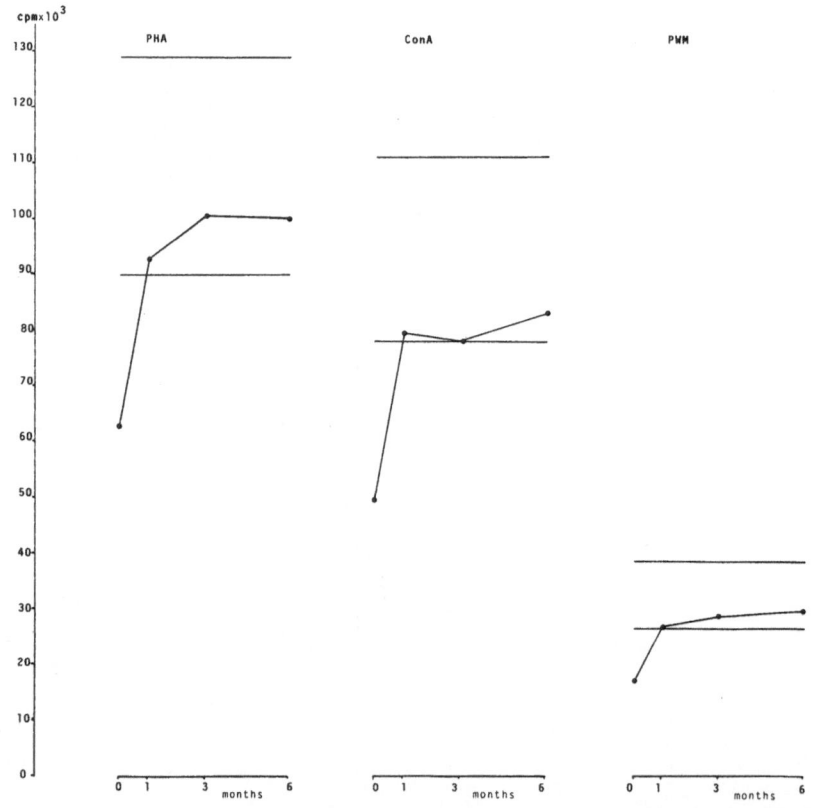

Fig. 4. Response to mitogens: functional tests showed a markedly reduced
   response to all of the mitogens (PHA, ConA, PWM). During the
   follow-up the lymphocyte function returned to normal.

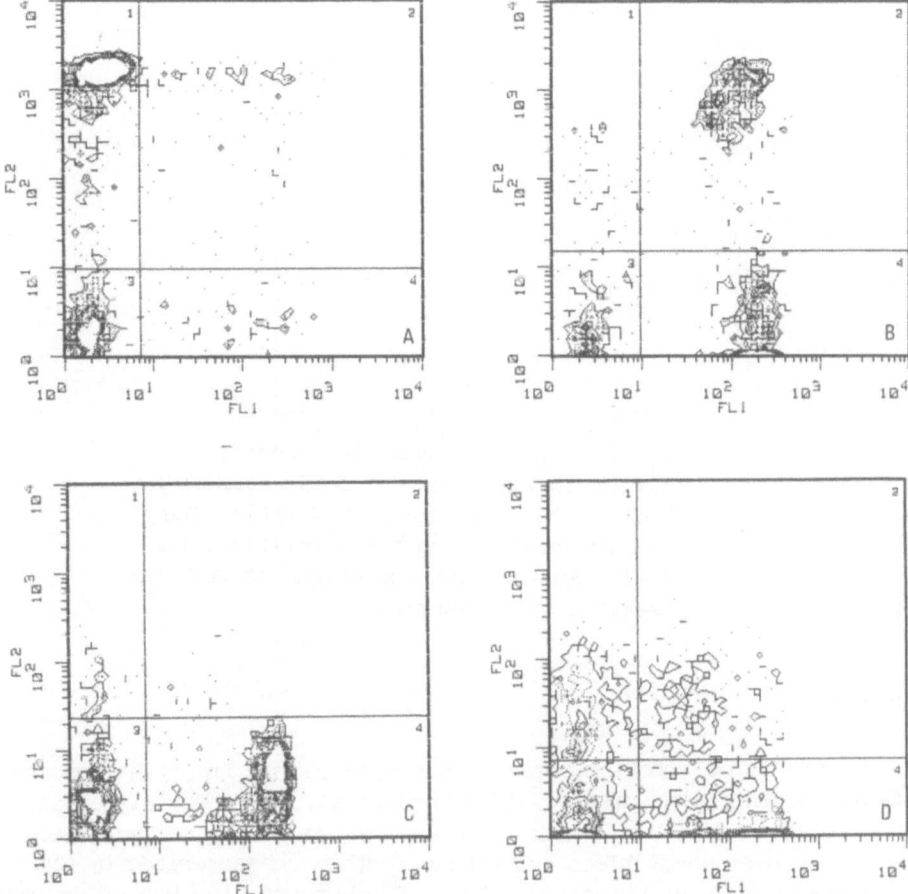

Fig. 5. Double labelling technique:
   A: phycoerythrin conjugated CD8 (FL2) plus fluorescein -
   conjugated Leu 7 (FL1)
   B: phycoerythrin - conjugated CD16 (FL2) plus fluorescein -
   conjugated CD8 (FL1)
   C: phycoerythrin - conjugated CD8 (FL2) plus fluorescein -
   conjugated CD3+ (FL1)
   D: phycoerythrin - conjugated leu 15 (FL2) plus fluorescein-
   conjugated CD8 (FL1)


During the acute stage, in some patients further studies concerning CD8
lymphocyte subset have been made, always employing double labelling technique
but making use of new monoclonal antibodies. The monoclonal antibody combinations employed were the following: Leu 2 phycoerythrin/Leu7 fluorescein,
Leu 11 phycoerythrin/Leu 2 fluorescein, Leu 15 phycoerythrin/Leu 2 fluorescein (on separated lymphocytes).
   The mainly represented phenotypical expression was as follows: CD8+
Leu7- CD16- CD3+ Leu15- (Fig. 5).
   CD8 lymphocytes stratified according to the intensity of their fluorescence in 2 distinct layers, dim and bright.The latter was more represented
in I.M. (Fig. 6).

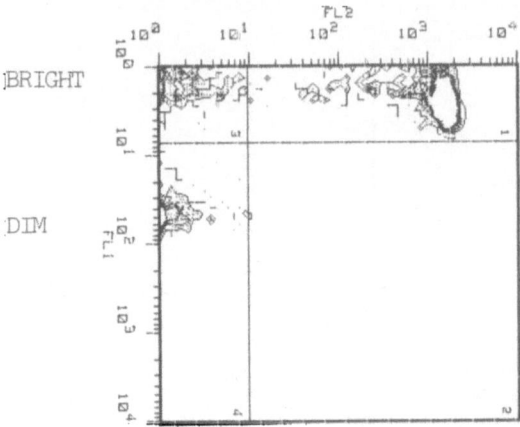

Fig. 6. Double labelling immunofluorescence:
phycoerythrin – conjugated CD8 (FL2) plus
fluorescein – conjugated CD4 (FL1). The
picture shows the double stratification
of CD8 cells in dim and bright layers and
CD4/CD8 ratio inversion.

DISCUSSION

CD8 phenotype subset expansion is considered an important response in
self-limitation of the disease. (Crawford et al.,1981; Dewaele et al.,1981).
Lymphocytes with HLA-DR/CD3 double receptors represent the "activated" lymphocytes of peripheral blood and appear further characterized by CD8 receptor of the suppressor-cytotoxic subset. Furthermore, the use of new monoclonal antibodies widens knowledge of cellular phenotypes.

It is extremely difficult to correlate phenotypes and functions.However
it is necessary to try to characterize more clearly lymphocyte phenotypes
also considering that, while bearing in mind the always more frequent reports concerning the possibility of co-expression of identical antigens on
cells with different functional activities (Gebel et al.1987; Lanier et al.
1983,; Forti and Ortolani, 1987), the CD4/CD8 ratio should, perhaps, be re-
examined more thoroughly.

A deficit of the T cellular response to B lymphocyte stimulation which
is set off by the virus, could be responsible for particularly serious or
protracted forms of I.M..

An increasing number of cases of "Chronic mononucleosis syndrome", clinically characterized by fatigue, myalgia, malaise, depression, headache and
lack of concentration are being reported. As regards the serological picture
these patients reveal constantly elevated titres of anti-VCA IgG and anti-EA,
together with low or even no titres of EBNA antibodies.

However, this syndrome is of difficult definition and it is essential
to exclude the presence of other diseases or disorders which could explain
the forementioned symptoms.

Following our investigation on the 4 patients, still suffering from protracted illness, it can be hypothesized that the scarse or absent proliferation of activated lymphocytes (HLA-DR/CD3) and of CD8 lymphocytes is responsible for this protracted duration. Serological findings of abnormally high

anti-VCA IgG titres and the absence of anti-EBNA, after 6 months, are suggestive of such evolution.

It would therefore be advisable to make an exhaustive study and pro-
long the follow-up period on these patients in order to reach a more accu-
rate prognostic definition.

REFERENCES

Baker, F.S., Kotchmar, G.S., Foshee, W.S. and Sumaya, C.V., 1983, Acute
    hemiplegia of childhood associated with Epstein-Barr virus infection,
    Pediatr. Infect. Dis. 2:136.
Crawford, D.H., Brickell, P., Tidman, N., Mc Connell, I., Hoffbrand, A.U.
    and Janossy, G., 1981, Increased numbers of cells with suppressor T
    cell phenotype in the peripheral blood of patients with infectious
    mononucleosis, Clin exp. Immunol, 43:291.
Crawford, D.H., Edwards, J.M.B., 1988, Virus di Epstein-Barr, in: "Virologia
    Clinica", A.J. Zuckerman, J.E. Banatvala, J.R. Pattison; Ed. Centro
    Scientifico Torinese, Torino.
De Waele, M., Thielemans, C. and Van Camp, B.K.G., 1981, Characterization of
    immunoregulatory T. cells in EBV-induced infectious mononucleosis by
    monoclonal antibodies, N. Engl J. Med 304:460.
Fleisher, G., Henle, W., Henle, G., Lennettes, E.T., Biggar, R.J., 1979,
    Primary infection with Epstein-Barr virus in infants in the United
    States. Clinical and serological observations, J. Infect. Dis., 139:
    553.
Forti, A., Ortolani, C., 1987, L'analisi multiparametrica nella citometria a
    flusso, Il patologo clinico, 8:70.
Gebel, H.M., Anderson, J.E., Gottschalk, L.R., Bray R.A., 1987, Dermina-
    tion of helper: suppressor T-cells ratio, N. Engl. J. Med., 316:113.
Horwitz, C.A., Henle, W., Henle, G., Rudnick H., Latts E., 1985, Long-term
    serological follow-up of patients for Epstein-Barr virus after recov-
    ery from infectious mononucleosis, J. Infect. Dis., 151:1150.
Jones, J.F., 1986, Chronic Epstein-Barr virus infection in children,
    Pediatr Infect. Dis., 5:503.
Kormaroff, A.L., 1987, Sindromi della "mononucleosi cronica", Minuti
    Menarini, 77:41.
Lanier,L.L., Lanier, L.L.,Le, A.M., Phillips, J.H., Warner, N.L., Bablock,
    G.F., 1983,Subpopulations of human natural killer cells defined by
    expression of Leu-7 (MNK-1) and Leu-11 (NK-15) antigens, J. Immunol.,
    131:1789.
Marklund, G., Ernberg, I., Lundberg, C., Henle, W. and Henle, G., 1986,
    Differences in EBV-Specific Antibody patterns at onset of Infectious
    Mononucleosis, Scand. J. Infect. Dis., 18:25.
Niederman, J.C., 1977, Infectious Mononucleosis, in: "Infectious Diseases",
    Paul D. Hoeprich, Harper and Row, London.
Pattengale, P.K., Smith, R.W., Perlin, E., 1974, Atypical lymphocytes in
    acute infectious mononucleosis: identification by multiple T and B
    lymphocyte markers, N. Engl. J. Med., 291:1145.
Sugden, B., 1982, Epstein-Barr virus: A human pathogen inducing lymphoproli-
    feration in vivo and in vitro, Rev. Infect. Dis., 4:1048.
Sumaya, C.V., 1986, Epstein-Barr virus serologic testing: diagnostic indica-
    tion and interpretation, Pediatr. Infect. Dis., 5:337.

Svedmyr, E., Ernberg, I, Seeley, J., Weiland, O., Masucci, G., Blomgren, H., Berthold, W., Henle, W., Klein, G., 1984, Virologic, Immunologic, and Clinical observation on a patient during the incubation, acute, and convalescent phase of infectious mononucleosis, Clin. Immunol. Immunopatol., 30:437.

Straus, S.E., 1988, The Chronic Mononucleosis Syndrome, J. Infect. Dis.,157: 405.

Tobi, M., Morag, A.,Ravid, Z., Chowers, I., Feldman-Weiss, V., Michaeli, Y., Ben-Chetrit, E., Shalit, M., Knobler, H., 1982, Prolonged atypical illness associated with serological evidence of persistent Epstein-Barr virus infection, Lancet 8263:61.

Tosato, G., Magrath, Z., Koski, I., Dooley, N. and Blaese M.,1979,Activation of suppressor T cells during Epstein-Barr-virus-induced infectious mononucleosis. N. Engl. J. Med., 301:1133.

Vella, S., Rocchi, G., 1988, Mononucleosi Infettiva.in"Infettivologia a confronto", F.De Rosa; Ed.CIC Edizioni Internazionali,Roma.

MULTIPLE SCLEROSIS AND ZOSTER ENCEPHALOMYELITIS:

A PROBABLE COMMON ETIOPATHOGENESIS

Riccardo Viel

Department of Neurology
General Hospital
33100 Udine, Italy

The involvement of herpesviruses[1] in the possible viral etiology of Multiple Sclerosis (MS) is radically opposed by several researches,[2] who attribute the high serum and CSF antibody titers to activation of the latent virus by prolonged corticosteroid therapy.

Personal findings and a review of the literature lead me to be skeptical of this conclusion. The type of case I would like to discuss[3] is that of the MS patient in whom Herpes Zoster develops in the course or, better still, at the outset of the illness and so cannot be dismissed as an isolated complication.

By way of example I cite the emblematic case presented by Alajouanine in 1931 of a 26 year old woman with MS.[4] It had all begin 9 years before with a serious attack of Zoster radiculomyelitis affecting L2 - L4 and the right lower limb, followed by successive exacerbation. On the basis of what was then known of the biology and clinical manifestations of herpesviruses, the author interpreted the initial episode as a symptomatic Zoster infection "which disclosed" - these are his words - "the MS virus anatomic localization to the level of the posterior roots." He did not know that "idiopathic or primary" Herpes Zoster does not give rise to serious eruptions or persistent neuralgias in young subjects[5] or that a Zoster encephalomyelitis (neuraxitis) may be spatiotemporally disseminated[6,7] with the same clinical pattern as MS. The case reported by Alajouanine is, in my view, important because it was a case of typical MS in which the first bout coincided with a Zoster radiculomyelititis. Increasingly refined techniques of virus isolation and identification may tell us in the future how many MS cases, especially those with recurrent exacerbations, fit the description of Zoster neuraxitis. Indeed, if one cannot - or does not - seek a candidate virus, one ends by mistaking a neuraxitis of probable herpetic origin for an exceptional , if not unique, case of MS in the elderly. This is what Filley et al.[8] did when they considered the episode that started the disease as important (rightly) but inexplicable (wrongly): their right $C_4$ radiculitis was, we think, a Zoster radiculoganglionitis without eruption. This obviously does not dispose of the problem of the etiopathogenesis of MS, which remains complex, related to a peculiar immunogenetic soil and to as yet unknown environmental cofactors. A primary Zoster neuraxitis, as MS might turn out to be, must be regarded as a bit peculiar: in its mild clinical expression, its non-exanthematous character and its frequent spontaneous remission. It must be supposed that the immune control, cellular and humoral, of the Varicella Zoster Virus (VZV) located, as long as latent, in the sensory ganglia of the brain and spinal cord, is temporarily and partially altered by some other infection caused by a "helper" virus, not necessarily neurotropic, such as Epstein Barr virus (EBV), as some recent personal studies would suggest.

Another nervous disease, in some respects antipodal to MS, with a controversial pathogenesis is sporadic Amyotrophic Lateral Sclerosis (ALS) or Charcot's disease. The viral hypothesis, despite laboratory failures, has some plausibility because of the following considerations:

1. ALS has a relatively broad span of age at onset, though preferring late, if not old, age, with the occurrence of small temporospatial clusters.[9,10]
2. The motoneuron degeneration begins in one or another of the following segments of the CNS: medullopontine, cervical or lumbar (of the spinal cord), the last being the pseudopolyneuritic or lumbar type of ALS.
3. The not infrequent complaints of sensory disturbances like painful paresthesias and dysesthesias confined to one or more nerve roots in the early stages of the subacute form of the disease.
4. The most recent pathological findings, which substantiate the involvement, marginal though it may be, of the posterior funiculi and, more constantly, of Clarke's nuclei with the spinocerebellar pathways,[11] and which, therefore, rule out a pure systematic first and second motoneuron degeneration, although this is the essential and preminent feature.

Our experience leads us to concentrate on points 2 and 3. The dermatomes corresponding to the aggregates of motoneurons initially affected by the disease are under the control of major sensory pathways: the trigeminal roots for the medullary forms, the brachial and lumbosacral plexuses for the spinal forms of ALS. These are the sensory areas and roots preferentially affected by Herpes Zoster (shingles). The sensory disturbances, in certain cases, are not purely subjective: they often leave zones of hypoesthesia with a radicular distribution or even present at onset as a subacute amyotrophy of the shoulder (as in a recent case in a Swiss neurological clinic, with liquor and gas myelography negative, now evolving to ALS).

All these points might fit, and be explained by, a viral etiology of ALS. According to a suggestive hypothesis, the painful paresthesias at the onset of the disease, not rarely present, are a warning that something is moving in the cranial and spinal ganglia, where herpesviruses are known to gather in the latent stage.[12,13] Just as they may migrate to the periphery, to the skin and mucosae, there is no reason to think that they may not take the centripetal path of the ganglionic T cell and infect the neurons of the cord and brainstem to which the root is anatomically and functionally related. Two completely different nervous diseases, one affecting mainly the young and the other maturity and old age, might thus have an identical pathogenesis and express a different immune response to the same viral agent.

Human pathology is not without examples of different anatomoclinical entities caused by the same pathogen. A case in point is neurosyphilis, which presents first as a meningovascular symphilis with endarteritic processes affecting the meninges and blood vessels (Heubner's endarteritis) and later as progressive or general paralysis due to a chronic encephalitis through invasion and destruction of neurons by Treponema pallidum.

REFERENCES

1. J.R. Martin, Herpes simplex virus type 1 and 2 and Multiple Sclerosis, Lancet ii:777 (1981).

2. R.T. Johnson, The possible viral etiology of Multiple Sclerosis. Adv. Neurol. 13:1 (1975).

3. R. Viel, Sclerosi Multipla: osservazioni cliniche a favore del Virus Varicella Zoster nell'etiopatogenesi della malattia. Archivio di Psicol., Neurol. e Psichiat. 2:335 (1985).

4. T.H. Alajouanine, B. Griffith, Sclèrose en plaques précédée d'un éruption zostérienne avec paralysis crurale. Rev. Neurol. 55:84 (1931).

5. R.E. Hope Simpson, The nature of Herpes Zoster; a long-term study and new hypothesis. Proc. Roy. Soc. Med. 58:9 (1965).

6. D. McAlpine, Y. Kuroiwa, Y. Toyokura, S. Araki, Acute demyelinating disease complicating Herpes Zoster. J. Neurol. Neurosurg. Psychiat. 22:120 (1959).

7. E.R. Hogan, M.R. Krigman, Herpes Zoster myelitis. Arch. Neurol. 29:309 (1973).

8. C.M. Filley, P.E. Sternberg, M.D. Nörberg, Neuromyelititis optica in the elderly. Arch. Neurol. 41:670 (1984).

9. C. Melmed, C.H. Krieger, A cluster of Amyotrophic Lateral Sclerosis Arch. Neurol. 39:595 (1982).

10. D. Chad, Conjugal motor neuron disease. Neurology (NY) 32:306 (1982).

11. P. Averback, P. Crocker, Regular involvement of Clarke's Nucleus in sporadic Amyotrophic Lateral Sclerosis. Arch. Neurol. 39:155 (1982).

12. K.G. Warren, Isolation of Herpes Simplex Virus from human trigeminal ganglia. Lancet ii:637 (1977).

13. M.M. Esiri, A.H. Tomlinson, Herpes Zoster. Demonstration of Virus in Trigeminal nerve and Ganglion by immunofluorescence and Electron Microscopy. J. Neurol. Sci. 15:48 (1972).

# LOSS OF SUPPRESSOR-INDUCER T-CELLS IN CHRONIC-PROGRESSIVE MULTIPLE SCLEROSIS: PRELIMINARY RESULTS

Mauro Zaffaroni, Silvano Rossini*, Rosalia Palma*,
Angelo Ghezzi, Stellio Marforio and Carlo L. Cazzullo

Centro Studi Sclerosi Multipla – Università di Milano –
Ospedale di Gallarate – Italy
*Centro Immunotrasfusionale – Ospedale L.Sacco – Milano

## INTRODUCTION

Multiple Sclerosis (MS) has long been postulated to be an autoimmune disease of the Central Nervous System with a possible viral origin and with relevant abnormalities in immunoregulation (1). The most important immuno-logical finding in MS is a defective T-suppressor activity which is proven by functional assays as by phenotypic characterization (2).
Several laboratories using monoclonal antibodies (mAb) techniques have reported that a decreased number of CD8+ve lymphocytes and an increased CD4/ /CD8 ratio are important features of the chronic-progressive form of the disease (3-6).
More recently a selective loss of the inducer of suppression T-cell sub-population (CD4+/2H4+ve cells) has been reported in chronic-progressive MS (7). In order to confirm this finding, we started a cross-sectional study in 54 consecutive MS patients and 24 healthy controls.

## SUBJECTS AND METHODS

We studied 54 patients with MS clinically definite according to the criteria of McDonald and Halliday. They were 20 males and 34 females with age ranging from 18 to 54 years (mean 33.4 years). Forthy had a relapsing-remitting and 14 had a chronic-progressive course.
Controls consisted of 24 healthy blood donors and laboratory personnel: 14 males and 20 females with a mean age of 30.5 years.
Blood samples were obtained by heparinized syringes and mononucleated cells were separated by Ficoll-Paque density gradient. Cells were finally label-led as follows:
Tube 1: control mouse IgG FITC- and PE-conjugated;
Tube 2: anti-CD4 FITC plus anti-2H4 PE mAbs;
Tube 3: anti-CD3 FITC plus anti-CD8 PE mAbs.
All mAbs were purchased from Coulter-Clone. Cells were analyzed by dual-color laser flow cytofluorimetry using a logarithmic scale on an EPICS-C cell sorter (Coulter).

RESULTS

The results are summarized in the table. We found a significant reduction of CD4+ve and of CD4+/2H4+ve lymphocytes in chronic-progressive
patients with respect to relapsing-remitting ones and with respect to healthy controls. It must be stressed that 9 out of 24 chronic progressive MS
patients had less than 14% CD4+/2H4+ve lymphocytes whilst only 7 out of 24
controls had low percentages of suppression-inducers.

Table 1. Lymphocyte subpopulations in our series. Mean percentages and
standard deviations are given.

| Groups | CD4+2H4+ | CD4+ | CD8+ | CD4/CD8 |
|---|---|---|---|---|
| Controls (n=24) | 19.0±8.2 | 44.3±9.6 | 24.3±8.0 | 2.1±0.9 |
| Chron.Progr.MS (n=14) | 11.4±4.6 | 36.9±10.7 | 22.5±8.8 | 1.9±0.9 |
| Relaps.Remit.MS(n=40) | 24.8±8.6 | 51.7±9.3 | 23.7±8.2 | 2.4±1.1 |
| AN.O.VA. (p) | < 0.001 | <0.001 | n.s | n.s. |

CONCLUSIONS

To our knowledge this is the first confirmation of the results of
Morimoto and co-workers (7).
We found a significant reduction of the inducer of suppression T-cell subset in chronic-progressive MS patients. In parallel we found that the
whole CD4+ve subset was reduced in the same class of patients.
In contrast with our previous reports (6,8) we did not find a selective
loss of CD8+ve cells in chronic-progressive patients: this may be explained
by the use of different mAbs and by the exponential scale of fluorescence
required by double-color analysis in contrast with the linear scale previously used for single-color analysis.
Moreover, low percentages of CD4+/2H4+ve cells unespectedly did not correlate with low percentages of CD8+ve cells or with increased CD4/CD8 ratios.
As a consequence we suggest that the selective loss of inducer of suppression T-cells may act on suppressor/cytotoxic T-lymphocytes at a functional
level rather than reducing the number of CD8+ve cells. This condition may
play a role in activating certain T or B cell clones reactive against an
unknown antigen of the central nervous system.
It is well known that an environmental aetiological factor for MS is
strongly suggested by epidemiological studies and that, more recently, a
retrovirus related to the family of human T-lymphotropic virus has been
claimed as the possible cause of MS (9). Similarly to the mechanism of
action of human T-lymphotropic retrovirus, a selective involvement of
CD4+/2H4+ve T-cell subpopulation could explain the immunological abnormalities found in MS.

REFERENCES

1) B.H. Waksman and W.E. Reynolds, Multiple Sclerosis as a disease of immunoregulation, Proc.Soc.Exp.Biol.Med. 175:282 (1984).

2) B.G.W. Arnason, Immunocyte abnormalities in multiple sclerosis, in:
   "Virology and immunology of multiple sclerosis: rationale for therapy"
   C.L. Cazzullo, D.Caputo, A.Ghezzi and M.Zaffaroni Eds., Springer-
   Verlag, Berlin (1988).
3) M.A. Bach et al., Deficit of suppressor T-cells in active multiple
   sclerosis, Lancet 2:1221 (1980).
4) E.L. Reinherz et al., Loss of suppressor T-cells in active multiple
   sclerosis: analysis with monoclonal antibodies, N.Engl.J.Med. 303:
   125 (1980).
5) A. Compston, Lymphocyte subpopulations in patients with multiple sclerosis, J.Neurol.Neurosurg.Psych. 46:105 (1983).
6) M.Zaffaroni et al., Monoclonal antibody analysis of blood T-cell subsets in multiple sclerosis, Ital.J.Neurol.Sci. 5:45 (1984).
7) C. Morimoto et al., Selective loss of the suppressor-inducer T-cell subset in progressive multiple sclerosis, N.Engl.J.Med. 316:67 (1987).
8) C.L. Cazzullo et al., The role of lymphocyte subset analysis in defining
   the clinical evolution of multiple sclerosis, Europ.Neurol. 27:5 (1987).
9) H. Koprowsky et al., Multiple sclerosis and human T-cell lymphotropic
   retroviruses, Nature 318:154 (1985).

# INDEX

Antigens (continued)
   of RSV, 14, 149-150
   T cell-dependent, 124
   T cell-independent, 124
   viral, 136-137
Antigen-antibody complexes, 157
Antigen-presenting cells (APC),
        237-245, 260
Antiprotease, 160
Aplastic anemia,
   and IFN production, 54
Arachidonic acid metabolites, 159
Arthritis,
   rheumatoid, 141
      and EB virus, 1
Atherosclerosis, 142
Atropine effect on IgA secretion,
        23
Autoantibodies, 140
   in hepatitis Delta virus, 273-
        275
   in HIV, 3
Autoimmune,
   disease and IFN production, 47,
        50
   reaction or response, 3, 138-140

Bacillus Calmette-Guerin (BCG),
        221-222
BALT (Bronchus-associated lymphoid
        tissue), see MALT
Basophil or mast-cell, 21-23, 159
B cells,
   activation, 176
      by HIV antigens, 3-4
   and dengue virus, 157
   determinants, 119
   in EBV infection, 3, 280
   EBV-transformed lines, 5-7
   in HIV infection, 3
   and IFN, 47-48, 53
   immunoglobulin synthesis, 19-23
   in LCM infection, 118-128
   in multiple sclerosis, 294
   neuropeptide receptors, 22
   proliferative response, 235-245
   specificity, 120
   in virus infection, 135
Beta$_2$ microglobulin, see Major his-
        tocompatibility complex
B lymphocytes see B cells,
Bronchial lavage, 14

Bronchitis and bronchiolitis,
   in RSV infections, 10, 14, 211-
        212
Burkitt's lymphoma,
   and EBV virus, 1, 137


Cancer,
   cervical and papillomavirus, 1,
        137, 172, 176
   viral etiology, 137-139, 143
CD3 monoclonal antibody, 239-242,
        281-286, 293
CD4, 3, 13, 32, 93, 96-103, 176,
        221, 281-283
   monoclonal antibodies, 4, 277-
        278, 293-294
   role in HIV-related disorders, 3-
        5, 221
   soluble, 6-7
CD8, 4-5, 13, 21, 32, 93, 96-103,
        109, 176, 212, 281-286
   monoclonal antibodies, 277-278,
        293-294
Cell (or cellular),
   damage in viral infection, 135
   hyperplasia, 179
   modification and transformation
        by IFN, 51
   transcription, 180
Central Nervous System (CNS),
   and autoimmune disease, 293-295
   and chronic disease, 138
   degeneration, 290
   and herpes virus infection, 187-
        189
   and immunity, 20
Charcot's disease, see Sclerosis
        lateral amyotrophic
Cholecystokinin (CKK), 23
Choriomeningitis lymphocytic (LCM),
        91-107, 135, 138
Chron's disease and viral/bacterial
        infections, 142
CKK, see Cholecystokinin
CMV, see Cytomegalovirus
Colitis ulcerative, 142
Colony stimulating factors (CSF),
        see Interleukins
Complement in dengue virus infec-
        tion, 157, 159
Concanavalin A (Con A), 22, 238,
        248-250, 253, 280

in the skin (SALT), 176
Mast-cells, see Basophils
M cells, 20
Measles, 2
  SSPE infectious, 136-137
  vaccine for, 1, 221
  virus antigen, 137, 142
Mediators,
  chemical in dengue virus, 157,
    159
MHC, see Major histocompatibility
    complex
Mitogen, 22
  IFN induction, 47-48, 50
Monocyte or Mononuclear phagocytes,
    see Macrophages
Mononucleosis infections, see
    Epstein-Barr virus
Mucosal immunity (see also MALT),
    19-26
  nervous, 21
  neuropeptides or neurotransmit-
    ters, 19-26
  role of IgA, 19-20
Multiple sclerosis, see Sclerosis
    multiple
Mumps virus,
  and juvenile diabetes, 141
  meningoencephalic complication,
    277
  parotitis, 277
Mx protein, 51
Mycoplasma,
  antibodies to, 136
  and arthritis, 141

Nasopharyngeal carcinoma, 137
Natural killer (NK),
  activity, 21-22, 27, 53
    in CMV infection, 27-36
    in dengue virus infection, 158
    in herpes simplex virus infec-
      tion, 257
    in influenza infection, 247-254
    in papillomavirus infection,
      177
    in RSV infection, 211-220
  cells and IFN production, 47-48
  cell lysis, 32
  "like"-cell, 218
  neuropeptide effects on, 21-24
Neuraxitis, 289-291

Neuroaminidase inhibiting antibody
    in influenza, 37
Neurological disease and viral
    infection, 138-140
Neurones,
  retinal and viral infection, 135
Neurosyphilis, see Syphilis
NK, see Natural killer
Noradrenalin, 21

OKT4, see $CD_4$ monoclonal antibody
OKT8, see $CD_8$ monoclonal antibody
Oligodendrocytes and viral infec-
    tion, 135

Paget's disease of bone, 142
Papilloma, 169-186
Papovaviruses, 136-137
  and cervical cancer, 1, 137, 170,
    172, 176
  and IFN, 55
  papillomavirus, 169-186
    antigenicity, 174-175
    cell-mediated immunity, 174-176
    characteristics, 169-171
    detection, 172-173
    immune response to, 174-176
    isolation, 170
    persistent infection, 173, 179-
      180
    replication, 170-171
    vaccine, 174
Parainfluenza,
  vaccine for, 1
Paramyxovirus, 9, 12
Pararotavirus, see Rotavirus
Parkinson's disease viral etiology,
    139
Parotitis, see Mumps virus
Parvo viruses, 137, 141
Paul-Bunnel reaction, see Epstein-
    Barr virus
Peroxidase anti-peroxidase (PAP)
    method, 226-227
Peyer's patches (see also MALT), 22
PHA, see Phytoemagglutinin
Phytoemagglutinin, 22, 257-261,
    277, 280
Pituitary cells, 135
Plasminogen activator, 160
Pneumonia, 211, 221
  haemorragic 14